A Review of Science and Technology During the 1973 School Year

Science Year

The World Book Science Annual

1974

Field Enterprises Educational Corporation

Chicago   London   Paris   Rome   Stuttgart   Sydney   Tokyo   Toronto

The publishers of *Science Year* gratefully
acknowledge the following for permission to use
copyrighted illustrations. A full listing of illustration
acknowledgments appears on pages 432 and 433.

258 Copyright © by National Geographic Society–
     Palomar Observatory Sky Survey; reproduced
     with permission from Hale Observatories
305 © 1972 World Magazine, Inc.
311 Drawing by Richter © 1973
     The New Yorker Magazine, Inc.
342 © *Punch*, London
367 Drawing by John Corcoran © 1973
     The New Yorker Magazine, Inc.
413 © National Geographic Society

The Cover: Detail from the painting by Herb Herrick for
the Special Report "X Rays from the Sky."
See page 70.

Preface

Publishing a review of science and technology over a period of several years has given us an opportunity to see new ideas, and new applications of old ideas, develop—sometimes in unexpected directions. It is fascinating, for example, to watch, through *Science Year* articles, as the molecular biologists gradually fit together the pieces of the puzzle of the cell. It is equally fascinating to watch the astronomers, as they find more and more quasars, find less and less agreement as to what and where these sources of energy are.

The broader changes that take place in science over a period of years—a sort of evolution of emphasis—has also been reflected in *Science Year*. In the beginning, we devoted more pages to space technology than any other field. Gradually this emphasis shifted to the environment; later to the biomedical sciences. This year, the energy crisis permeates both the Special Reports and the Science File.

Yet the scope of *Science Year* has been limited. Heretofore the editors felt that to adequately cover the life, earth, physical, and space sciences it was necessary to exclude the social and behavioral sciences. Thus we erected fences within fields. For example, in anthropology, we would report on physical but not cultural research; in psychology, we would cover experimental but not clinical work.

Because these fields are not that easily divided, and because of increasing reader interest, we have continuously moved those fences. This year we leveled them—particularly with the Special Report: "The Ins and Outs of Mind-Body Energy." Psychologists Elmer and Alyce Green describe how a person can be trained to control his autonomic nervous system. To better understand these processes, the scientists have studied persons with unusual powers to control not only their own bodies, but outside objects as well.

The Greens' work is part of a growing trend to take a serious scientific look at unexplained but persistent phenomena, sometimes associated with Eastern philosophy, that has little or no scientific basis from the point of view of Western cultures. Perhaps the best known studies are on the painkilling effects of acupuncture. According to accepted theories of pain, acupuncture should not work. But apparently it does for many people. Extrasensory perception (ESP) has no physical explanation either. Yet there is evidence that some people can predict the turn of cards, beyond probability, while others can manipulate objects without touching them. These studies in parapsychology are being given the "respectability" of the research laboratory, where physical data are recorded and analyzed. And organizations such as the American Association for the Advancement of Science are now scheduling sessions on ESP at their annual meetings.

The blending of Eastern and Western philosophies—of physics and parapsychology—portends some major changes in scientific thinking. It seems appropriate to note this in the year that celebrates the 500th anniversary of the birth of Copernicus.　　　　　[Arthur G. Tressler]

Contents

10 **Apollo—The End of the Beginning** by Darlene R. Stille
Seven visits, in less than four years, to the surface of the moon have widened
not only our vision of the moon, but also of the entire solar system.

24 **Taking Sexism out of Science** by Betsy Ancker-Johnson
A physicist who has personally faced sex prejudice in universities and
industry suggests the problem can be solved by removing sex stereotypes.

33 **Special Reports**
Fifteen articles, plus a *Science Year* record album, give in-depth
treatment to significant and timely subjects in science and technology.

34 **Talking with Chimpanzees** by Roger S. Fouts
The ability of chimpanzees to learn and use a human language indicates that
nonhuman primates may be more capable of abstract thought than we had realized.

50 **The Power Above, the Power Below** by John F. Henahan
In the face of an increasing shortage of conventional fuels, exotic energy
sources—the sun and the earth's heat—are being investigated.

64 **X Rays from the Sky** by Lee Edson
Through the eye of orbiting satellite Uhuru, astronomers are gathering
a new spectrum of data on the violent events in the universe.

78 **The Case of the Missing Neutrinos** by William A. Fowler
A leading astrophysicist analyzes the reasons for the scarcity of detectable
solar neutrinos and proposes that our picture of the sun's interior may be wrong.

92 **A Summer's Search for a Buried Past** by Donald L. Wolberg
A young paleontologist recounts his experiences on an archaeological expedition
that has recruited scientists from a number of disciplines.

108 **New Insights into Immunity** by Jacques M. Chiller
Two specialized lymphocytes, working together, may be responsible
for much of the success of the immunological system.

120 **Lengthening Our Lives** by Bernard L. Strehler
Scientists are looking for ways to slow or stop the disarrangement
of the molecules within cells that occurs as we age.

136 **The Ins and Outs of Mind-Body Energy** by Elmer and Alyce Green
Medicine, physics, and psychology are being combined in studies that may
show the powers of mysticism are not so mystical after all.

148 **The Bees That Got Away** by Michael Sheldrick
A genetic experiment to improve honey production in the Western
Hemisphere has loosed a horde of ferocious bees on the land.

160 **It's A Dog's Life** by Michael W. Fox
A veterinarian and psychologist views with sadness the fate of the dog,
whose instincts are being blunted by our modern urban society.

175 **How Canids Communicate**—A *Science Year* Special Feature
A two-record album—"The Music of the Canids"—is combined with
a picture story—"Decoding a Dog's Body Language"—
to provide insight into what dogs and their cousins are trying to say.

184 **Getting Around the Rush Hour** by Samy E. G. Elias
A personal rapid transit system that is fast, dependable, and inexpensive
may be the answer to the problems of intracity transportation.

198 **Lasers Today: Still Not in Full Focus** by James L. Tuck
A variety of lasers have been put to a variety of uses,
but not all the visions of a decade ago have been fulfilled.

212 **Engineering the Environment** by Michael Reed
An engineering project that took a quarter-century to complete has changed
the face and future of a large part of eastern Australia.

220 **The New Rainmakers** by Charles L. Hosler
Continuing attempts to modify the weather are hampered by our incomplete
knowledge of the behavior of clouds.

232 **Science Parts the Iron Curtain** by William J. Cromie
The summit agreements of May, 1972, have opened the floodgates of
scientific cooperation between the United States and the Soviet Union.

243 **Science File**
A total of 44 articles, alphabetically arranged by subject matter, plus
8 *Science Year* Close-Ups, report on the year's work in science and technology.

371 **Men and Women of Science**
This section takes the reader behind the headlines of science and into the
research laboratories to focus on the scientists who make the news.

372 **Robert Wilson** by Robert H. March
Putting aside his natural instincts as a loner, this physicist pulled together
people and materials to build the largest particle accelerator to date.

388 **Jonas Salk** by John Barbour
His polio vaccine made him a national hero, and now, in the twilight of that
fame, his goal is to bring a new comprehensiveness to all the biological sciences.

404 **Awards and Prizes**
Science Year lists the winners of major scientific awards and describes
the scientists honored, their work, and the awards they have won.

413 **Deaths**
A list of noted scientists and engineers who died during the past year.

415 **Index**

Frank D. Drake is director of the National Astronomy and Ionosphere Center, Arecibo, Puerto Rico; associate director of the Center for Radiophysics and Space Research, Ithaca, N. Y.; and professor of astronomy at Cornell University. He received the B.E.E. degree from Cornell in 1952, and the M.A. degree in 1956 and Ph.D. degree in 1958 in astronomy from Harvard University. Dr. Drake is chairman of the Division for Planetary Sciences of the American Astronomical Society and a member of the visiting committees for the Kitt Peak National Observatory and the Cerro Tololo Inter-American Observatory.

Dr. Walsh McDermott is Livingston Farrand Professor of Public Health and Chairman of the Department of Public Health at the Cornell University Medical College. He received his B.A. degree from Princeton University in 1930 and his M.D. degree from Columbia University's College of Physicians and Surgeons in 1934. In 1967, Dr. McDermott was named chairman of the Board of Medicine of the National Academy of Sciences. He is editor of the *American Review of Respiratory Diseases*.

Roger Revelle is Richard Saltonstall Professor of Population Policy and director of the Center for Population Studies at Harvard University. He received a B.A. degree from Pomona College in 1929 and a Ph.D. degree from the University of California in 1936. He is president-elect of the American Association for the Advancement of Science. Much of his current work involves studies of the interactions between population growth and change, development of natural resources, and the environment.

Alvin M. Weinberg is director of the AEC's Oak Ridge National Laboratory in Tennessee. He is a graduate of the University of Chicago, 1935, and earned his doctorate in physics there in 1939. After three years of researching and teaching mathematical biophysics at Chicago, he joined the Metallurgical Laboratory (part of the Manhattan Project) in 1941. He moved to Oak Ridge in 1945, becoming director of the Physics Division in 1947, research director in 1948, and director of the laboratory in 1955.

Contributors

Ancker-Johnson, Betsy, Ph.D.
Assistant Secretary of Commerce
for Science and Technology
Taking Sexism out of Science

Araujo, Paul E., Ph.D.
Research Fellow in Nutrition
Harvard University School of
Public Health
Nutrition

Auerbach, Stanley I., Ph.D.
Director, Environmental Sciences Division
Oak Ridge National Laboratory
Ecology

Barbour, John, B.A.
Science Writer
Associated Press
Jonas Salk

Belton, Michael J. S., Ph.D.
Astronomer, Kitt Peak
National Observatory
Astronomy, Planetary

Boffey, Philip M., B.A.
Managing Editor
Science and Government Report
Science Support

Bromley, D. Allan, Ph.D.
Henry Ford II Professor and Chairman
Department of Physics
Yale University
Physics, Nuclear

Buchsbaum, S. J., Ph.D.
Executive Director
Research, Communications Sciences
Division Bell Telephone Laboratories
Communications

Budnick, Joseph I., Ph.D.
Professor of Physics
Fordham University
Physics, Solid State

Burton, Charles, M.D.
Associate Professor of Neurosurgery
Temple University Health Science Center
Close-Up, Medicine, Surgery

Burwell, Calvin C., M.S.
Planning Group, Research Engineer
Oak Ridge National Laboratory
Energy

Chesher, Richard H., Ph.D.
Vice-President
Marine Research Foundation
Oceanography

Chiller, Jacques M., Ph.D.
Associate, Department of
Experimental Pathology
Scripps Clinic and Research Foundation
New Insights into Immunity

Cromie, William J., B.S.
President and Editor
Universal Science News, Inc.
Science Parts the Iron Curtain

Crow, James F., Ph.D.
Professor of Genetics
University of Wisconsin
Genetics

Deffeyes, Kenneth S., Ph.D.
Associate Professor of Geology
Princeton University
Geoscience, Geology

Diener, Theodor O., Dr. sc. nat.
Research Plant Pathologist
U.S. Department of Agriculture
Agriculture Research Service
Close-Up, Microbiology

Drake, Charles L., Ph.D.
Professor of Earth Sciences
Dartmouth College
Geoscience, Geophysics

Edson, Lee, B.S.
Free-Lance Writer
X Rays from the Sky

Elias, Samy E. G., Ph.D.
Chairman, Department of
Industrial Engineering
West Virginia University
Getting Around the Rush Hour

Ensign, Jerald C., Ph.D.
Professor of Bacteriology
University of Wisconsin
Microbiology

Evans, Earl A., Ph.D.
Professor and Chairman
Department of Biochemistry
University of Chicago
Biochemistry

Fouts, Roger S., Ph.D.
Research Associate
Institute for Primate Studies
Department of Psychology
University of Oklahoma
Talking with Chimpanzees

Fowler, William A., Ph.D.
Institute Professor of Physics
California Institute of Technology
The Case of the Missing Neutrinos

Fox, M. W., Ph.D.
Associate Professor of Psychology
Washington University
It's A Dog's Life

Gillette, Robert, B.S.
Staff Writer, *Science*
Close-Up, Environment

Goss, Richard J., Ph.D.
Professor of Biology
Division of Biological and Medical
Sciences Brown University
Zoology

Gray, Ernest P., Ph.D.
Chief, Theoretical Plasma Physics
Applied Physics Laboratory
Johns Hopkins University
Physics, Plasma

Green, Alyce, B.A.
Research Psychologist
Menninger Foundation
The Ins and Outs of Mind-Body Energy

Green, Elmer, Ph.D.
Head, Voluntary Controls and
Psychophysiology Laboratory
Menninger Foundation
The Ins and Outs of Mind-Body Energy

Griffin, James B., Ph.D.
Director, Museum of Anthropology
University of Michigan
Archaeology, New World

Hawthorne, M. Frederick, Ph.D.
Professor of Chemistry
University of California, Los Angeles
Chemistry, Synthesis

Hayes, Arthur H., Jr., M.D.
Chief, Division of Clinical Pharmacology
Milton S. Hershey Medical Center
Drugs

Henahan, John F., B.S.
Free-Lance Science Writer
The Power Above, the Power Below

Hilton, Robert J., D.D.S.
Associate Professor Operative Dentistry
Northwestern University Dental School
Medicine, Dentistry

Hines, William
Science Correspondent
Chicago Sun-Times
Space Exploration

Hosler, Charles L., Ph.D.
Dean, College of Earth and
Mineral Sciences
Pennsylvania State University
The New Rainmakers

Irvin, Robert W.
Automotive Writer
The Detroit News
Close-Up, Transportation

Isaacson, Robert L., Ph.D.
Professor of Psychology
University of Florida
Psychology

Kellerman, Kenneth I., Ph.D.
Scientist
National Radio Astronomy Observatory
Close-Up, Astronomy, Cosmology

Kessler, Karl G., Ph.D.
Chief, Optical Physics Division
National Bureau of Standards
Physics, Atomic and Molecular

Kristian, Jerome, Ph.D.
Staff Member, Hale Observatories
Astronomy, Cosmology

Maglio, Vincent J., Ph.D.
Assistant Professor of Paleobiology
Princeton University
Geoscience, Paleontology

Maran, Stephen P., Ph.D.
Head, Advanced Systems and
Ground Observations Branch
Goddard Space Flight Center
Astronomy, Stellar

March, Robert H., Ph.D.
Professor of Physics
University of Wisconsin
Robert Wilson
Physics, Elementary Particles

Marsh, Richard E., Ph.D.
Research Associate
California Institute of Technology
Chemistry, Structural

Mayer, Jean, Ph.D.
Professor of Nutrition
Harvard University School
of Public Health
Close-Up, Nutrition

Meikle, Thomas H., Jr., M.D.
Dean, Cornell University Graduate
School of Medical Sciences
Neurology

Merbs, Charles F., Ph.D.
Chairman, Department of Anthropology
Arizona State University
Anthropology

Novick, Sheldon
Publisher, *Environment*
Environment

Price, Frederick C., B.S.
Managing Editor
Chemical Engineering
Chemical Technology

Price, John B., Jr., M.D.
Associate Professor of Surgery
College of Physicians
and Surgeons
Columbia University
Medicine, Surgery

Rodden, Judith, M.Litt.
Research Archaeologist
Archaeology, Old World

Romualdi, James P., Ph.D.
Professor of Civil Engineering
Carnegie-Mellon University
Transportation

Shank, Russell, D.L.S.
Director of Libraries
Smithsonian Institution
Books of Science

Sheldrick, Michael, B.S.
Coordinator, Environmental Field Program
Washington University
The Bees That Got Away

Spar, Jerome, Ph.D.
Professor of Meteorology
New York University
Meteorology

Steere, William Campbell, Ph.D.
President Emeritus and Senior Scientist
New York Botanical Garden
Botany

Strehler, Bernard L., Ph.D.
Professor of Biology
University of Southern California
Lengthening Our Lives

Tilton, George R., Ph.D.
Professor of Geochemistry
University of California, Santa Barbara
Geoscience, Geochemistry

Treuting, Theodore F., M.D.
Professor of Medicine
Tulane University School of Medicine
Medicine, Internal

Tuck, James L., M.A.
Associate Physics Division Leader,
Los Alamos Scientific Laboratory
Lasers Today: Still Not in Full Focus

Weber, Samuel, B.S.E.E.
Executive Editor
Electronics Magazine
Electronics

Wittwer, Sylvan H., Ph.D.
Director, Michigan Agricultural
Experiment Station
Michigan State University
Agriculture

Wolberg, Donald L., B.A.
Teaching Associate
Department of Geology
University of Minnesota
A Summer's Search for a Buried Past

Woltjer, Lodewyk, Ph.D.
Rutherfurd Professor of Astronomy
Columbia University
Astronomy, High Energy

Wright, James E., Ph.D.
Research Entomologist
U.S. Department of Agriculture
Close-Up, Agriculture

Zare, Richard N., Ph.D.
Professor of Chemistry
Columbia University
Chemistry, Dynamics

Contributors not listed on
these pages are members of
the *Science Year* editorial staff.

Apollo–
The End
Of the
Beginning

By Darlene R. Stille

**The moon landings not only provided
priceless scientific data, but also freed
man to explore worlds beyond his own**

On Dec. 19, 1972, the Apollo 17 space capsule
bearing three astronauts splashed down in the
South Pacific Ocean, marking the end of the historic Apollo space program that had landed the
first men on the moon. In its three-and-a-half-year
lifetime, the Apollo program's moon-landing phase
opened a new era in the history of mankind.

On July 20, 1969, astronauts Neil A. Armstrong
and Edwin E. Aldrin, Jr., stepped out of their
Apollo 11 lunar module (LM), Eagle, and walked
on the surface of the moon. These pioneers returned
to the earth with a bounty of 48.4 pounds of rock
and powdery soil samples taken from the desolate
lunar area known as the Sea of Tranquility. Then

The tracks of a two-wheeled pull cart used by the
Apollo 14 astronauts trail off into the lunar distance.

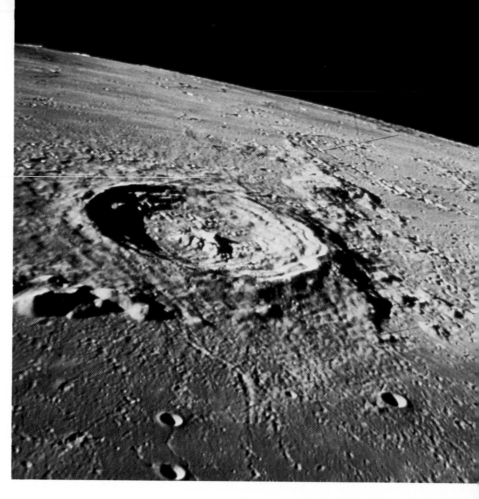

Before landing on the moon, astronauts took photographs from their spacecraft, orbiting as low as 60 miles above the lunar surface. This gave scientists an increasingly more accurate view of the moon's cratered terrain.

The author:
Darlene R. Stille is a senior editor for *Science Year.*

followed a series of five more Apollo missions carrying men to the moon. These explorers set up supersensitive seismometers to detect moonquakes and other geologic activity, magnetometers to measure magnetic forces on the moon, and solar-wind instruments to capture the charged particles of energy that speed out from the sun and bombard the lunar surface.

Each of the Apollo moon landings was targeted for a different geological area. The Apollo 12 LM set down in the crater-pocked Ocean of Storms on Nov. 19, 1969. The Apollo 14 astronauts in January, 1971, explored the rugged, boulder-strewn Fra Mauro region, venturing out more than half a mile from their landing site and pulling their tools and instruments around in a two-wheeled cart.

The Apollo 15 mission in July, 1971, brought motorized transport to the moon in the form of the lunar rover. This moon-adapted dune buggy carried Apollo 15 astronauts to three different topographical areas: Hadley Rille, a gorge; the Marsh of Decay, a plain; and the slopes of the Apennine Mountains. The Apollo 16 lunar module touched down in the Descartes region of the lunar highlands on April 20, 1972. Again using a lunar rover, the astronauts explored huge craters and a mountain about 3 miles from their landing site.

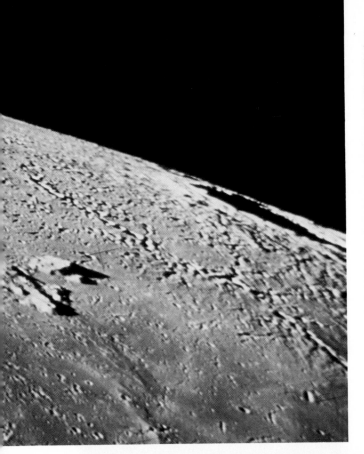

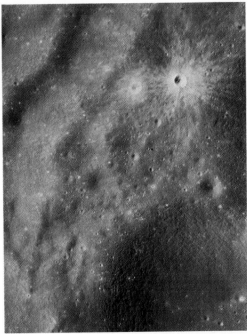

A small, bright crater, *above,* was photographed during the Apollo 8 lunar orbital mission. The Apollo 10 crew photographed the far side of the moon, *below,* which cannot be seen from earth.

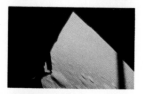

The last Saturn 5 rocket bearing an Apollo spacecraft roared off its Cape Kennedy, Fla., launch pad shortly after midnight on Dec. 7, 1972, carrying the Apollo 17 astronauts toward their landing target in the Taurus-Littrow Valley. At the Taurus-Littrow site, the astronauts drove their lunar rover over an area covered with boulders and marked by shallow depressions. One place they visited was a record 4.7 miles from the lunar module. They harvested a richly varied crop of rocks and core samples, including a glassy, orange soil.

The data collected on all six Apollo moon landings will keep the scientists who are investigating the moon busy for years. Much about the moon remains a mystery, but careful examination of the rock samples and analysis of other information already indicate past volcanic activity and a period when objects from space rained heavily down on the moon, gouging out the craters that scar its surface. These events, which took place on the moon billions of years ago, are probably similar to events that occurred on the earth. But, unlike the earth, the moon has no wind or water to erase the record of its distant past. So scientists hope they can use the information gathered by the Apollo astronauts to piece together a more complete picture of the sun, the earth, and the other planets in our solar system.

Unquestionably, the Apollo program provided a wealth of invaluable information. But perhaps more important, it taught us that we can break the bonds that tie us to earth and soar off into space to explore other worlds. "No one can know where this exploration will finally take us," said historian Arthur Schlesinger, Jr. "But the pursuit of knowledge and understanding has been humankind's most abiding quest, and to have confined this quest to our own small planet would surely have been a betrayal of man's innermost nature."

On July 20, 1969, the Apollo 11 lunar module touched down on the Sea of Tranquility. About 8 hours later, Neil A. Armstrong climbed down a ladder to take man's historic first step on the moon.

14

Edwin E. Aldrin, Jr.,
set up a moonquake
detector about 30 feet
from the Apollo 11
lunar landing craft,
above. One of man's
first footprints in the
moon's powdery soil,
left, is preserved in
the eerie stillness.

Apollo 12 astronauts
Alan L. Bean and
Charles Conrad ventured
out a quarter of a mile
from the lunar module
to explore areas of
the Ocean of Storms
and collect samples
of lunar soil and rock.

For a seismic test in the Fra Mauro uplands, Apollo 14 astronaut
Edgar Mitchell created shock waves by setting off explosive charges.

On the Apollo 15 mission to the Apennine
Mountain area, David R. Scott drilled holes
to implant heat sensors near the base of
Mount Hadley, *above.* James B. Irwin checked
out the lunar rover, *right,* a moon-adapted dune
buggy first used by the Apollo 15 astronauts,
which permitted the explorers to travel miles
from the safety of their home on the moon.

With a camera along to record the rock in its natural setting, Apollo 16 astronaut Charles M. Duke examined a large boulder, looking for promising samples. At the Descartes landing site, John W. Young used a lunar rake to sift some walnut-sized rocks out of the powdery soil.

At the rim of Plum crater, formed by the impact of a meteor,
Duke collected rocks that had sprayed out when the meteor struck.

In the Taurus-Littrow valley, target of the scientifically rewarding Apollo 17 mission, geologist-astronaut Harrison H. Schmitt described a huge boulder that had bounced down the slope of a hill and split in two. Among the other discoveries the astronauts made was a mysterious, glassy, orange soil, *right.*

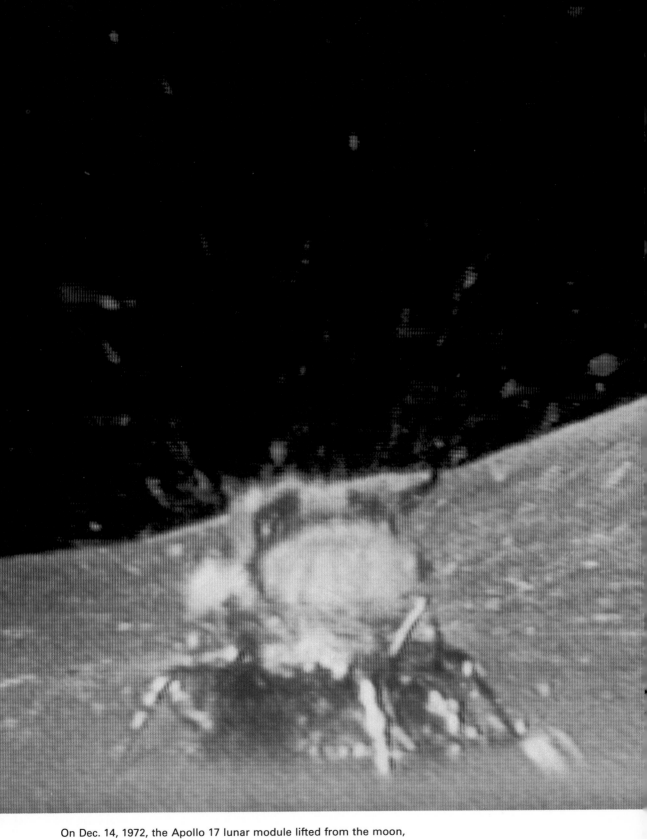

On Dec. 14, 1972, the Apollo 17 lunar module lifted from the moon, ending what was probably man's last lunar visit in this century.

Taking Sexism out Of Science

By Betsy Ancker-Johnson

The stereotypes of "masculine" and "feminine" must be erased from our culture before women can pursue careers in science unhindered

Why would any woman want to be a scientist or engineer? To succeed in these "masculine" professions she obviously has to be aggressive, competitive, analytical–hardly traditional characteristics of the feminine sex. Therefore she must be some kind of freak.

It is extremely difficult to disprove this picture of a woman scientist because there are so few of us. When you spread us out over government, industry, and the universities, we have properties something like an electron cloud around a nucleus: If you do track one down, what you learn about her characteristics, other than where she is located, has extremely large errors.

According to the 1970 National Register of Scientific and Technical Personnel, the percentage of women among American scientists ranged from 19 per cent in anthropology to less than 2 per cent in the atmospheric and space sciences. In my own field of physics, the figure was 3.7 per cent, and only 2.1 per cent had a Ph.D. degree. Women

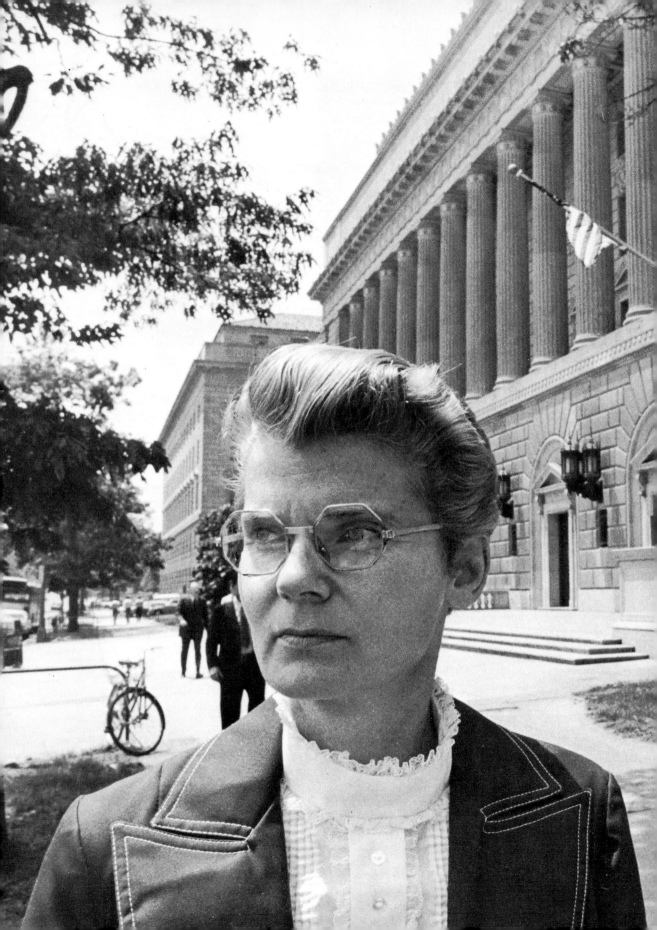

"...the 'conspiracy' of discouragement begins long before she seeks her first job...."

physicists are even less visible in management, representing less than 1 per cent of those promoted to higher jobs. They are more visible, unhappily, in the unemployment ranks. In 1970, 20 per cent did not have jobs, compared to only 5 per cent of the male physicists.

Although these figures must be very bad news to any young woman contemplating a scientific career, the "conspiracy" of discouragement begins long before she seeks her first job. As science students, women are rarely taken seriously for what they are—students and not husband-hunters—and what they aspire to be—scientists (and probably also wives). A woman science student, frequently the only woman in her classes, is almost always lonely and isolated. Both professors and classmates contribute to this. During my first year of graduate work at Tubingen University in West Germany, my (male) colleagues did not believe I was really there to learn physics. Women cannot think analytically and therefore I must be there to find a husband. For the same reason, I was excluded from the small informal discussion groups (again male) that are so helpful in early graduate work. This isolation amplified my sense of appearing stupid and the fear of failure, worries common to all graduate students. But the experience had one advantage. I became determined to succeed in research, probably twice as determined as most men.

Determination is equally imperative when women scientists become job hunters. The devastating effects of discrimination can be seen in the National Research Council Doctorate File for the 1958 to 1970 period. Of those scientists who got postdoctoral education, 60 per cent of the men, but only 40 per cent of the women, had definite job commitments after they were interviewed. Even more discouraging, while only 15 per cent of the men had no job prospects at all, over 30 per cent of the women were in this uncomfortable situation.

There is an informal network, composed of professors and employers, that is very effective in finding teaching and research jobs for new graduates. But women seldom benefit from this placement service. Even with a professor's help, women jobseekers can be foiled. A member of an outstanding science faculty told me that he could not find suitable employment for the two women Ph.D. students he had supervised. But he had had no trouble in placing men of the same caliber in highly desirable jobs.

A woman is always considered a riskier employee than a man. If she is single, the prevalent attitude is that she'll get married and quit. If she is married, she won't take her job seriously and will leave if her husband moves to a new job. Statistics from the U.S. Department of

The author:
Betsy Ancker-Johnson is assistant secretary of commerce for science and technology. She was formerly a manager in the research and engineering division of the Boeing Company, and also an affiliate professor of electrical engineering at the University of Washington.

"...deduction...that a mother is unlikely to achieve her potential as a scientist..."

Labor show, to the contrary, that the work-life expectancy is essentially the same for women and men in relatively the same circumstances. For example, at age 35 both a man and a woman have a work-life expectancy of 28 more years. Moreover, while 11 per cent of men make a change in their jobs at least once during their careers, only 9 per cent of women do. Yet the job-changing practices of men don't keep them from being employed.

If a competent woman scientist somehow hurdles all of these obstacles and gets a good job, are her troubles over? Probably not. If she is single, she is unlikely to be accepted socially, which dims her chances for promotion. If she marries and becomes a mother she encounters a different set of difficulties. It is commonly believed that a scientist's most productive years are those during which women usually are bearing children. The deduction then is that a mother is unlikely to achieve her potential as a scientist or to compete effectively with men.

As a result, in a college or university our woman scientist is almost certainly to be found in the lowest ranks and at the less-esteemed institutions. Data from the 1970 National Register shows that 6 per cent of the physicists, for example, employed in colleges without graduate programs are women. This is noticeably higher than the percentage of women Ph.D.s in physics (2.1 per cent), and even the percentage holding bachelor's degrees (3.7 per cent). On the other hand, only 2.5 per cent of the physicists employed in colleges with master's programs and only 1.7 per cent of those working in the universities are women. In the 10 universities with the best graduate programs in physics in 1970, less than 1 per cent of the faculty were women.

Industry is the most inhospitable, or perhaps the least attractive, employer of women scientists. A *Management Review* survey published in November, 1970, found that nearly 60 per cent of the companies surveyed hired women only for clerical jobs. In nearly three-fourths of these companies, women held less than 5 per cent of the professional or technical jobs. About 40 per cent of the firms had no women in management and only about 10 per cent had as much as 5 per cent in that category. Thus women are effectively discouraged from leaving the academic world. In the case of physicists, 30 per cent of the men work in industry and an equal percentage in colleges and universities whereas 9 per cent of the women physicists work in industry and 58 per cent in academia. A mere 1.2 per cent of all industrial physicists are women and only 0.7 per cent hold managerial positions.

The story is boringly repetitious in government. Only 2.8 per cent of physicists holding government jobs are women, again less than the 3.7

per cent of all women physicists. And, as is the case in industry, only 0.7 per cent are in supervisory roles.

Sexual discrimination comes full circle with the salary structure. The National Register shows that women scientists consistently draw lower salaries than men who are in comparable jobs. The discrepancies are smallest in government-sponsored jobs and greatest in management positions in industry. But on the average, women's salaries are only 75 per cent those of men.

Apart from salary and promotions, a woman often encounters other barriers to her professional growth. She finds it difficult to advance from the apprentice role of research assistant. This means that work on a project results only in gracious acknowledgments and not co-authorship of the resulting paper. If she is permitted to graduate from this role and organize some kind of research, or a study on her own, she is unlikely to get much help from the institution or her associates in obtaining the necessary grants or fellowships.

A woman is often isolated, perhaps unintentionally, by her male colleagues from fruitful, casual "shop talk." For example, lunch periods can be times of profitable scientific exchange but frequently the lone woman is not invited to join the group. For years, at an eastern university it was the custom for the physics faculty to take visiting speakers to lunch at a restaurant that did not admit women. It did not matter whether the visitor was a scientist or not. Thus the exclusion from the scientific fraternity that she first met in college continues.

Although a woman scientist is isolated, she is also far too visible and aware of it. I call this the "sore-thumb syndrome." Everywhere, she is looked upon as different. Harmless remarks become trials. For example, a former classmate who had not pursued a career invited me to a party. Her husband, and most of the male guests, were also physicists. As the guests arrived, she introduced me, adding "and she's a physicist, too." Eventually she saw me wince, and stopped it.

As a student, a woman needs to learn how to handle the egotistical men who think she is hunting for a husband. Later, she must learn how to deal with their wives. Even if a wife is not disturbed about your sharing a lab with her husband, she may still inadvertently make upsetting remarks. One time the wives of some colleagues were helping me clear the table after a party my husband and I had given. During the party, our baby was being looked after by the young woman who regularly cared for her during my absences from home. One wife remarked, "I should have pursued an interesting career like you so I could have avoided changing diapers." I know this woman did not really think I studied physics to avoid changing diapers. Such remarks, nevertheless, will pop out and cause those who aspire to excel both as scientists and mothers considerable anguish.

All of these problems stem from the stereotypes our society has created regarding what feminine and masculine mean. "Feminine" usually is associated with the characteristics of passivity, gentleness, intuition, and physical weakness, and "masculine" with the characteristics of aggression, hardness, analytical ability, and physical strength.

It begins in the crib with the color of the blanket. A child's toys are chosen according to sex—girls have to play with dolls, boys with erector sets. During times of stress, girls can cry, but boys are sissies if they do. And heaven help the boy who prefers girls as playmates!

A boy probably suffers more than a girl if he resists being forced into a stereotype. I know from experience that a 10-year-old girl who would rather take her bike apart than ride it, much less play with dolls, will be tolerated. We call her a "tomboy," implying that "she'll grow out of it as she becomes a woman." What happens, however, to the boy who dares to ask for permission to take sewing or cooking in junior high school instead of shop? Very few of us know the extent to which such boys suffer when their worth as individuals is assailed because of their honest interests.

The stereotyping continues through high school. Boys are urged to take science and mathematics, while girls major in English or commercial courses. Parents, friends, teachers, and worst of all, guidance counselors, cement career directions by encouraging boys to pursue science or engineering in college while urging girls to avoid them.

Even women who overcome the stereotype barrier and persist in a scientific career probably started out as little girls who were deprived of mechanical toys and discouraged from fooling around with Dad's tools. Hence, it is not surprising that those who do become scientists predominantly choose the nonlaboratory fields, such as mathematics and anthropology and tend to avoid the more laboratory-oriented fields—physics, chemistry, the biological sciences, the earth and marine sciences, and the atmospheric and space sciences.

Why do so many people want boys and girls—and men and women —to conform to established patterns? One reason, I think, is the fear of change—a sense of threat to the familiar life style. I think this feeling is irrational because it is totally inconsistent with the development of human history, particularly with that of our American heritage. The Pilgrims were certainly not afraid of change. They were, in fact, an adventuresome group of people and were rather experimentally oriented. The Declaration of Independence was born out of deep desires for change. But even then there were those who were afraid of it. Many people argued against independence, preferring to continue under the protection of King George III.

A modern version of the colonial Tories, in my opinion, are the men and women who oppose the Equal Rights Amendment. They argue that it will force women into combat duty, or lose them legal protection from long and hard physical labor or child custody in divorce cases. But in a civilized society, few men or women would choose combat, although both sexes have performed such duty when it was necessary. In my judgment, protective laws that are not specifically based on biological differences are simply not needed. Since I weigh 115 pounds, I'm not interested in a job heaving 50-pound cement sacks, but then I would not be even if I weighed 250 pounds. On the other hand, any woman who can handle such a job should have it if she wants it. As for child custody, I wonder how women who fear the loss of legal favoritism can be so sure that they are better parents than the fathers of their children.

I had an experience that shows women are as threatened as men by a challenge to the stereotype. For about a year, at an industrial research laboratory, I collaborated with a theorist, a man, on high-temperature electron-hole plasmas. He had to thoroughly understand my experimental results and I had to grasp the meaning of his equations. We developed theories to explain the measurements, and further experiments to test the theories. A secretary in our group complained to the personnel department that we spent hours at a time together with the office door closed. Apparently, while we were struggling to understand each other's work and do new physics, she was picturing us as romantically involved. Presumably she could not imagine a professional relationship with a member of the opposite sex. I did not know what was happening until after she had been transferred. I both felt sorry for her and dismayed that I had to live with the ugly aftermath. The tack I took then is one I have taken several times—simply carry on and eventually enough data will accumulate to establish the truth. In this case, the data consisted of several co-authored papers and the co-authors' two thriving and intact marriages.

I suspect that these fears are not the real issues, even to those who vehemently express them. A more basic fear seems to be that if we were to have sexual equality and abandon the culturally accepted definitions of feminine or masculine, we would somehow evolve into a unisex. If there were any danger of this, I would certainly have some fears myself. I'm convinced, however, that the human race is permanently divided into two clearly distinguishable sexes.

My conviction is based on a study I have been pursuing for about a decade. It is, admittedly, founded on a rather small statistical sample. Our daughters–Ruth, 13, and Martha, 9–and sons–David, 12, and Paul, 10–are definitely growing up in a climate of sex equality. Both sexes do dishes, wash cars, empty garbage, cook, help repair bikes, learn to play musical instruments, backpack, ski, kayak, or practice Aikido. Ruth and Paul prefer verbal subjects to math. David and Martha are taller than average for their age groups and Ruth and Paul are shorter than average. David and Martha might be called more "aggressive," and Paul and Ruth more "passive." These characteristics can, and probably will, change with time. But the point is– there is absolutely no confusion about who is a girl and who is a boy.

The question does not arise in this family about the husband's role or the wife's. Each has a satisfying career in science. And our equal sharing of parental activities and responsibilities has been no threat to his masculinity. In fact, of all the couples we know where the wives are capable career women, the husbands are invariably interesting, happy, capable career men.

Many problems stem, I think, from the nature of sex. Women and men live together in a special intimate relationship from which I see no desirable retreat or substitute. Racial, religious, or ethnic minorities can find a haven from the battle for equality with each other at home, or in social or religious activities. But the home could become a battleground in the drive for sex equality. Therefore, men and women must work together to understand the essence of masculinity and femininity, to define equality, and to strive for it.

A definition of sex equality should depict the essence of femininity and masculinity in biological terms only. Men are usually physically stronger than women and each sex has a specific role in procreation. Everything else is a cultural manifestation.

In fact, masculine or feminine "characteristics" comprise a spectrum. It is something like the element helium that has three prominent lines in the blue. The wave length at 4,471 Å is more "blue" than those at 4,388 Å or 4,713 Å, but most of us would see all three as blue. A person in the spectrum ranging from intuitive to analytical (compared with red to violet) may be on the yellow or green side of blue. But is this a useful test of masculinity and femininity? Some of the best scientists I know are so intuitive they can jump over several logical steps and guess an answer. They always go back carefully through all the logic, of course, to find whether their intuitive jumps are justified. An experienced scientist often develops a "good intuition," and it is an extremely useful tool. No one would dream, however, of concluding that this person has become more "feminine."

Every person should be free to exhibit various parts of the spectrum at any time. Abandoning sex stereotypes would, I believe, allow men to be more genuinely masculine and women more fully feminine. And they could enjoy each other and themselves more completely, because no one would be forced to *act* masculine or feminine.

Many people's life style would not change one iota if we abandoned sex stereotypes and had instant equality. Given a genuine choice, however, I think many people would choose different paths. Men who grew up fitting naturally into the cultural expectations would presumably fit just as easily without these imposed expectations. Women who are happy as wives and mothers exclusively would presumably be just as happy if they had chosen their current roles rather than having to assume them. And that is the point—let us have a choice.

Both women and men must work at it. A mother can proceed with a demanding career such as science only if the father supports her completely. This means more than just sharing parental responsibilities equally. It means sharing the joys and disappointments of each other's workday, both in and out of the home, and having a mutual pride in each other's achievements.

It also means that families must become highly inventive in arranging genuinely adequate child care so that the mother and father are completely at peace about the children's welfare and each can concentrate fully on his or her work. Sometimes a mother and father can arrange their schedules so that one is always at home. Hopefully, both universities and industries will develop programs to provide adequate care of preschool children whose parents have careers.

Sexual equality in science will take time to develop. To bring it about more quickly, currently successful women scientists accept their role as models. Hopefully they will be willing to demonstrate that it is possible to succeed simultaneously in career and family, contribute to neighborhood and "extracurricular" activities, and be happy, interesting people in the process. Perhaps then, more young women will be encouraged to seek the challenges and excitement of science, and more young men will be ready to support them.

I can hardly wait for the day when young readers of this or a similar essay will comment, "How quaint!" The sooner the reason for writing on this subject disappears, the better.

Special Reports

The special reports and the exclusive *Science Year* record album give in-depth treatment to the major advances in science. The subjects were chosen for their current importance and lasting interest.

34   **Talking with Chimpanzees** by Roger S. Fouts

50   **The Power Above, the Power Below** by John F. Henahan

64   **X Rays from the Sky** by Lee Edson

78   **The Case of the Missing Neutrinos** by William A. Fowler

92   **A Summer's Search for a Buried Past** by Donald L. Wolberg

108   **New Insights into Immunity** by Jacques M. Chiller

120   **Lengthening Our Lives** by Bernard L. Strehler

136   **The Ins and Outs of Mind-Body Energy** by Elmer and Alyce Green

148   **The Bees That Got Away** by Michael Sheldrick

160   **It's A Dog's Life** by Michael W. Fox

      175   **How Canids Communicate**
           A *Science Year* Special Feature

184   **Getting Around the Rush Hour** by Samy E. G. Elias

198   **Lasers Today: Still Not in Full Focus** by James L. Tuck

212   **Engineering the Environment** by Michael Reed

220   **The New Rainmakers** by Charles L. Hosler

232   **Science Parts the Iron Curtain** by William J. Cromie

Talking with Chimpanzees

By Roger S. Fouts

"Come   hug   me."

**A young scientist teaching sign language to chimps
finds they can create and use language as humans do**

Whheen I was a boy growing up on a farm in central California, I remember stepping off the school bus in the afternoon and greeting my dog Pal waiting by the roadside. His excited barks and wagging tail seemed to be saying, "I've missed you and I'm very glad that you're home." Then as he ran ahead of me toward the house, he stopped from time to time with an impatient look that said, "Are you coming?" or "Hurry up!"

Anyone who has ever established a relationship with another animal can probably remember wishing many times: "If only he could speak, I'm sure he'd tell me what he's thinking." Countless popular myths and stories from *Aesop's Fables* to *Alice's Adventures in Wonderland* reflect this desire. Yet, very few people think that we could ever de-

"Fruit"

When Fouts asks, "Who am I?" during a lesson in sign language, Washoe quickly tugs her left ear, "R," for "Roger." In addition to naming her friends, the 8-year-old chimp also makes signs for nouns and verbs. Washoe uses more than 160 signs and can combine them into phrases.

The author:
Roger S. Fouts is a research associate at the Institute for Primate Studies of the University of Oklahoma in Norman. He has been teaching chimpanzees the American Sign Language for six years.

velop Dr. Dolittle's ability to "talk to the animals." My research is aimed at determining if a species other than our own—chimpanzees—can learn a human language and use it as we do.

Many people believe language is unique to human beings, but some scientists think otherwise. For more than 40 years, some of them have been trying to teach nonhuman primates to speak. The most successful attempt was made during the late 1940s and early 1950s by psychologists Keith and Cathy Hayes of the Yerkes Laboratories of Primate Biology, then in Orange Park, Fla. They took a baby chimpanzee, Viki, into their home and raised her as a member of the family. After six years of intensive training, Viki had learned to say four words—mama, papa, cup, and up—but she uttered them in a very heavy chimpanzee accent. The word "mama" was, for the most part, voiceless, although she could make the proper lip movement required to produce the "m" sound. She sometimes confused the words, however, and used them incorrectly. Like other chimpanzees, Viki never tried to imitate human speech sounds on her own as all normal children do.

The Hayes's research suggests two possible conclusions: Either chimps do not have the mental ability to learn a human language, or they are physically incapable of vocal speech. However, they may be able to learn to communicate in some other way.

Philip Lieberman, professor of linguistics and electrical engineering at the University of Connecticut, has since shown that the chimpan-

"Hat"   **"Listen"**   **"Baby"**   **"Smile"**

zee's vocal apparatus is quite different from that of an adult human. He claims that chimpanzees cannot change the shape of their vocal tract by moving the back of the tongue as human beings can. As a result, they cannot produce the important human vowels "a," "i," and "u" and the other finely regulated sounds of human speech.

Psychologists R. Allen Gardner and Beatrice T. Gardner of the University of Nevada, Reno, found a way around this limitation. They knew that several scientists have observed chimpanzees, both in the wild and in captivity, using gestures such as holding a hand out with the palm down to appease another chimpanzee, as well as hugging and kissing while greeting their companions. Adrian Kortlandt, an ethologist at the University of Amsterdam, who has studied the behavior of chimpanzees in the wild, said in 1967 that particular gestures, such as one meaning "Come with me," may differ from one group to another. In other words, chimpanzees living in different areas—forests and grasslands, for example—may even develop "dialects." Thus, the Gardners' assumption that they could be taught a human language based on gestures was a reasonable one.

In June, 1966, they started a project that eventually shook the foundations of scientific thought about what a chimpanzee can accomplish in terms of human language. They began teaching American Sign Language (ASL)—a set of hand gestures used by the deaf in North America—to Washoe, a 10-month-old female chimpanzee.

The signs of ASL are comparable to words in a spoken language. Each individual sign is composed of a particular hand configuration and movement in addition to a specific position where the sign begins and ends in relation to the signer's body. The sign meaning "hat," for example, is made by patting the top of the head with an open hand; and the sign for "pencil" is made by drawing the extended index finger of the right hand across the open palm of the other hand.

When I met Washoe in September, 1967, she was almost 2 years old and I was a graduate student just beginning my doctoral studies under Allen Gardner. My first encounter with Washoe is still engraved on my memory. Gardner took me to a children's playground at the university, where Washoe was playing. As we approached the chain link fence, a small, dark object streaked across the playground, bounded over the fence, leaped into my arms, and gave me a kiss. This was the beginning of my long friendship not only with Washoe, but also with the chimpanzee as a species.

For the next three years, my time was divided between assisting the Gardners and their other graduate students with Washoe's education and studying for a Ph.D. degree in experimental psychology. The project was carried out in the Gardners' back yard, where Washoe lived in a 24-foot house trailer that had all the conveniences of home, such as chairs and tables, mirrors, cupboards, water faucets, and a refrigerator, bathroom, and bed. There was also a large yard for her to play in, and she had many of the same toys, played the same games, and had the same experiences that a human child might have.

All members of our research team were fluent in ASL and became her close, affectionate friends. At least one of us was with her during all her waking hours. Although we did not use speech in Washoe's presence, we did laugh and imitate the sounds that she made. Washoe was totally immersed in a sign-language environment: We signed to her constantly—at breakfast, lunch, and dinner, during hide-and-seek games and automobile rides, when we washed the dishes, and while changing her dirty diapers.

Washoe learned some signs by imitating us. But the most efficient way to teach her a sign intentionally was by guiding her hands into the correct position. We first drew her attention to an object—a tree, for example—by pointing to it. Then we bent her left arm at the elbow so that her hand pointed up and placed her right hand under the bent elbow. After we did this a few times, she seemed to grasp the idea that the sign meant "tree"—not just that tree, but all trees.

Washoe proved to be a good student and learned to make ASL signs skillfully. In addition to signs for such objects as hats, cats, and flowers, Washoe also learned signs for such actions as in, out, and run. Her vocabulary grew to include numerous other signs for nouns, pronouns, proper nouns, verbs, direct objects, prepositions, and adjectives. And she proved that she understood the meanings of the signs by using them in situations unrelated to those in which she learned them. For example, after she learned to use the "open" sign for a particular door, she then used the sign for cupboard doors, car doors, boxes, and even

Lucy's List of Signs
By December 26, 1972, 7-year-old Lucy was using almost 90 signs. She could name herself and friends, request forks and kisses, and apologize for mischief.

baby/doll	me
ball	mine
banana	mirror
barrette	more
berry	no
blanket	nut
blow	oil
book	open
bowl	out
brush	pants
candy	pen/write
car	pick/groom
cat	pipe
catch	please
clean	purse
coat	run
cold	shoe
comb	smell
come/gimme	smile
cry	smoke
cup	sorry
dirty	spoon
dog	string
drink	swallow
eat	tea
enough	telephone
flower	there
fork	this/that
fruit	tickle
go	want
hankie	what
hat	yes
hug	you
hurry	yours
hurt	
in	Jack
key	Janet
kiss	Lucy
leash	Maury
lipstick	Roger
listen	Steve
look	Sue

water faucets and capped pop bottles. She also combined her signs into meaningful small phrases, such as "Roger tickle Washoe," a plea for her favorite game, and "Open key food," when she wanted the refrigerator in her house trailer unlocked.

Previously, most scientists believed that the ability to name things in the environment is unique to human beings and that only human beings can assemble words into new sentences. Since one of the essential characteristics of human language is combining words in a grammatically correct order, some scientists have also assumed that only human beings are able to do this. If this were true, Washoe would have combined her signs in a random manner. But more than 90 per cent of the time, she used the pronoun "you" before verbs or the pronoun "me," as in signing "You out me," when she wanted to go outside. She seemed to understand syntax, or sentence structure, for when I signed, "Tickle Washoe," she would get ready to be tickled; but if I signed, "Washoe tickle," she would tickle me.

Washoe can also make relatively long combinations that seem to have syntax. She once pestered me for a cigarette I was smoking with a flurry of signs: "Give me smoke," "Smoke Washoe," "Hurry give smoke," and "You give me smoke." Since I do not approve of her smoking, I denied her requests. Finally, I signed to her, "Ask politely." She quickly responded with, "Please give me that hot smoke."

When we tested Washoe's vocabulary, which now includes more than 160 signs, we made sure no one accidentally gave her hints to the answers. First, one researcher put an object such as a lollipop into a box with a clear Plexiglas side and covered that side with a piece of fiberboard. Then another person came into the room or area and sat down behind the box where he or she could not see what was inside. With Washoe sitting in front of the box, he uncovered the Plexiglas, asked Washoe what was in the box, and recorded her answers. Initially, the test was a battle of wills in trying to get her to sit where we wanted her to. Often, in frustration, she would refuse to pay attention, run up a tree, or even steal the test item without naming it. But once

When Washoe wants another car, boat, or piggyback ride, she requests, "More Roger ride me." Her ability to combine words into sentences has convinced most scientists that what she has achieved is really language.

"More   **Roger**

we learned to wait until she was interested and allowed her to uncover the box, she did very well. In other words, we turned the test into something she enjoyed doing, rather than forcing her to do it.

In October, 1970, the Gardners decided to end the project because several members of the research team were graduating and it might be difficult to replace so many skilled people. I had received my degree by then, so Washoe and I went to the Institute for Primate Studies near Norman, Okla., at the invitation of William B. Lemmon, its director and a psychology professor at the University of Oklahoma.

The Institute for Primate Studies is a primate colony housed on Lemmon's farm. There, 24 chimpanzees and many kinds of monkeys live in several buildings and outside cages. On fine days we take several chimpanzees by rowboat to one of the small islands on a large pond near Lemmon's house. They can play there among the bushes and trees that they have stripped bare. In addition to the chimpanzees living at the institute, I am also studying three chimpanzees that live in private homes in the area, where they are raised as much as possible as human children are raised. Lemmon is examining the innate characteristics of chimpanzees by comparing the behavior of these chimpanzees, who have never seen another chimp, with that of chimps raised by their natural mothers at the institute.

I am teaching sign language to the chimpanzees in both situations. At the institute, I work with chimpanzees of different ages and sex, often with as many as three at a time. I also continue to work with Washoe. When she came to the institute, she saw chimpanzees for the first time since she was an infant. At first she called them bugs, thus classifying them as rather despicable animals. Since that time, how-

ride   **me.''**

ever, she has adjusted very well and often refers to the young chimpanzees with the "baby" sign.

Besides teaching Washoe signs, I am continuing to examine the way in which she uses signs in new ways. One day while we were in a rowboat near the chimp island, I signed to Washoe, "What that?" referring to some swans swimming nearby. Previously I had always identified the swans with the "duck" sign. This time her answer astounded me. She called them "water bird." Perhaps she is more correct in describing swans as "water birds," for they are waterfowl. But her response also reveals how she classifies swans.

Washoe's great skill with ASL had led some scientists to wonder if Washoe was unique in this ability. So one of the first things I did at the institute was to find out whether other chimpanzees could also learn signs. I started with four young chimpanzees from the chimpanzee colony—males Bruno and Booee, and females Cindy and Thelma. More than 30 student volunteers that I instructed in ASL helped me to teach them 10 signs by physically molding the chimpanzees' hands into the correct sign form and then giving them a raisin when they made the sign correctly. They learned some signs quickly, taking as little as two minutes of training, but mastered others quite slowly, taking almost eight hours.

Of course, some were difficult to learn because they were similar in the way they were formed or in the concept they represented. But the chimpanzees' personalities accounted for most of the differences. Booee, for example, did quite well and seemed eager to learn and please. (Also, he is willing to sell his soul for a raisin.) Bruno was at the other extreme. When I tried to teach him his first sign, "hat," he

Little Salome and her foster mother, Susie Blakey, greet Fouts when he arrives for a weekly lesson. Only 21 months old, Salome already knows 15 words.

As soon as her diaper is changed, Salome makes her favorite sign, "tickle," by pulling a finger across the back of her right hand.

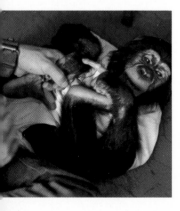

would merely sit passively, refusing to take an active part in the training session. In hopes of overcoming his stubbornness, I changed his reward from raisins to apple slices, to banana slices, and finally to a sweet soft drink. But he still would not cooperate. Finally I threatened him with a frown and harsh tone of voice, and immediately he put his hand on top of his head—the sign for "hat."

After teaching the chimpanzees all 10 signs, we tested them. Some made 90 per cent of the signs correctly, while others only made 26 per cent correctly. But they all performed at a level they could not have attained merely by chance. In any event, I was now sure that chimpanzees can learn and use signs.

Since that study, the chimpanzees have enlarged their vocabularies. I have been gathering data on how they combine signs to form phrases and how they use signs with each other. I first tried to promote such an exchange between Booee and Bruno when they each had a vocabulary of 35 signs. One day when they were together, I gave Bruno a piece of fruit. Booee then spontaneously signed: "Give me fruit" or "Fruit Booee." Bruno, however, promptly ran off with his treasure when he saw Booee ask for it. Once I became aware of how chimpanzees feel about sharing fruit, I encouraged signing in other circumstances with success. For example, in situations of mutual cooperation, such as when one chimpanzee asks another for a tickle, they will now sign back and forth in this manner: Booee signs "Tickle," Bruno replies "Booee hurry come," and Bruno gets tickled. But on one occasion, when Bruno was eating some raisins from a graduate student's hand, Booee approached Bruno and signed, "Tickle Booee." Bruno replied: "Booee me food"—possibly meaning, "Booee, I'm eating"—and quickly went back to his raisin eating.

Washoe signs quite readily to other chimpanzees. Unfortunately, many of them still do not have large enough vocabularies to understand her or to respond. One day Washoe, Bruno, Cindy, and Thelma were having an outing on the chimpanzee island. Cindy and Thelma were playing at the east end, and Washoe was grooming a lounging Bruno in the middle of the island. Suddenly, Cindy and Thelma began giving "oo-oo" alarm calls—probably after seeing a snake—and walked rapidly toward the other end of the island. Washoe noticed their reactions, stood upright, and hesitantly began to move in the same direction. Her naïve young friend, however, remained oblivious to the excitement. When Washoe was a few feet away from him she turned, saw him casually lounging in the sun, and signed to him, "Come hug. Come hug,"—so she could carry him to safety. But Bruno did not understand, so Washoe finally gave up signing, hurried back, and dragged him out of danger.

I have also seen that a chimpanzee can give the original meaning of a sign a new significance. Washoe did this with the sign "dirty," which she initially used only to refer to soiled items and feces. One day I took Washoe into a building where monkeys live in order to teach her a new sign, "monkey." While I prepared a data sheet for the training, Washoe began actively threatening a macaque, and the monkey responded in kind. As soon as the data sheet was ready, I stopped the fight and had Washoe turn around to face a siamang in another cage. I asked her in sign language what it was and she did not respond. After I molded her hands into a "monkey" sign three times, she began to refer to the siamang as "monkey." I interspersed questions about other objects—shoes, pencils, and dolls—with questions about the monkey in order to make sure she had learned the new sign. Then, I had her turn to a cage with some squirrel monkeys in it, and she readily

Lucy pays rapt attention to teacher Fouts during a morning break for tea. Later, they will continue her lesson, with breaks for romps and tickles.

More agile than any child, Salome joyfully leaps from her stroller, *opposite page*, and races away to hug Farkle, the family dog. Salome likes to play choo-choo train with her 15-month-old "sister," Robin, and share a hugfest with their mutual "mama."

called them "monkey." After she responded several times to other questions, I pointed to the macaque she had fought with earlier and asked her, "What that?" Her answer was "dirty monkey." Since that time, Washoe has also used the "dirty" sign as an adjective for people who do not do what she wants them to, and it seems to fit the situation very well. When I refused "Out me," a request from Washoe to be taken off the island, she called me "dirty Roger." On another occasion, she called me "dirty Roger" when she asked me for some fruit and I told her I did not have any.

When I teach sign language to the institute chimpanzees that are living in human homes, it is amusing to observe not only how these puckish rain-forest dwellers must adjust to *Homo sapien* surroundings, but also how the surroundings must sometimes be adjusted. Everything in a chimpanzee's environment seems to hold no purpose other than being a potential toy or interesting object to explore or take apart. To a chimpanzee, for example, curtains are not objects for screening harsh sunlight, but rather excellent material to climb and swing from. As a result, the chimp's human home usually has no curtains, drapes, or other potential "branches." And there are many locks on doors and other places that must be protected from the chimpanzee's inquisitive nature.

Excluding adjustments for the chimpanzee's character, however, the homes are very similar to those of most human infants. There are also the same concerns with toilet training, table manners, eating the proper food, brushing teeth, and discipline. The chimpanzees sleep in the same room with their human parents, often eat their meals with the family, go on outings with them, and, of course, play children's games with their foster brothers and sisters.

I spend several hours a week with Lucy, a bright pupil. She was born on Jan. 18, 1966, and has been living in a country home with

Ally offers to share a bite with Talbot, the family cat. The 3½-year-old chimp eats meals at the table like a human child.

Maurice and Jane Temerlin and their teen-age son Steve for the past seven years. I started teaching her sign language when she was 4 years and 8 months old. Every weekday at about 9 A.M., I arrive at Lucy's house for a two-hour lesson. After greeting her with tickles, I get out several objects for which she knows signs—dolls, strings, brushes, combs, shoes, pipes, beads, and plastic flowers—and we talk about them for a while in sign language. Next we go to the kitchen, where I make each of us a cup of tea and chat with her about it. When we return to the living room, we play games she enjoys and that she will make the signs for. These include tickle games or games where she gives me commands to do such things as comb my hair ("Comb Roger") or pretend to swallow a pencil ("Swallow Roger"). Once a week I drill her on all the signs in her vocabulary.

When she feels particularly mischievous or energetic or becomes bored with a lesson, a short tussling or tickling session is in order. Then we resume her education. After about an hour, Lucy will ask to go outside ("Out Lucy"). She usually wants an automobile ride ("Hurry open car"), during which she particularly enjoys seeing dogs. After the ride, we return for some outdoor play—tree swinging, tickling, and chasing—and then go back into the house for more tickling or such quiet activities as grooming each other's hair, during which she signs, "Pick there." Before I leave, I oil her feet and put her in her room.

During the session, I record all the signs she makes and the situation in which she makes them. I also note her errors and the signs she does not make correctly—sloppily formed signs, which are comparable to mispronounced words. She now has a vocabulary of nearly 90 signs, and, like the other chimpanzees, she combines them into meaningful phrases. Lucy has even invented a sign for the leash we use when we take her for walks. She formed the sign by placing her hooked index finger on the side of her neck. I found no sign for leash in an ASL dictionary, so I accepted hers. Lucy's invention might be comparable to a child inventing a new word.

Late in 1972, I completed a study with Lemmon and psychologist Roger Mellgren that revealed Lucy's conceptual ability. We gave her 24 different fruits and vegetables that she was free to handle or, if she liked, eat, and asked her what they were. The situation was similar to playing a naming game with a young child. She already had five food-related signs in her vocabulary: "food," "fruit," "banana," "drink," and "candy." During the first four days of tests, we gave her the foods to see how she would refer to them. Then, on the fifth day, we taught her the "berry" sign, using a cherry as a stimulus. We wanted to see if she would keep the berry sign specific to cherries, in the same way she uses the banana sign to label only bananas. Or would she generalize the berry sign to other fruits in the way she uses the fruit sign for all kinds of fruit? The berry sign turned out to be highly specific to cherries; and she did not apply it to strawberries, blackberries, radishes, cherry tomatoes, grapes, kernels of corn, peas, or raisins, as we thought she might. When we tried to train her to call a blackberry a "berry," she resisted. Apparently she thought that only cherries belonged to a special class labeled "berry."

Her responses to other fruits and vegetables, however, enabled us to begin to understand how she sees the world. For example, she called a piece of watermelon "candy drink" or "drink fruit," and signed "Lucy smell drink food" when she saw a cherry tomato. She often labeled the citrus fruits (grapefruits, lemons, limes, and oranges) as "smell fruits," probably because of their pungent odor. When she bit into an old radish, which she had been calling "food" or "fruit," she suddenly changed its name to "cry hurt food." Sweet pickles and celery often elicited the "pipe" sign from her, perhaps because they are eaten by inserting an end into the mouth and removing the remaining portion in much the same manner as a smoker handles a pipe. Throughout the study, she referred to vegetables as "food" more often than as "fruit," and to fruit as "fruit" more often than as "food." This study revealed that chimpanzees can classify things in their environment as opposed to simply labeling specific objects. It also appears that they can combine various signs in their vocabulary to describe a novel object, such as calling a radish "cry hurt food."

As Washoe nuzzles her doll, scientists wonder if she will try to teach her future baby to use sign language.

The youngest chimpanzee I work with, Salome, is being raised in the home of Churchill and Susie Blakey and their infant daughter Robin in Oklahoma City. Salome was born on July 7, 1971, and learned her first sign, "food," when she was 4 months old. By April, 1973, she already knew 15 words and was combining them in short phrases such as "Give food," "Give drink," "Tickle me," and "You tickle me." Since Salome is an infant, we are obtaining data on how an infant chimpanzee learns sign language at various developmental stages.

Ally, my third home-raised pupil, was born on Oct. 15, 1969, and lives with Sherri Roush in Norman. I have been teaching him sign language since he was a little more than a year old, and his vocabulary now includes more than 70 signs. In January, 1973, William Chown and Lawrence T. Goodin, graduate students in psychology, and I examined the relationship between his understanding of vocal words and gestural signs because he, like Lucy and Salome, constantly hears spoken English in his home. Initially, we used the vocal word "spoon" to refer to a spoon. Then we removed the spoon and taught him to respond to the vocal word for spoon with its sign, one he did not previously have in his vocabulary. One of us would tell him, "Sign spoon," or would sign "Sign" and then say the word "spoon." We helped Ally learn the new sign by saying "spoon" while molding his hands into the proper "spoon" sign. Later, another person who did not know what sign we had taught Ally asked him in sign language to give the signs for several objects that were in the room, including a spoon. Ally made the correct sign for spoon. Eventually, he learned 10 signs for corresponding vocal words and then spontaneously transferred these signs to the objects that the vocal words represented.

These results indicate that Ally can understand spoken words and translate them into signs when the objects they represent are not visible. This is contrary to a popular belief that only human beings can understand spoken words and that other animals respond only to cues in the environment, not the actual words. Ally's learning to make signs for vocal words bears a strong resemblance to the manner in which human beings learn a foreign language in school. For example, a student in Spanish class is taught that the Spanish word "sombrero" and the English word "hat" refer to the same object. If a student learns this, he or she can label a hat "sombrero."

New research is revealing that other apes can also learn ASL. In an exploratory study with Tommy, an infant orang-utan, I have found that he can label objects with signs and combine these signs into two-sign combinations. As of May, 1973, Francine Patterson, a graduate student in psychology at Stanford University, had taught signs to Koko, a 19-month-old female gorilla being raised in a zoo, and Koko was using them in short combinations. Once scientists have thoroughly studied how apes use sign language, we may well find the rudiments of language in lower primates, such as monkeys.

Now that some dreams and fantasies about talking to animals have been realized scientifically, what does the future hold? Research at the institute will continue in the same general vein, but we hope that ultimately it will go far beyond this. For example, if we can teach sign language to a whole colony of chimpanzees and they use it among themselves, we will be able to move into the study of culture and perhaps find answers to a whole series of questions: Will these chimpanzees teach sign language to other chimpanzees? Will they do this by imitation or by tutoring their infants? Will this result in any changes in their behavior? Will they begin to approach human culture by passing knowledge from one generation to the next? The possibilities are endless. All that is required of the researchers is an inquisitive attitude that matches that of the chimpanzees.

For further reading:

Bronowski, J., and Bellugi, Ursula, "Language, Name, and Concept," *Science*, May 8, 1970.

DeVore, Irven, "The Ways of the Primates," *Science Year,* 1971.

Fouts, Roger S., "Acquisition and Testing of Gestural Signs in Four Young Chimpanzees," *Science,* June 1, 1973.

Gardner, R. Allen, and Gardner, Beatrice T., "Teaching Sign Language to a Chimpanzee," *Science,* July 15, 1969.

Hahn, Emily, "On the Side of the Apes," *New Yorker,* April 17, 1971, and April 24, 1971.

Hahn, Emily, "Washoese," *New Yorker,* Dec. 11, 1971.

Hewes, Gordon W., "Primate Communication and the Gestural Origin of Language," *Current Anthropology,* February-April, 1973.

Premack, Ann James, and Premack, David, "Teaching Language to an Ape," *Scientific American*, October, 1972.

The Power Above,
The Power Below

By John F. Henahan

**With traditional sources of energy rapidly dwindling,
scientists have begun to explore two promising options**

At the southern end of the Rocky Mountains, about 20 miles west
of Los Alamos, N.Mex., engineers are drilling deep into rocks at the
edge of an extinct volcano. Although they are using Geiger counters
and oil-drilling equipment, they are not seeking metals or petroleum.
Instead, they are looking for sources of heat that they can mine to boil
water that will drive steam turbines.

In northern Arizona, optical physicists are testing strange mirror
devices that focus the rays of the sun, again, as a heat source to drive
turbines. Both groups are after the same thing—to develop new sources
of electric power that will help solve the growing energy crisis that is
gradually affecting almost every part of the world.

The traditional fuels—natural gas, fuel oil, and gasoline—are in
short supply. Our rising utility and service station bills verify this, as
do the fuel suppliers themselves with unprecedented advertising that
tells us how to use less fuel. Worse yet, during 1973 there was a threat
of widespread gasoline rationing.

Large quantities of coal are still available, but coal is a serious polluter, and the development of coal gasification plants, in which coal is converted into clean-burning gas, lies some distance in the future. The construction of nuclear-power plants, once seen as the panacea for all our power demands, has lagged because of the public's concern with the danger of radioactive pollution. The most promising of such plants—fusion reactors, which would make use of the fusion of hydrogen nuclei, and breeder reactors, which would produce their own radioactive fuel—are not yet technically feasible. However, because we lack other immediate energy alternatives, it now appears that nuclear fuels and coal must provide most of the energy that will be needed for the next 25 years.

Accordingly, most of the $662 million listed in the 1973 federal budget for energy research and development has gone into research on nuclear power and fossil fuels. However, $7 million has been dispensed primarily for research and development on what many experts consider to be two other prime energy options for the future—geothermal energy trapped below the earth's surface, and energy from the sun.

Geothermal energy is, literally, the earth's heat. It is produced when molten rock, heated by the decay of radioactive elements, squeezes up from the earth's mantle and heats a layer of porous rock saturated with water that has seeped down from the surface. Because such porous rock is capped by layers of less-permeable rock, the water is under high pressure and cannot boil, even though its temperature may be from 200 to 300° F. higher than the boiling point of water at the earth's surface (212° F.). Wherever there are natural crevasses, or openings, the water converts into steam, usually mixed with hot water, and rushes to the surface as geysers and hot springs.

Trapped geothermal energy is tapped by drilling a well through the less-permeable rock into the porous rock layer or a deep natural crevasse just above the porous rock. The well releases the pressure on the underground water, and steam or steam and hot water spurts to the surface at velocities of up to 740 miles an hour. At the surface, the steam is fed into conventional power turbines to generate electricity.

Geothermal power is already used in various parts of the world. In Russia and New Zealand, it produces electricity for both domestic heating and air conditioning. In Italy and Japan, it heats greenhouses and water reservoirs on fish farms, and has various industrial uses. In northern Chile, the government has drawn up plans for a geothermal power plant at the Tatio Geysers in the Andes Mountains. It will use the geysers' steam to generate electricity and then condense the steam to make fresh drinking water. Chile also plans to extract useful minerals, such as salt, from the geyser water.

In 1972, geothermal steam accounted for an electrical capacity of about a million kilowatts throughout the world, enough to supply the industrial and household needs of a city of 4 million persons. Geothermal power contractors expect that the capacity will be quadrupled by

The author:
John F. Henahan is a science writer and former editor of *Chemical and Engineering News.*

The World's Geothermal Resources

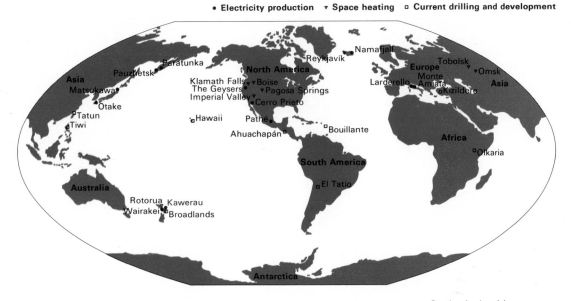

Geological evidence shows many potential areas around the world where the heat of the earth can be tapped for power, though few have yet been utilized.

1985. That would place geothermal energy production at the present level of nuclear energy. However, the potential is great. For example, Donald E. White, a geologist with the U.S. Geological Survey Laboratories in Menlo Park, Calif., estimates that if only 10 per cent of the geothermal energy available in the upper 2 miles of the earth's crust could be tapped, it would provide almost 60 million kilowatts of electricity for the next 50 years, more than enough to meet the energy needs of all the people in the United States.

Geothermal tapping is possible only where the interior heat of the earth has welled up near the earth's surface. On a worldwide basis, such geothermal hot spots lie in areas where earthquakes and volcanoes occur. Geological prospectors use various techniques to locate these hot spots. Usually they dig test wells and determine how rapidly the heat increases with depth. For example, a vast geothermal field was found under the Imperial Valley east of San Diego, Calif., in this way. Geologists found that the temperature in some wells drilled there may increase more than 10° F. per hundred feet of depth, compared to an increase of only 1 to 2° F. in nongeothermal areas. Geothermal zones can also be spotted from the air with infrared photography; the hot spots appear as white areas on a dark-gray photograph.

Power companies consider dry steam—steam that emerges from the earth unmixed with hot water—to be the most economical and easy-to-use form of geothermal energy. Dry stream from the Larderello geothermal wells in west-central Italy has supplied several small towns with electricity since 1904. Union Oil Company, which operates dry-steam wells about 75 miles north of San Francisco, has increased the area's output of electricity 200 per cent since 1960. New wells drilled

between now and 1980 should increase the electrical capacity there from the present 247,000 kilowatts to more than 1 million kilowatts.

One disadvantage of geothermal steam is that it loses pressure as it rises to the earth's surface. The steam exerts from 750 to 1,000 pounds per square inch (psi) at the surface, while the steam produced by boiling water in conventional power plants exerts 3,000 psi. Geothermal steam cannot produce as much energy, but since the steam is free, power company officials say that geothermal-power plants can be built and operated at about half the cost of nuclear-power plants and about three-fourths the cost of fossil-fuel plants.

Dry-steam fields are a relatively rare geological phenomenon. However, there are 20 times as many wet-steam fields, where hot water rises with the steam. Proponents of geothermal energy believe that these wet-steam fields may ultimately be the real energy bonanza. Robert Rex, geologist and vice-president for exploration of Pacific Energy, Inc., Los Angeles, estimates that the wet-steam fields under the Imperial Valley contain enough steam to produce a constant supply of from 20 to 30 million kilowatts. This would meet the electric power needs of the population in the American Southwest for two or three more centuries.

Several leading oil companies and utilities are now drilling exploratory wells in the Imperial Valley in search of steam. A Mexican plant is producing electricity with steam from 19 wells in what is believed to be an extension of the valley's geothermal reservoir. This small, 75,-000-kilowatt plant at Cerro Prieto, near Mexicali, will eventually produce electric power for all the northern section of the Baja California peninsula, which has a population of about 1 million.

Progress north of the border in the Imperial Valley has been hampered by the high salt content of the geothermal fluids, which causes corrosion, as well as environmental and handling problems. This salinity has not been much of a problem at Cerro Prieto because the hot brines that rise with the steam there are merely dumped into ditches that drain into the Gulf of California. Such a practice would damage valuable adjacent croplands in the Imperial Valley, so the salty water will be pumped back into the earth through injection wells. Once returned, the brines may be reheated by hot subterranean rock and the steam that is produced will be used again. Reinjection might also help to prevent sections of the valley from sinking because of the removal of the water, as has occurred near Cerro Prieto.

The San Diego Gas and Electric Company has drilled four exploration wells near Niland, which lies at the south edge of the Imperial Valley's Salton Sea. If these wells prove promising, they and other wells will feed a small 10,000-kilowatt power plant that is equipped with newly developed techniques for coping with the highly corrosive brines. Instead of separating the steam from the hot water, the plant

High pressure geothermal steam rushes from a safety valve in the pipe that carries it to a power plant turbine at Cerro Prieto, Mexico.

Steam from Wairakai geothermal fields on North Island, New Zealand, is used to generate electric power for a paper-manufacturing plant.

Three Power Sources

Hot magma deep in the earth heats water that seeps down from the surface and collects in the permeable rocks below. Hot water and steam under pressure then rise to the surface through natural cracks and form geysers. Where there is no natural water source, cold water may be pumped down through a hole drilled into the earth. Steam then rises through a second hole drilled nearby.

Hot Water

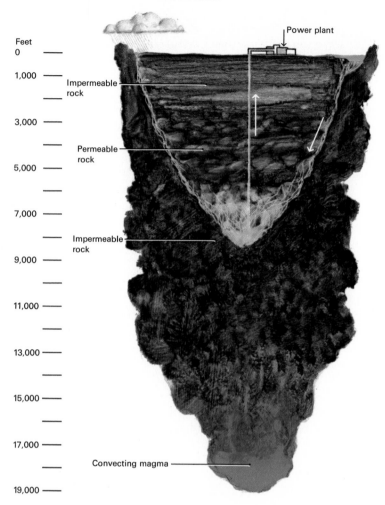

Feet
0

1,000

3,000

5,000

7,000

9,000

11,000

13,000

15,000

17,000

19,000

Power plant

Impermeable rock

Permeable rock

Impermeable rock

Convecting magma

will keep the brines under pressure and pass them in pipes through a heat exchanger made of a corrosion-resistant titanium alloy. Heat from the exchanger will vaporize a low boiling fluid, such as isobutane. This vapor, instead of steam, will drive a turbine.

The engineers working near Los Alamos are drilling into a geothermal reservoir that contains neither steam nor hot water, only hot rock. These researchers, from the Atomic Energy Commission's Los Alamos Scientific Laboratory, plan to produce steam by pumping cold water down a 25,000-foot well into the hot, dry rock area. There it is expected the cold water will create a "fracture zone," a crisscrossed network of cracks and fissures about 3,000 feet in diameter. The researchers believe the water will spread out through the cracks and absorb heat from the rocks. Still under pressure, the hot water would then be forced to the surface through a second well drilled into the top edge of the fracture zone. Once a suitable temperature differential is

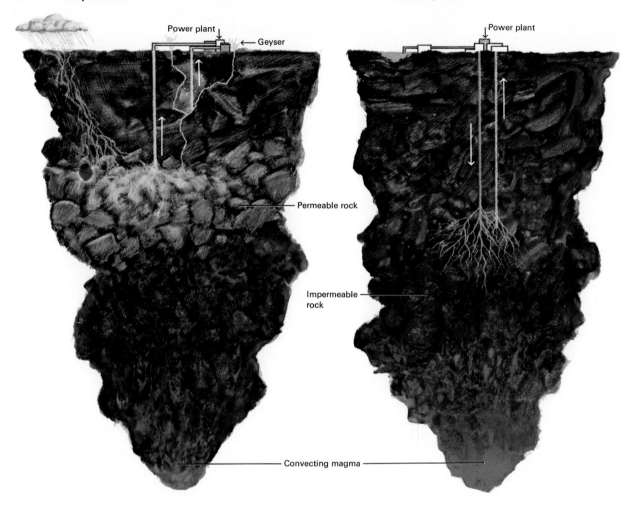

Dry Steam

Power plant

← Geyser

Permeable rock

Impermeable rock

Convecting magma

Hot Dry Rock

Power plant

created between the cold water entering the system and the hot water leaving it, the engineers believe that the pump could be turned off and the cycle would run itself by natural convection.

The first experiments, to see if the water would actually create a fracture zone, were carried out successfully in April, 1973. They produced a fracture zone about 140 feet in diameter at a depth of 2,500 feet, where the rock temperature is only about 210° F. The researchers are confident that they will also get good results at the much deeper levels in the earth where the higher temperatures required to generate electricity are found.

Preliminary economic estimates "make us feel optimistic," says Morton Smith, a metallurgist at the Los Alamos Scientific Laboratory. He points out that drilling into dry rock could make geothermal energy available in many areas where geothermal hot water or steam do not exist. Such areas include the uranium-heated granites beneath

A solar furnace built in the Pyrenees Mountains by French scientists has 63 mirrors that collect the sun's rays. It can produce temperatures up to 3800° C. (6872° F.). The aluminum roof of a house in a Washington, D.C., suburb traps the sun's rays and converts them to heat the house.

the White Mountains of New Hampshire and the volcanic rock beneath Hawaii. It could prove exorbitantly expensive to drill deep wells in some of these areas. However, Rex says that the heat available from rock that is close enough to the surface to be tapped economically by present methods could satisfy energy needs in the United States for "at least several thousand years."

To spur further geothermal development in the United States, a panel of the National Science Foundation headed by former Secretary of the Interior Walter J. Hickel has recommended a federal expenditure of $684.7 million to raise production of geothermal energy from 300,000 kilowatts in 1973 to 395 million kilowatts by the year 2000. For its part, the federal government is prepared to lease 100,000 acres of federal land throughout the country to geothermal developers if legal questions—such as rights for mineral development—at the state, local, and national levels can be resolved.

However, most power and utilities companies approach geothermal energy cautiously. They point out that geothermal-power plants must be built in areas where geothermal hot spots occur, and the loss of electricity during transmission is always a problem. Other power plants can be built where the electricity is needed. In addition, power company executives still view geothermal energy as an unproved resource. They seem to be counting on nuclear, and possibly coal, plants and new methods of recovering natural gas and petroleum to get us through the energy crisis. In addition, geothermal energy is not environmentally pure. Geothermal wells release sulfur dioxide from the earth, one of the same atmospheric pollutants given off by coal-burning power plants. At the same time, no one really knows whether taking fluids from the earth and then pumping them back in will trigger earthquakes in areas that are already earthquake prone. Pumping fluid wastes into deep holes near Denver in the 1960s apparently caused minor quakes in that area.

The deep earth as a potential energy source is rivaled by the sun. In fact, solar energy has an advantage over geothermal energy—it is pollution free, and it offers an almost unlimited power potential. The thermal energy from the sun's rays falling onto the earth is about 5,000 times greater than the world's usable geothermal capacity. However, the sun's energy is so diffuse that it must first be collected and concentrated before it can be used. In addition, the solar energy available at any one time is limited by the weather and the length of the day. Nevertheless, solar energy could help greatly to ease the energy crisis if it could be converted into electricity and ways could be found to store it on an unprecedented scale.

It has already been used on a small scale to heat and distill water, to cook food, and to heat homes. In addition, millions of photovoltaic converters, or solar cells, that transform sunlight directly into an electric current have been used in the space program because they provide a reliable source of electricity for many years. Unfortunately, these

Solar Power Collector

Solar device on the roof of a building at the University of Arizona was built by Aden and Marjorie Meinel. It heats compressed air within a steel pipe to 900° F.

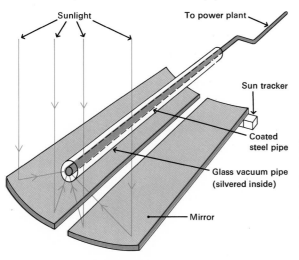

Solar farm of the future would use thousands of Meinel devices spread over vast areas of the desert and semidesert cattle-grazing lands.

devices, usually made of thin slices of highly purified silicon, are prohibitively expensive for wide use at the present time.

On a large scale, water could be heated to its boiling point by concentrated solar energy and then fed as steam into a power turbine. This approach is suggested by Aden Meinel, director of the Optical Science Center at the University of Arizona in Tucson, and his wife and co-worker, Marjorie. They propose to build huge "solar farms" to harvest electricity from the sunlight that falls on marginal grazing land and semideserts in the American Southwest.

The Meinels say that if they could cover 5,000 square miles of land in the Southwest with panels to collect energy from the sun's rays, the complex would produce a billion kilowatts of electricity—enough to

meet all of this country's energy needs by the year 2000. The area, of course, is large—equal to several counties in some states. But Meinel says that this area could be distributed over the six to eight states that now have abundant sunshine. He adds that strip-mining operations used to service coal-burning power plants would ruin a much greater area during the same period.

The Meinels also suggest that heat left after electricity has been generated would be used to remove the salt from the area's brackish water, providing large quantities of potable water. As an added bonus, the rain water and dew that would run off the huge panels could be collected and used to irrigate the adjacent lands, converting normally desert areas into grazing land for livestock.

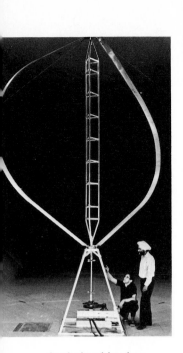

A wind turbine has been developed by two engineers in Canada. Unlike traditional windmills, it rotates around a vertical axis. Large models may be used to produce electricity for lighting and heating cottages in the Arctic. At Rance River estuary in France, ocean tides drive generators that provide electric power for many nearby towns.

To prove that their idea is feasible, the Meinels have erected what they call a "credibility model" on the roof of the university's Optical Sciences building. The model has two long mirrors placed parallel to each other to form a channel. An evacuated glass tube containing a steel pipe passes through the channel. The glass tube is silvered on the inside to reflect the light as does the mirror. Sunlight, concentrated by the mirrors, enters the glass tube through narrow, unsilvered lines along its sides. The light is then reflected by the silvered walls of the tube into the steel pipe, heating compressed air that flows through the pipe. The air can be heated to about 900° F. by the concentrated sun's rays, and presumably could boil water to produce steam. The Meinels hope to improve their device until it can heat the contents of the pipe to 1000° F., which is the standard temperature to which steam is raised in a high-pressure turbine.

One way to do this is to trap the infrared part of the sunlight that now escapes. To do this, Meinel is experimenting with ultrathin coatings about 1/100,000 of an inch thick. He is particularly interested in coatings composed of a layer of silicon backed by a metal, such as gold, silver, or aluminum, that absorbs infrared radiation. These coatings are placed around the steel pipe with the silicon side out. Silicon absorbs most of the sunlight, converting it into heat, while the infrared rays, which pass through the silicon, are absorbed and turned into heat by the metal part of the coating on the opposite wall of the tube. Meinel uses compressed air in the pipe in his model because it is readily available and easy to handle. In the large-scale solar farms he proposes, molten sodium nitrate would be used because it is a more efficient heat exchanger and holds the heat much longer, allowing it to be stored for use at night and on cloudy days.

Meinel estimates that it would cost $1 billion annually for research and construction during the next 10 years to perfect his idea and build solar farms that could provide the bulk of our electric power. This compares favorably, he says, to the $17 billion that the utilities now invest in other types of power plants every year. Meinel admits that the cost of generating electricity from solar energy would probably boost electricity costs about 30 per cent higher than they are now. However, he believes the economic disadvantage would not be as serious as it appears because the cost of electricity produced by conventional methods is certain to rise sharply in coming years.

Many scientists have doubts about solar energy, geothermal energy, and other "exotic" energy options. They suggest that the research money and time should be spent on more practical problems, such as extracting gas and gasoline from coal. Ralph Lapp, who is a nuclear physicist and board member of Quadri-Science, Incorporated, says that the continuing increase in our demands for energy, coupled with the uncertainties in geothermal and solar technology, make it improbable that any major development of these resources could occur before the year 2000. "It is just economically not feasible to put our research and development funds into uncertain sources of energy when our immediate needs are so staggering," he says.

Alvin M. Weinberg, director of the Atomic Energy Commission's Oak Ridge (Tenn.) National Laboratory, says that the proponents of the exotic-energy schemes are much too optimistic about getting the costs down. He believes that the eventual cost per kilowatt-hour may be five times what the solar power experts have figured. Nevertheless, Weinberg believes that the experiments to develop these sources, especially geothermal energy, should be continued.

Perhaps the work of the engineers and physicists who favor the innovations in energy development will someday lead to significant sources of power. In any case, because of the time it takes to develop any new energy option, we must pursue such studies more seriously if we hope to tap the power of the earth or sun before most of the fuels we are now using disappear.

For further reading:

Fallermayer, E., "Energy Joyride Is Over," *Fortune,* September, 1972.

Henahan, John, "Full Steam Ahead for Geothermal Energy," *New Scientist,* Jan. 4, 1973.

Hoke, John, *Solar Energy,* Watts, 1968.

Lapp, Ralph E., *The Logarithmic Century,* Prentice Hall, 1973.

Thirring, Hans, *Energy for Man: From Windmills to Nuclear Power,* Torchbooks, 1962.

Weaver, K. F., "Search for Tomorrow's Power," *National Geographic,* November, 1972.

X Rays
From the Sky

By Lee Edson

**The Uhuru satellite's historic sweep through space
has gleaned remarkable discoveries about the universe**

On the morning of Dec. 12, 1970, as the blood-red sun beat down on the ancient Kenyan fishing village of Ngimini, natives went about the same daily chores their forebears had performed for many centuries. Men swished dead branches at the backsides of reluctant, scrawny cows being led to pasture. Women balanced the family wash on their heads and clutched babies wrapped in colorful blankets. Naked children played noisily in front of thatched-roof huts.

But there was a new element in this placid picture on that December morning. Three miles out in the Indian Ocean, there was a rumble in the air, and a cushion of flame flared out from a steel tower. As the African villagers paused to watch, a Scout rocket belonging to the National Aeronautics and Space Administration (NASA) soared gracefully into the sky. It carried the world's first X-ray satellite into an orbit about 300 miles above the equator. Launched on the seventh anniversary of Kenya's independence, the satellite was named Uhuru,

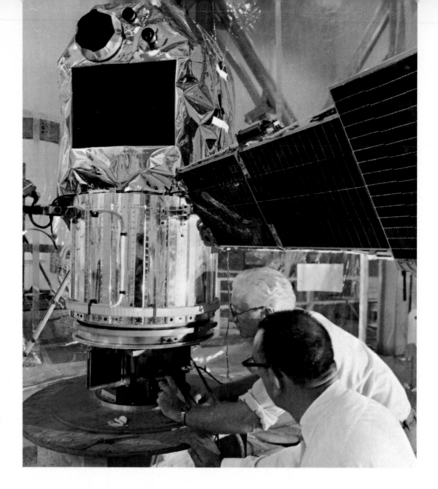

Technicians complete final adjustments during magnetic checks on the Uhuru satellite at NASA's Goddard Space Flight Center.

The author:
Lee Edson is a science writer whose articles have appeared in many magazines and books. He wrote the article "The Search for a Human Cancer Virus" in the 1973 edition of *Science Year.*

the Swahili word for freedom. It was to become one of the most significant instruments in recent scientific history.

More than two years later, Uhuru–officially known as SAS-A (small astronomy satellite A)–was still orbiting the earth, collecting and measuring X rays from outer space, and searching out violent events as far away as the edge of the universe. "Uhuru has dramatically opened up a new way to tell us things about the universe that were unavailable to the relatively limited instruments of earth," says astronomer Riccardo Giacconi of American Science & Engineering, Incorporated (ASE), Cambridge, Mass. He conceived and directed the Uhuru project.

"Until rockets began to probe outer space, the secrets of the sky could be read only by optical and radio telescopes," Giacconi said. "Most of what we know about the dynamics of the universe–the birth and death of stars and galaxies–came to us only through visual and radio waves. Of course, we knew theoretically that the cosmos radiates electromagnetic energy in all wave lengths, from short gamma rays and X rays on one end of the spectrum to the long radio waves on the other end. But the earth's atmosphere absorbed or reflected most of the waves. A good thing, too, because if X rays and ultraviolet rays could penetrate the atmosphere and reach the earth's surface, they would tear up living molecules and destroy life. But while the atmosphere has saved our lives, it has also thwarted much of our understanding of outer space.

"Rockets have changed all that," Giacconi continued. "Operating beyond the atmosphere, they have enriched our knowledge of galaxies we cannot see by other means, of hot gases at temperatures of millions of degrees, of giant explosions in remote regions where quasars roam. We have a whole new catalog of stars we did not know existed before. There is no doubt of it—Uhuru has ushered in a new age of deep-space astronomy based on X rays."

Despite their exotic origins, space X rays belong to the same family of high-frequency radiations that serve us in familiar medical applications. Both are usually the result of swift changes in the velocity of fast-moving electrons. The sudden jar causes the tiny charged electrons to send out radiation in the form of an X ray, in much the same way that a hammer striking an anvil sends out a sound wave.

In space, X rays can be generated in a number of ways. One is thermal generation. Matter becomes highly agitated when it is heated to extraordinarily high temperatures, and the fast-moving electrons lose energy in the form of X rays. On earth, only a nuclear explosion generates enough heat to produce X rays. Thermal generation is also known as black body radiation. A unique thermal mechanism of X-ray production that exists only in space occurs whenever electrically charged particles fall into the deep gravitational wells created when stars exhaust their fuel, die, and collapse into themselves. X rays can also be generated nonthermally by whipping electrons to velocities approaching the speed of light and jetting them through the turning forces of a magnetic field. This process is known as synchrotron radiation. It occurs in space and also on earth in betatrons, synchrotrons, and other circular high-energy accelerators.

The first American spacecraft launched by another country, Uhuru heads for space from an Italian mobile launch platform off the coast of Kenya. It circles the globe 335 miles above the equator once every 96 minutes.

Scientists' first practical inkling that the sky was shedding X rays came in 1949, when Herbert Friedman of the Naval Research Laboratory (NRL) in Washington, D.C., widely regarded as the dean of rocket astronomy, sent V-2 rockets aloft in experiments dealing with communications research. One of these rockets climbed about 50 miles and gathered data showing that the sun was a powerful emitter of X rays. This information surprised Friedman and his colleagues. They knew the sun's surface was hot enough, 5500°C. (about 10,000°F.), to vaporize any metal but not nearly hot enough to produce energy in the X-ray region. So a small mystery developed that was not resolved until the researchers turned their attention to the sun's corona, a blazing halo that stretches and fumes for millions of miles along the sun's circumference.

The corona shoots out tongues called flares, and Friedman decided to use a rockoon—a combination of a rocket and a balloon—to take a close look at this phenomenon. In 1956, he and his colleagues lofted the first rockoon from a ship in the Pacific Ocean. They triggered the rocket when colleagues, watching the sun through a solar telescope at the High Altitude Observatory at Boulder, Colo., signaled the beginnings of a flare. Freed from the balloon, the rocket streaked above the earth's atmosphere, capturing X rays through an ultrathin metal foil window. Mounted behind the window was a Geiger counter that recorded X-ray energy and radioed the data back to the earth. The data showed that a jump in X-ray activity could be tied to the solar flare. The scientists concluded that the sun's corona must reach a temperature as high as 50 million°C. This temperature is a measure of the energy of plasma particles in the corona. The X rays are produced when these energetic particles collide.

To Friedman, however, the data showed something more–that high-energy X rays were also being emitted in the absence of a solar flare. These might be coming from somewhere beyond the sun.

But from where? Part of the answer came in June, 1962, when Giacconi and his colleagues at ASE and Bruno B. Rossi, then of the Massachusetts Institute of Technology (M.I.T.), lofted an Aerobee sounding rocket with a powerful X-ray detector 130 miles into space. They were looking at the fluorescence on the moon, which is caused by solar X rays, as a means of analyzing the chemistry of the lunar surface. But during its four minutes of flight, the rocket had a full view of other regions in the sky and startled the researchers by reporting a far more powerful source of X rays than they had anticipated. The source seemed to lie in a cluster of stars along the galactic plane, the line that cuts across the heart of the 100 billion stars that make up our Galaxy, the Milky Way. Here was the first strong evidence of X rays beyond

The Rotating Observatory
Uhuru rotates slowly as it orbits, and detectors sweep a band 5 degrees wide in the sky. The closely stacked slats of the collimator act like a Venetian blind to limit the view. X rays that enter produce electrical pulses in the proportional counter and these are relayed to computers on the earth.

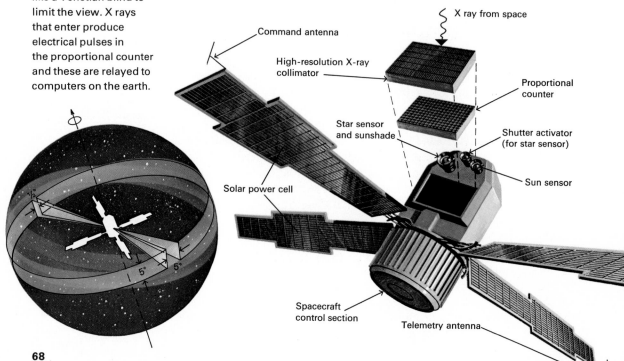

our solar system. Giacconi presented his findings to an X-ray analysis conference at Stanford University in August, 1962. This talk marked the quiet, almost obscure birth of X-ray astronomy.

Back at the NRL, Friedman probed the sky with rocket-mounted detectors 10 times more sensitive than those he had used earlier, and pinpointed two X-ray sources. One was Scorpius X-1—one of the brightest X-ray objects in our Galaxy. No indication of a bright visible star was found in the area, however. But in June, 1966, thanks to earlier brilliant detective work by ASE astrophysicist Herbert Gursky, an optical counterpart—a faint blue star of the 13th magnitude—was found by astronomers at Tokyo University and confirmed at the Palomar Observatory in California a few nights later.

Not too surprisingly, Friedman also found another strong source of X rays, the Crab Nebula. This great, lacy cloud of gas in the constellation Taurus, roughly 6,000 light-years from the earth, contains the remnants of a massive supernova explosion so enormous that it was seen in daylight in A.D. 1054, long before the invention of the telescope. This highly studied cloud embraces virtually all the puzzles of the cosmos in its glowing body, and from earth is one of the brightest radio transmitters in the heavens. It had been found to generate other kinds of radiation, so why not X rays? Friedman studied the Crab during the five minutes that the moon eclipsed it in July, 1964, an event that occurs every nine years. He found that the X rays were generated in an extended region one light-year long—one-sixth the total length of the nebula. This was the first experiment that linked an X-ray source with a previously known celestial body. "We were elated," Friedman recalls. "We felt we were helping in establishing the beginnings of a new science."

With interest in X rays burgeoning among astronomers and with NASA approval and funding, Giacconi and his associates at ASE began to design and construct a satellite that could observe the sky far longer than the few minutes eked out by the Aerobee and other sounding rockets. That satellite is Uhuru.

To see how Uhuru was getting along, I visited its earthly corporate home at ASE headquarters, a mile or so from M.I.T. On the surface, the efficient, modern ASE offices look like they might be processing life insurance. But the walls display charts and pictures of astronomical phenomena and the air is full of talk about X-ray astronomy. There is also a full-sized prototype of the X-ray-detecting satellite that is now orbiting the earth.

Harvey Tananbaum, an astrophysicist who has shared a number of X-ray discoveries with Giacconi, showed me the ever-growing stack of computer data received from Uhuru. The data are received at Quito, Ecuador, transmitted to Goddard Space Flight Center in Greenbelt, Md., and then sent to Cambridge for analysis. Physicists and other scientists pore over the computer readout charts, searching among the peaks and valleys for telltale patterns.

Three Kinds of Cosmic X Rays

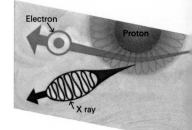

Thermal Bremsstrahlung occurs when hot plasma particles make speeding electrons change their directions. They give off energy as X-ray photons.

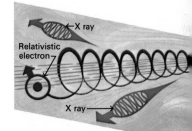

Synchrotron radiation requires a magnetic field that forces relativistic electrons into spiral paths. The spiraling electrons emit X rays.

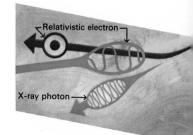

Inverse compton scattering occurs when photons below X-ray range gain enough energy from relativistic electrons to reach X-ray frequencies.

X Rays from an Evolving Binary System

An eclipsing binary system develops in five stages. The more massive of the two orbiting stars is the first to exhaust its nuclear material and swell. Some of its gas falls onto its smaller companion. The massive star's core collapses to a neutron star or black hole and blows off more matter to its companion. When the companion in turn exhausts its fuel, it swells, transferring gas back to the collapsed star. The streaming cloud collapses to a disk, *below,* from frictional forces in the gas.

While all this bustle takes place on earth, the 300-pound Uhuru continues to spin slowly on its axis, completing one full rotation every 12 minutes, as it moves through its 300-mile-high equatorial orbit. This orbit was chosen to avoid the interfering radiation in the Van Allen belts, two rings of charged particles that lie above the earth's atmosphere. As the satellite rotates, X-ray detectors sweep a 5-degree-wide band in the sky. Uhuru's spin and orientation can be controlled by radio command from the ground. Once a day, scientists change the spin axis so that the detector covers a different band of the sky.

As the satellite orbits, X rays strike a detecting device known as a proportional counter, a box filled with the rare gas argon and covered on one side with a grilled beryllium window. When an X ray penetrates the window, it is absorbed in the gas and produces an electron that carries off most of the X-ray energy. The electron smashes its way among the atoms in the gas and yields another electron each time there is a collision. The electrons accumulate at the anode wire in the box and are counted. Their number is proportional to the energy of the X ray. From such data, scientists can calculate the flux, or flow, of the X rays. At the same time the X rays are being measured, sun and star sensors in the satellite pinpoint the location of the X-ray sources. In this way they can later be checked against known optical and radio sources in the same part of the sky.

This continuous flow of data makes ASE a primary source of information for astronomers throughout the world. "The phone never stops ringing," says Tananbaum. While I was there, a call came in from astronomers in England who wanted the latest data from Uhuru, so they could reset their telescopes.

Uhuru's odyssey has produced some remarkable information about the X-ray sky. In the first few months, for example, Uhuru pinpointed many objects in areas where the star catalogs showed blanks. Yet, some of these sources—60 of which were found to lie in clumps along

The Collapsed Sources
Binary X-ray emissions from neutron stars and black holes are different. Hercules X-1, *below left,* a neutron star, rotates with a strong magnetic field. The hot gas in the disk can fall to the star only along the magnetic field lines. The X rays appear to pulse as the magnetic poles whirl about. Cygnus X-1, *below,* is a black hole. It has no magnetic field, so we see no pulsations. Gas streaming toward the black hole heats by friction and emits variable X rays.

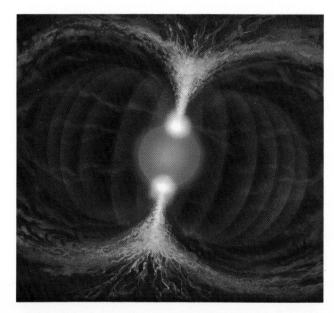

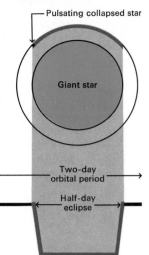

Pulsating collapsed star

Giant star

Two-day orbital period

Half-day eclipse

X-ray pulses from the collapsed star virtually vanish regularly for half-day periods, indicating that the smaller star has been eclipsed by its giant companion.

the horizontal plane of our own Galaxy—are from 10,000 to 100,000 times brighter in terms of X-ray emission than the sun. In fact, they are among the most energetic objects in our Galaxy.

Such high energies are linked to processes that occur after great explosions within the Galaxy, such as the supernova explosion that produced the Crab Nebula. A rapidly spinning neutron star, which is only about 10 miles in diameter, has provided the Crab's X-ray energy for 920 years after the explosion. Scientists agree that the outpouring of X rays from the hot plasma of the nebula seems to be produced by the synchrotron radiation process. That is, the X rays apparently are caused by electrons racing through a strong magnetic field. However, scientists do not agree on how the more recently discovered pulsing X rays from the center of the Crab are produced.

One speculation as to the origin of the pulsar's radiations, including X rays, was proposed by Thomas Gold of Cornell University, and called the "cosmic slingshot model." Because of the enormous compression of a neutron star, its magnetic field's strength may be increased by a factor of 10 billion. Gold suggests that far out from the spinning star, this rotating field would move almost at the speed of light. Particles thrown off by the neutron star and swept into the magnetic field spiral along the field lines, going faster and faster as they approach the outer parts of the spinning field. There they will give off beams of radio waves, light, and X rays. The spinning beams sweep the cosmos, causing us to see regular pulses if the sweeping beam happens to pass by us.

For three years, the Crab pulsar, known as NP-0532, pulsating 30 times per second, was the only one of the 89 known pulsars that had been found to emit X rays. In 1972, however, ASE scientist Paul Gorenstein detected an X-ray pulsar in the constellation Vela, which is closer to us than the Crab—1,200 light-years away. Gorenstein points out that the Crab pulsar is relatively young, less than 1,000 years old, while the Vela pulsar may be as much as 10,000 years old and is still producing great quantities of high-energy particles.

Discovery of the Vela pulsar has helped astronomers fill out the story of how neutron stars evolve in a supernova explosion. The neutron star is now generally accepted as an object whose nuclear fires have been banked to such an extent that there is not enough internal pressure from the heat of nuclear reactions to counteract the gravitational forces that draw atoms of matter to one another. As a result, the star collapses so far that its electrons combine with protons to create neutrons. At this stage, the mass of the neutron star is still about that of the sun, even though the star is only about 10 miles in diameter. It has grown so dense through gravitational collapse that a cubic inch of its material weighs an incredible 1,000 billion pounds.

A neutron star then slowly dies, not all at once, but in stages. First it spins rapidly, even madly, giving off light, radio waves, and X rays. As it ages, it loses energy and slows down, rapidly at first, but at a steadily

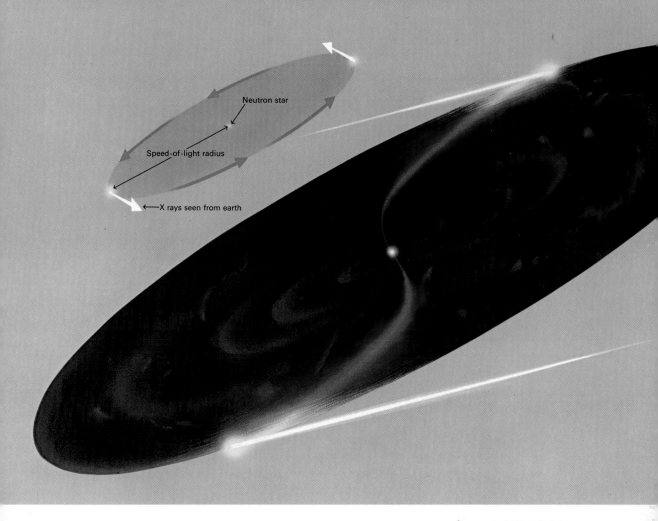

Neutron star

Speed-of-light radius

← X rays seen from earth

decreasing rate. X rays and light waves shortly disappear, but radio emission can last for millions of years.

Pulses of X rays have also been found emanating from a few of the many binary, or two-star, systems, which comprise perhaps half of the star groupings in the known universe. The X-ray-emitting binary is an astronomically odd couple—a giant star waltzing with a gravitationally collapsed celestial partner that may be either a white dwarf, a neutron star, or a black hole.

The first X-ray binary discovered was Centaurus X-3, which Uhuru found in 1971. It showed up as a pulsating X-ray source with a regular, slowly shifting period. Tananbaum recalls that the Uhuru data excited them so much that he and Giacconi could not wait for a computer analysis and began to plot the data by hand. They found that Centaurus X-3 pulsates every 4.8 seconds and is eclipsed by its giant companion every three days. The mass of the X-ray object is one-tenth that of the sun. An optical counterpart of the companion was found in 1973 by William Liller of Harvard College Observatory.

A more revealing X-ray-eclipsing binary, whose pulsations were also discovered by Uhuru in 1971, is Hercules X-1. This neutron star

An X-Ray Pulsar
A solitary star may collapse and become a spinning neutron star that emits X rays. In this model, electrons from the star, whipped out along magnetic field lines at nearly the speed of light, emit synchrotron X rays.

pulses rapidly and regularly every 1.2 seconds. The star's X-ray flux rises to a peak, then drops, and sharply returns to its peak every 1.7 days. The alteration in flux occurs when the X-ray source slips behind its partner in an eclipse. However, astronomers have also recorded a mysterious 35-day cycle, in which the X rays are switched on for 11 days and off for 24 days. The cause of this cyclic behavior is uncertain. Friedman suggests that, in a transfer of gas from one star to the other, the gravitational well of the giant star fills up, waves are set in motion, and the gas sloshes over to the neutron star in the same way that water splashes out of a full washtub. The neutron star pulls the gas down into its deep gravitational well at enormous velocity. In the process, the gas rises in temperature to 50 million degrees C. and radiates X rays. An alternative theory proposed by Kenneth Brecher of M.I.T. suggests that the cause lies in the precession of the neutron star, which makes it wobble around like a spinning top.

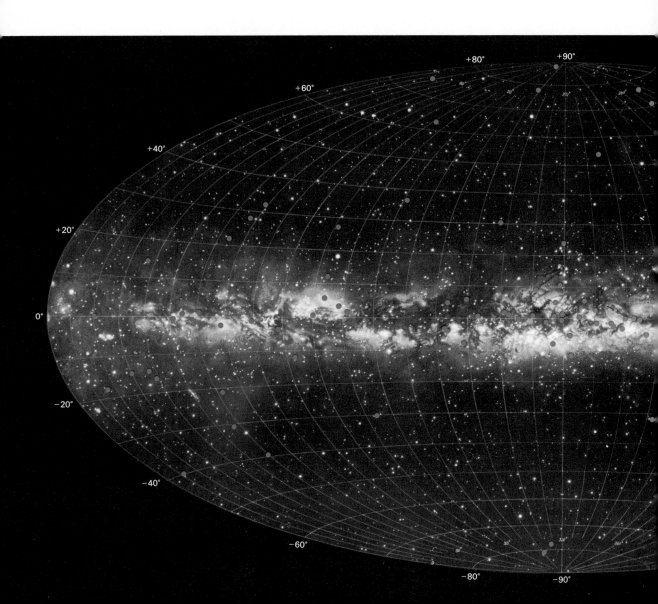

Hercules X-1 has been a continuous source of excitement. Astronomers have discovered that it is also associated with a variable star called HZ Herculis, whose light grows and dims in a cycle. Until recently this optical feature was a puzzle. It now appears that the trigger is likely to be the variable X-ray flux, which heats up one side of the optical star and makes it glow.

Still another impressive finding of Uhuru is the recent discovery of rapidly pulsing X rays coming from Cygnus X-1. This binary X-ray source may actually be the first identifiable example of that mysterious and elusive stellar object known as a black hole. The evidence for this depends on a process of elimination. The fast variations of pulse observed in Cygnus X-1 could only result from a small compact star, so it must be a white dwarf star, a neutron star, or a black hole. But observers have determined that the object's mass is greater than 10 solar masses, so it cannot be a white dwarf or a neutron star, which are less

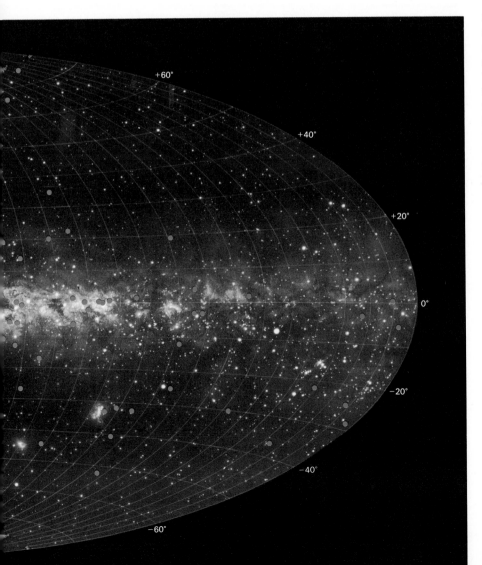

Uhuru discovered 125 new X-ray sources during the first 70 days of its space odyssey. Two-thirds of these lie within 20 degrees of the galactic plane. Some found at higher latitudes have been identified with known extragalactic objects.

massive. Hence, it must be a black hole. With such a large collapsed mass, the star, in effect, disappears from the universe. Not even a slender ray of light can escape from the power of its gravitational grip.

Still, not every scientist is convinced that Cygnus X-1 is a black hole, mostly because the source is so elusive.

Casting its sights beyond our Galaxy, Uhuru has uncovered, at last count, some 60 extragalactic X-ray sources, covering all known types of galaxies from those like the familiar Milky Way to the hyperactive and peculiar Seyfert galaxies. Even quasars, those highly energetic celestial objects at the very edge of the visible universe, emit X rays.

What is the source of their overwhelming X-ray energy? Giacconi says: "In our own Galaxy, or in other normal galaxies such as the Great Nebula in Andromeda, we could explain it by adding up the flux from the individual stellar X-ray sources. But in the very violent galaxies, the X-ray emission is so great that it cannot be accounted for by simple addition. There must be something else."

Several clusters of galaxies have also turned out to be especially strong sources of X rays. One such cluster is the Virgo Cluster, which contains about 1,000 visible galaxies and emits strong radio waves at its core. Gursky and Edwin Kellogg of ASE have found X rays not just at the center of Virgo, but throughout the cluster—a region ranging up to millions of light-years in size. Similar findings have been obtained for the Coma Berenices and Perseus clusters, two of the largest in the known universe, with more than 1,000 galaxies each. In fact, the Perseus Cluster is so violent that some scientists believe it may be wracked by a chain of galactic explosions.

Looking between the clusters, Uhuru has confirmed a phenomenon even more mysterious—a rich X-ray flux throughout the entire sky. The flux comes from all directions. It is of extraordinary strength, suggesting that it is more than the stormy, violent behavior of galaxies and clusters of galaxies seen out to the farthest reaches of the universe. It may be man's first sight of a vast intergalactic cloud that fills the spaces between the clusters. The X-ray energy emitted is what might be expected of a very hot diluted gas at a temperature of 100 million degrees. But how, for example, is such a gas heated? The latest suggestion, offered independently by two groups of astronomers—Sabatino Sofia and his associates at the University of South Florida in Tampa and Allan Solinger and Wallace Tucker of ASE—suggests that the heat comes from the excitation of the gas by cosmic rays from one of the galaxies. Not all scientists appear willing to accept this theory yet.

Assuming the existence of a hot intergalactic gas has one immediate cosmological consequence: The universe is more massive than we thought—massive enough, in fact, to gladden the hearts of those cosmologists who believe that the universe is a closed system, constrained to stay within some kind of limits. For years, astronomers have been tantalized by the "mystery of the missing mass of the universe." If you count the galaxies and determine their total mass, the answer does not match the mass required to generate the gravitational force needed to halt the expansion of the universe. Is this missing mass the tenuous, fiercely hot plasma between the galaxies? Only one particle per cubic centimeter—an amount invisible to optical and radio telescopes—would add up to a tremendous mass on the galactic scales. The present conclusion is a cautious, "maybe." Indeed, some marginal evidence exists that the X-ray flux in the sky might come from relatively nearby sources, such as the halo of the Milky Way. If, on the other hand, further experiments show that there is an intergalactic mass, it would "close the universe." That is, it would suggest that the expansion of the universe from the original fireball, or big bang, that began billions of years ago is slowly coming to a halt. This in turn would encourage the theory that the universe neither remains in a steady state nor expands infinitely, but expands and contracts, perhaps endlessly, like a great cosmic heart.

The wealth of Uhuru's cosmological findings has helped to brighten the future of X-ray astronomy. ASE now has a continuing program of rocket investigations lined up at White Sands, N.Mex., and M.I.T. scientists have developed a new satellite with greater resolution than Uhuru which will be launched in 1975. Scientists have also prepared a substantial program of experimental study of solar X rays to be conducted from Skylab, the first U.S. orbiting space station.

The most striking X-ray space probe in the works is NASA's High Energy Astronomy Observatory (HEAO), the largest X-ray spacecraft yet designed. Three stories high, it contains 20 square feet of counter area compared to the 2 square feet available in Uhuru. When lofted, HEAO will go 180 miles into space, where it will spin lengthwise between two and six revolutions per hour, picking up X rays over a wide range of energies. Within six months, HEAO should see and identify X-ray stars fainter than man has spotted before.

Recent federal budget cuts have postponed the launch of HEAO to 1977 and reduced the number of its assigned experiments, but that has not dampened the astrophysicists' enthusiasm for the project. In fact, Giacconi and Friedman feel that the revamped and restructured observatory, costing half as much as before and with a third of the weight, is a good one. Others may not agree, but most scientists feel that they will make the new HEAO the instrument that will give man his richest and most complete view of the universe.

For further reading:

Metz, William D., "X-ray Astronomy: Observations of new Phenomena," *Science,* Jan. 28, 1972.

Metz, William D., "X-ray Astronomy (II): A New Breed of Pulsars," *Science,* March 9, 1973.

Schnopper, Herbert W., and Delvaille, John P., "X-ray Sky," *Scientific American,* July, 1972.

Thomsen, Dietrick, "Intergalactic Space: Something's Making X rays," *Science News,* June 10, 1972.

The Case of the Missing Neutrinos

By William A. Fowler

**Our picture of the sun, the source and nurture of
life on earth, is in question because of baffling results
from a most unusual astronomical observatory**

After six years of observations almost a mile deep in the Home-
stake Gold Mine at Lead, S.Dak., chemist Raymond Davis, Jr., of
Brookhaven National Laboratory has discovered that something mysterious is happening. There is a severe shortage of high-energy neutrinos arriving at the earth. At most, only about one-fifth as many of
these particles are pouring from nuclear reactions at the center of the
sun as physicists have predicted.

Where are the missing neutrinos? All of the detectives' clichés must
be advanced—"The case is still open" and "We have a number of
promising leads." Some of the leads challenge our belief in the stability of the sun and the stars. Others, such as the possibility that neutrinos are unstable and decay to other particles, would produce a fundamental breakthrough in elementary particle physics.

Neutrinos carry energy and momentum, as do all other elementary
particles. In fact, it was to balance just these quantities in certain
nuclear reactions that Austrian physicist Wolfgang Pauli postulated
neutrinos in 1932 and Italian physicist Enrico Fermi described in
detail the role they should play. However, neutrinos are thought to
lack almost all the other properties that elementary particles have.

The great neutrino detector—a giant steel tank filled with 100,000
gallons of cleaning fluid—lies a mile beneath the earth's surface.

Four forces act on elementary particles—gravity, electromagnetism, and the strong and weak nuclear forces. Because neutrinos have no mass at all, they travel at the speed of light. Gravity acts on neutrinos, but only through the equivalent in mass of their energy of motion. Neutrinos have no electric charge and no magnetic moment. In fact, neutrinos cannot take part in any electrical or magnetic process. Neutrinos are not affected by the strong nuclear force, either. They participate only in weak-force interactions, which are 10-billion-billion times weaker than electromagnetism.

Neutrinos are downright unsociable and, as a consequence, those born in the nuclear furnace deep within the sun should travel out without being absorbed or even scattered by the solar material. All but 1 in 100 million neutrinos pass out through this huge body as if it did not exist. Traveling at the speed of light, they journey to and through the earth in only 8 minutes.

Sunlight also gets its energy from nuclear processes at the center of the sun, but the nuclear energy is transformed immediately into thermal energy, which—scattered and absorbed electromagnetically—must diffuse from the center out to the surface in a random-scattering process that takes roughly 10 million years. Only then is this energy transformed into sunshine. Thus, neutrinos should bring us important information about the nuclear processes occurring at the center of the sun in only 8 minutes. Sunlight bears a message that is in some ways 10-million years old.

The case of the missing neutrinos is more than 25 years old, and I have worked on it for more than 15 years. It all started just after World War II, when Bruno M. Pontecorvo, an Italian-born theoretical physicist, suggested a way of detecting the neutrino. Pontecorvo predicted that chlorine 37, the heaviest of the two stable isotopes of the element chlorine, would occasionally absorb a high-energy neutrino by a weak-force interaction. The chlorine nucleus would promptly emit an electron and become argon 37 (Ar-37), a radioactive isotope of the rare gas argon. Detect the radioactive argon, said Pontecorvo, and that reaction—and thus the neutrino—would be found. In 1948, Luis W. Alvarez, an American physicist, discovered this approach independently, and a year later he proposed an experiment capable of detecting the extremely low rate of absorption that was predicted for neutrinos. Davis took up Alvarez' challenge at the Brookhaven National Laboratory in 1955.

Neutrinos come in four kinds—electron neutrinos, electron antineutrinos, muon neutrinos, and muon antineutrinos. The sun emits electron neutrinos. Nuclear fission reactors radiate electron antineutrinos in great quantities, and these were first detected in 1953 by their absorption in hydrogen nuclei, or protons, which produced detectable neutrons and positive electrons, or positrons. Muon neutrinos and antineutrinos are emitted along with muons when pions produced in high-energy accelerators decay, and these have also been detected.

The author:
William A. Fowler is institute professor of physics at California Institute of Technology. Not only has he made numerous contributions to nuclear physics and astrophysics, but to the earth sciences as well.

Scientists believe that the fusion of hydrogen into helium is the prime source of power in the sun and similar stars. A helium nucleus has slightly less mass than the four hydrogen nuclei that fused to make it up. By Albert Einstein's famous $E=\triangle mc^2$ law, this small mass difference ($\triangle m$) multiplied by the square of the velocity of light (c^2) yields the energy (E) released in the sun.

The main energy-producing sequence in the sun is called the proton-proton (pp) chain. The sun consists mostly of hydrogen, and the high temperature of its interior breaks down hydrogen atoms into protons and negative electrons, all in rapid motion. In violent collisions, two protons can fuse to form a deuteron, a nucleus of heavy hydrogen. A positive electron, or positron, and a neutrino are usually emitted. These pp-neutrinos range in energy from zero to 0.42-million electron volts (MeV). Once in 400 fusions, a negative electron (e) is captured instead of a positron being emitted. However, a neutrino is still emitted—a pep-neutrino with an energy of roughly 1.44 MeV.

Researchers have deduced these neutrino energies by carefully measuring the masses and energies of the other particles in the reaction, the electron, positron, proton, and deuteron. The ratio of the pep- to pp-neutrinos, 1 in 400, cannot be determined in the laboratory, but we can calculate it on theoretical grounds that are known to correctly predict the ratio of electron capture to positron emission in measurable weak interactions.

The deuteron eventually collides with a proton and forms a helium nucleus with a mass of three (He-3) and a photon of gamma radiation —more energy for the solar furnace. A collision between pairs of He-3 nuclei produces a nucleus of ordinary helium with a mass of four units and releases two energetic protons. Six protons have interacted to form the He-4 nucleus and two of the protons were released. In the process, two neutrinos are emitted.

Painstaking terrestrial measurements show that direct fusion by the pp-chain usually occurs in the sun, but not always. An He-3 nucleus occasionally should fuse with an He-4 nucleus that was previously produced. This produces the element beryllium (Be-7), again releasing gamma radiation. When we produce Be-7 in the laboratory, it decays radioactively by capturing an electron and emitting a neutrino. Its average lifetime is 77 days. Ninety per cent of the Be-7 neutrinos have an energy of 0.861 MeV, and 10 per cent have 0.383 MeV.

I pointed out in 1958 that there is still another fusion reaction that can occur—in the sun, but not on the earth. Be-7 may capture a proton before it captures an electron. Proton capture produces a nucleus of boron (B) with eight units of mass. This B-8 decays rapidly, emitting a positron and a neutrino. B-8 neutrinos range in energy from zero to the relatively large value of 14 MeV.

How many of these various kinds of neutrinos should rain on the earth, through each square centimeter facing the sun, every second? We know the energy of sunlight is about 1/30 calorie per square cen-

Deep in a South Dakota gold mine, chemist Raymond Davis, Jr., draws liquid nitrogen from a small tank. At a temperature of −320°F., the nitrogen will freeze out any neutrino-caused argon flushed from the larger observatory tank.

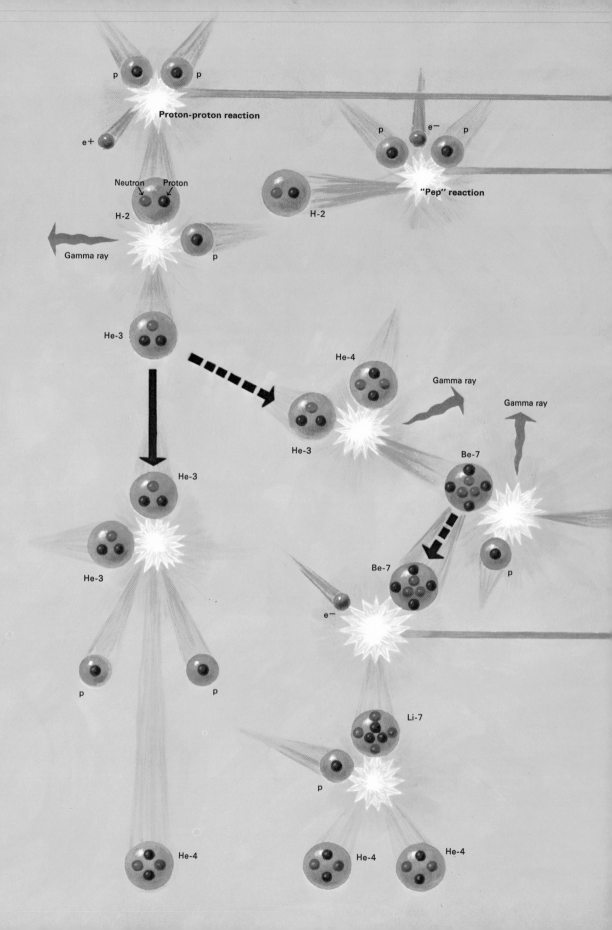

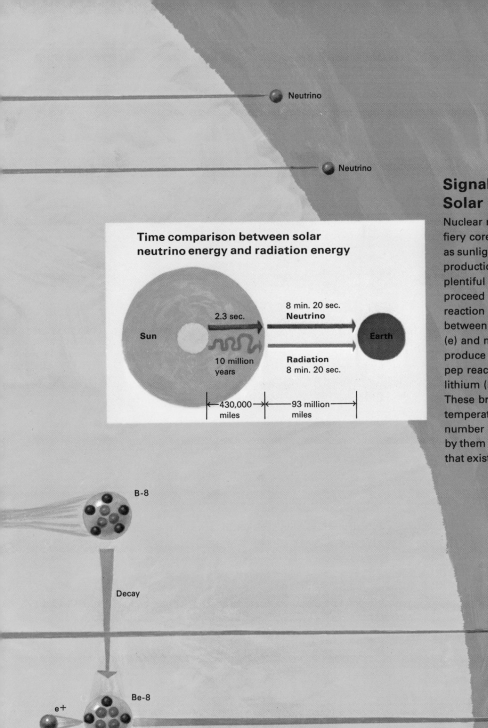

Neutrino

Neutrino

Signals from a Solar Furnace

Nuclear reactions in the sun's fiery core send neutrinos as well as sunlight toward earth. Most production of helium (He) from plentiful hydrogen 1 (p) should proceed along the proton-proton reaction pathway. Side reactions between fast-moving electrons (e) and nuclei also occasionally produce deuterium (H-2), by the pep reaction, and beryllium (Be), lithium (Li), and boron (B). These branching reactions are temperature sensitive, and the number of neutrinos produced by them will reveal the conditions that exist at the sun's center.

Time comparison between solar neutrino energy and radiation energy

Sun

2.3 sec.

8 min. 20 sec.
Neutrino

Earth

10 million years

Radiation
8 min. 20 sec.

|←430,000 miles→|←93 million miles→|

B-8

Decay

Neutrino

Be-8

e+

Neutrino

Decay

He-4

He-4

How to Capture Neutrinos

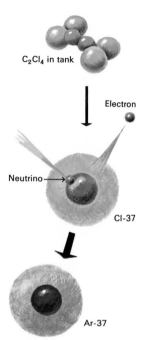

C₂Cl₄ in tank

Electron

Neutrino →

Cl-37

Ar-37

One-fourth of all the chlorine on earth—thus one atom in a C_2Cl_4 molecule—is Cl-37, and its nucleus can absorb neutrinos that have more than 0.814 MeV of energy. The nucleus expels an electron, transforming it into chemically unreactive but radioactive argon 37.

Neutrinos from the solar reactions have different energies. The B-8 neutrinos should be most readily absorbed by Cl-37 nuclei.

timeter per second. We also know that 1 million-millionth of a calorie —one-half that much for each neutrino—is released when four protons fuse into helium. Dividing the radiant energy of sunlight by the energy released per neutrino, we find that roughly 60 billion pp-neutrinos must pass through a square centimeter on the earth every second. We need only divide by 400 to find that 150 million pep-neutrinos arrive each second. Our calculations neglected the solar energy carried by the neutrinos themselves, but the laboratory measurements show that neutrinos carry only a few per cent of the total.

The numbers of these neutrinos were based on the known energy from the sun's surface, and are independent of the specific temperature and density at the interior of the sun. However, the Be-7 and B-8 neutrinos do not share this temperature independence. How much Be-7 forms depends on how often an He-3 nucleus fuses with an He-4 nucleus rather than another He-3 nucleus. The ratio of these competing processes is called a branching ratio. B-8 production also depends on this ratio, and on how often a Be-7 nucleus captures a proton rather than how often an electron.

Laboratory measurements show that both these branching ratios depend critically on the temperature at the sun's core. The measurements were carried out during the last 15 years, primarily at the California Institute of Technology (Caltech) by Ralph W. Kavanagh and Thomas A. Tombrello. They accelerated protons, for example, to high energies, and caused these to strike Be-7 nuclei to produce B-8,

Detecting Neutrinos from the Sun

Neutrinos arriving on earth
(per square centimeter per second per MeV)

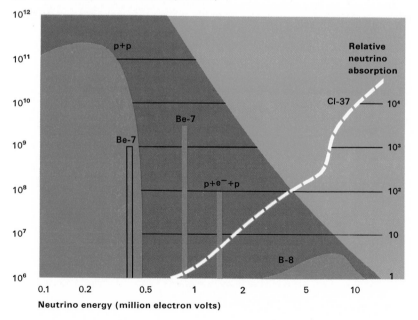

just as in collisions in the sun. They detected the rate at which B-8 forms by counting its decay products, a positron and two He-4 nuclei.

These laboratory reactions, however, occur so infrequently at the energies we think occur in the sun that their rates could only be measured at higher energies. The researchers must then extrapolate the rates down to solar energies to determine the branching ratios.

This might mislead the experimenter because many reactions occur more often at certain temperatures. These speed-ups are called resonances. If the He-3 plus He-3 reaction has a resonance at solar temperatures, it will consume more He-3 than we expect and "short circuit" production of Be-7 and B-8 and create the neutrino shortage. In 1970, Mirmira R. Dwarakanath repeated the accelerator measurements at Caltech using more sensitive equipment. He found that the critical He-3 plus He-3 reaction rate is still incredibly small, and matches the estimated rate without resonance at temperatures very close to those in the sun. So a resonance seems now to be extremely unlikely to be the culprit.

It may seem presumptuous to claim that physicists are able to estimate the central temperature of the sun accurately enough to get the exact branching ratios. However, theoretical calculations, using astronomical observations combined with the laws of physics, can be very accurate. A typical solar model—the theorist's "picture" of the inside of the sun—for instance, must exactly match the sun's present mass, radius, and luminosity. The sun neither expands nor contracts at a detectable rate, so the model must be stable. It is almost in perfect hydrostatic equilibrium—the outward pressure at every point in the sun must support the weight of the overlying material. The pressure, caused by thermal motions of nuclei and electrons, is proportional to the temperature and density of the material. And the densities at each point must average to the known average solar density. These constraints fix the range of temperature at each point within narrow limits.

Furthermore, if the sun is in a steady state, the nuclear energy produced in the sun's interior must match the energy radiated at the surface; this, added to the other constraints, narrowly limits the temperature at the core. Standard models for the sun agree that the sun's temperature builds to a maximum very close to 15 million degrees Kelvin (27 million degrees F.) at the center.

The branching ratios, when extrapolated to the range of temperatures the model gives for the sun's interior, indicate that 3 billion Be-7 neutrinos and 3 million of the more energetic B-8 neutrinos should arrive at a square centimeter on the earth each second.

Solar neutrinos should far outnumber those reaching the earth from all the other stars. It is much like the difference between night and day. The light from the stars at night shows how little of their luminous energy reaches us compared to that from the sun. Neutrino "brightness" should be much the same, except that the earth's bulk does not block neutrinos. The sun's neutrinos shine on us 24 hours a

How to Count Neutrinos

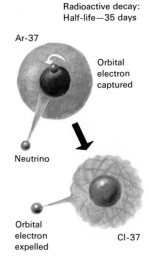

Radioactive decay: Half-life—35 days

Ar-37

Orbital electron captured

Neutrino

Orbital electron expelled   Cl-37

Argon 37 atom decays radioactively when an inner electron in its atomic shell is captured by the nucleus. This reverses the original neutrino capture process, and a neutrino is emitted. The excited Cl-37 atom quickly expels an electron that can be detected in a sensitive counter.

The observatory tank is flushed with helium about every 3 months to remove argon that neutrinos produce. Davis watches, *above,* as the helium passes through a trap filled with liquid nitrogen to collect the argon, which is then placed in a small proportional counter, *above right.* Shielded from natural radiation, it is then electronically monitored, *far right,* to detect its very weak radioactivity.

day. If we are wrong, and the stars outshine the sun, then the case of the missing neutrinos is even more serious.

From his station in a Black Hills mine, Davis looks for the high-energy solar neutrinos. Over the last six years, this chemist-cum-detective has painstakingly developed the search—one might call him the chief inspector. He has installed an enormous steel tank about a mile deep in the mine to shield against cosmic rays and other extraterrestrial radiation. It is filled with 100,000 gallons of the cleaning fluid perchloroethylene, C_2Cl_4. Davis calls the tank and its auxiliary apparatus his "Brookhaven Neutrino Observatory."

Davis has 2.2 million-trillion-trillion C_2Cl_4 molecules in his tank. On the average, one of the four chlorine atoms in each molecule is Cl-37. This is the natural ratio in chlorine on the earth—one part Cl-37 to three parts Cl-35. Cl-35 does not capture neutrinos. Laboratory mass-energy measurements show, too, that a neutrino having less than 0.814-MeV energy cannot transform a Cl-37 atom into Ar-37. Thus, none of the pp-neutrinos and only 10 per cent of the Be-7 neutrinos can be detected. Theorists have shown, however, that Cl-37 becomes increasingly more able to intercept a neutrino as the energy of the neutrino rises above 0.814 MeV. Thus, B-8 neutrinos, with their energy reaching up to 14 MeV, are intercepted much more easily than the pep or Be-7 neutrinos. Even so, the target size of a single Cl-37 nucleus to an average B-8 neutrino is extremely small, only 1.35 million-trillion-trillion-trillionth of a square centimeter. Thus, even the

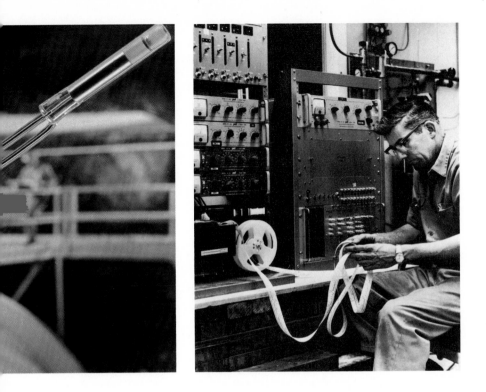

3 million B-8 neutrinos we expect to arrive at each square centimeter each second should produce slightly less than one argon atom a day in the huge tank. Adding the pep and Be-7 neutrinos brings the total production up to one a day.

The Ar-37 atoms produced are radioactive, and they begin to decay as soon as they are produced. Since the average lifetime of an Ar-37 atom is 50 days, Davis should have a maximum of 50 Ar-37 atoms in his tank, even after an extremely long exposure.

Radioactive Ar-37 is a noble gas, which means that it does not combine chemically with the C_2Cl_4 molecules. After about three months' exposure, Davis flushes it out of the tank by bubbling helium, also a noble gas, through the liquid. The helium-argon mixture passes through a cold charcoal trap, which freezes out the argon.

Davis places the concentrated argon gas in a small, well-shielded counter. Occasionally, an argon-37 nucleus captures one of the electrons from its atomic shell, and this re-forms the Cl-37 atom in an excited electronic state that decays rapidly to normal Cl-37 by ejecting an electron. Davis' electronic counters detect the ionization that each electron produces.

To make a long story short, Davis finds an average of fewer than 10 Ar-37 atoms in his tank. This result comes from six three-month runs, with most of the Ar-37 found in one of the runs.

How can we explain the missing neutrinos? We expect 50 and find at most 10. Is the technique Davis uses incomplete or incorrect? Is our

Possible Solutions to The Mystery

The missing neutrinos make scientists question their knowledge of both neutrinos and the sun. Does a neutrino have a small magnetic moment and lose energy as it passes out through the sun? Does it have a small mass and change to an antineutrino during its 8-minute trip to earth? Or does it decay to a muon neutrino? Unobserved internal properties of the sun, such as lower opacity, higher rotation, or a strong magnetic field, would produce a lower core temperature and fewer neutrinos. Is the sun changing? It might now have a lower core temperature in its temporary stage today than it would need in an equilibrium state.

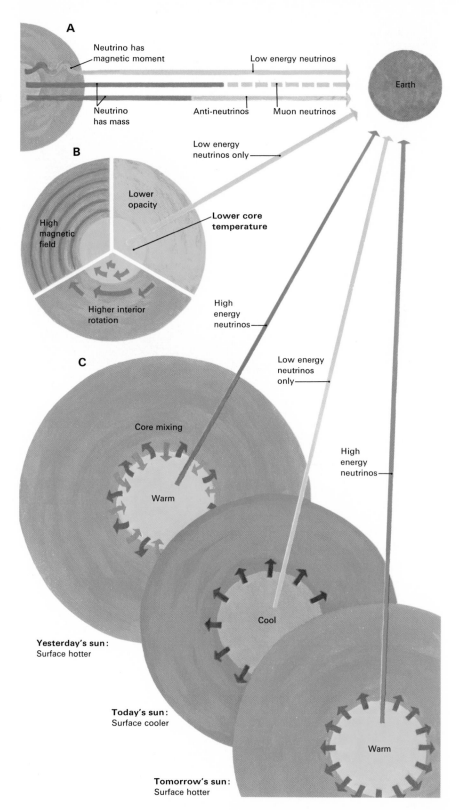

A

Neutrino has magnetic moment

Low energy neutrinos

Earth

Neutrino has mass

Anti-neutrinos

Muon neutrinos

B

Low energy neutrinos only

Lower opacity

Lower core temperature

High magnetic field

High energy neutrinos

Higher interior rotation

C

Low energy neutrinos only

Core mixing

High energy neutrinos

Warm

Yesterday's sun: Surface hotter

Cool

Today's sun: Surface cooler

Warm

Tomorrow's sun: Surface hotter

knowledge of elementary particles incomplete or incorrect? Or is it, instead, our knowledge of solar structure, evolution, and stability that is incomplete or incorrect?

Davis has checked his experimental technique. As a test to make sure he can extract the argon, he has added artificially produced Ar-37 into the tank and was able to flush 95 per cent of it out with helium. He has also learned to distinguish Ar-37 counts from those due to sources outside his counters.

He has flooded the mine chamber in which his tank is located with water to reduce the Ar-37 produced in the tank by neutrons from natural radioactivity in the mine's rock walls. Water slows down and absorbs neutrons. Cosmic rays also produce particles, muons, and neutrinos, that can penetrate even through the mile-thick blanket of rock above the tank. Some of the Ar-37 atoms Davis finds are almost certainly produced by these. But since Davis cannot be sure what produced them, he can only say that all of them are possibly due to solar neutrinos. Even if all 10 are from solar neutrinos, however, they are still only one-fifth of the number expected.

There are other tests that can be made. A University of Washington group has suggested that Davis build an accelerator that will produce B-8 alongside the tank, and see if he can detect its neutrinos. Alvarez has suggested that an extremely strong source of radioactive zinc be produced and lowered into the mine to irradiate the tank with the energetic neutrinos that it emits. Davis himself plans to expose perchloroethylene to the neutrino beam of the Los Alamos (N.Mex.) Meson Facility when it goes into operation in 1974.

In the meantime, Davis is attacking what seems to be the one loophole in his technique: Do the Ar-37 atoms produced when Cl-37 is hit by an energetic neutrino always recoil out of the C_2Cl_4 molecule? Can they somehow remain bound in the molecule and not be collected by the helium flush? Expert chemists insist that the argon produced will be collected. However, Davis plans to introduce C_2Cl_4 containing a small amount of radioactive chlorine 36 into a test tank. The Cl-36 decays to Ar-36 with a smaller recoil than Ar-37 has. If Davis can collect the Ar-36 he will close this loophole.

The missing neutrinos might, however, be telling us that something is wrong with our knowledge of elementary particles. Certain fractionally charged particles called quarks may exist in the free state, although there is no experimental evidence of this. But if they exist, they might catalyze pp-reactions in the sun and decrease Be-7 and B-8 neutrino production. Most experts believe that this is not the solution.

Pontecorvo and others suggest that neutrinos become antineutrinos on their way from the sun to the earth. A neutrino's velocity and angular momentum, or spin, are related like the advance of a left-handed screw, an antineutrino's like a right-handed screw. Because of this difference, antineutrinos cannot change Cl-37 into Ar-37. Davis has shown this by exposing perchloroethylene to the enormous antineutrino flux of the Savannah River nuclear reactor in Georgia. But according to theory, the transformation of neutrinos into antineutrinos is never completed. Even if it occurs, the theorists say that more argon should be produced than Davis finds.

Another suggestion is that electron neutrinos have a small mass and thus can decay into inert particles, such as muon neutrinos, before reaching earth. This is also unlikely but cannot be ruled out. If neutrinos have a slight magnetic moment, they would interact magnetically with electrons as they speed out through the sun, and this resistance could reduce their energy below the 0.814 MeV Cl-37 threshold. But there is no experimental evidence that neutrinos have a magnetic moment. All these elementary particle solutions to the solar neutrino shortage are so radical, so contrary to what we believe is correct, that we must question how much we really know about the sun.

Davis' results disagree with those predicted by the standard solar models. Be-7 and B-8 production are so temperature sensitive, however, that there is no disagreement if the temperature of the sun's core is only 10 per cent lower than we think. The core temperature depends on how transparent the outer gases are to the energy flow from the core. The more transparent the gas is, the lower the temperature of the core, and vice versa. If the sun is more transparent than scientists believe it is, there would be fewer Be-7 and B-8 neutrinos. The chemical elements that make the sun opaque are not its major constituents, hydrogen and helium, but the small amounts of heavier elements, such as carbon, oxygen, magnesium, silicon, and iron. Spectroscopists find about one atom in a thousand of these heavier elements at the surface of the sun. The composition of the surface material should be the same as the gaseous nebula from which the sun was formed, and thus the same as the entire sun. However, it is possible that the interior of the sun lacks heavier elements. If so, the opacity of the solar material would be slightly less and central temperature could be lower, and this would account for the small number of neutrinos that Davis finds.

Some of the scientific detectives question whether the pressure supporting the sun is entirely thermal. Could the sun contain a strong internal magnetic field—a billion gauss compared to the few gauss observed at the surface? Such a field would exert an outward pressure and thus decrease the temperature at the sun's center. Be-7 and B-8 neutrinos would then fall below detectable limits. A counterargument suggests, however, that large magnetic fields will quickly "bubble" to the sun's surface and disappear. And once lost, such huge fields could never be regenerated.

Could the sun be rotating much faster inside than its slow surface rotation of one revolution in 28 days? If it is, the outward centrifugal force would help support the sun against gravity, and, again, the central temperature would be lower. However, such rapid rotation should flatten the sun at the poles, and it would not appear perfectly circular. Robert H. Dicke and his collaborators at Princeton University claim that the sun is flattened very slightly. But some other physicists question Dicke's observations, or his interpretation of them.

I proposed another solution to the mystery in 1972—that the sun is in a transient state. What would happen if the interior of the sun suddenly became turbulent and new hydrogen fuel was swept into the central region where the fuel had been exhausted? At first, the nuclear fire would burn more brightly, but then the central region would expand and cool. The cooling would be so marked that it would practically stop the nuclear reactions. The sun's luminosity would decrease. After a long time, the core would shrink, nuclear reactions would begin again, and the sun would regain its original brightness. This decrease and recovery of luminosity should take 10 million years, the same time required for energy to reach the sun's surface. Over most of this period, neutrino emission from the sun should be well below that observed by Davis.

We can check this solution against the earth's geological record. Geologists note that the world's climate over the past few million years, the Pleistocene Epoch, is most unusual because of the Ice Ages, when many glacial and interglacial periods occurred. We may still live in an interglacial or postglacial stage. The extraordinary spread of the continental ice sheets required a fundamental change in climatic conditions, especially lower temperatures.

Did this decrease follow a decrease in the sun's luminosity? If mixing occurred in the sun a million years ago, the swollen core would produce relatively few neutrinos even today. Moreover, the sun will not recover its higher, normal brightness for almost 10 million years.

Other major Ice Ages have occurred in the earth's past history. One was in the Permian-Carboniferous Period, about 300 million years ago; another during the Cambrian, about 600 million years ago; and still another in the Precambrian, about 1.1 billion years ago. Each of these Ice Ages lasted about 10 million years. The sun may well have alternated many mixing stages with long intervals of stability. The geological record does not rule out this solution to the mystery.

The case of the solar neutrinos remains a puzzle, not only for astronomers, chemists, and physicists, but for geologists, as well. It may have an obvious solution that all the scientific detectives have so far overlooked—some simple defect in our observations, experiments, or theories. In a way, that would be a pity. All scientists hope the solution turns out to be a fundamental breakthrough in our knowledge of nature—perhaps in elementary particle theory, or in the theory of stellar structure and stability. That is the exciting prospect, and for that reason we cannot let the case rest. The search must go on.

For further reading:

Bahcall, John N., "Neutrinos from the Sun," *Scientific American*, July, 1969.

Fowler, William A., "What Cooks with Solar Neutrinos?," *Nature*, July 7, 1972.

A Summer's Search for A Buried Past

By Donald L. Wolberg

A graduate student tells how archaeologists are using the varied tools of modern science to reconstruct everyday life in Bronze Age Greece

I felt strange leaving New York City early in June, 1971. This was to be my first season with the Minnesota Messenia Expedition (MME), and my first journey to Europe. I was a young paleontologist and archaeologist, en route to a remote Greek hilltop called Nichoria Ridge. As my plane took off from John F. Kennedy International Airport, I could feel the incredible gap that now separated me from the Bronx slum only a few miles away where I grew up. I wondered what the kids I grew up with would say if they knew what I was doing. I was excited and sad, and a little sorry that there was nobody with whom I could share my feelings.

I had always wanted to study paleontology and archaeology. Reconstructing the way ancient animals or people lived, or what their world looked like, held a strange fascination for me. I had pursued that interest as a teen-ager by reading and amateur collecting, then attended New York University (NYU), and did graduate work in paleontology and archaeology there. Later, I took part in archaeological survey work and digs in the northeastern United States and did research on primates at the American Museum of Natural History in New York City. At the annual meeting of vertebrate paleontologists in 1970, I met paleontologist Robert E. Sloan of the University of Min-

Mission to Messenia

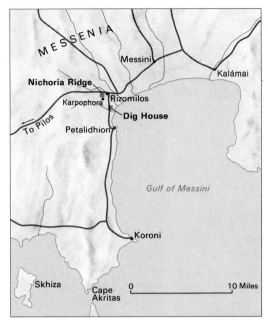

After studying several sites in Messenia, the expedition began to focus its efforts on Nichoria, near the ancient trade route along the gulf coast.

The author:
Donald L. Wolberg is a teaching associate in the geology department at the University of Minnesota. He joined the MME at Nichoria in 1971 and 1972.

nesota, who offered to become my adviser at the university and opened the way for me to become the MME project's paleontologist.

MME has been at work since 1969 on Nichoria, a ridge in Messenia Province on the Peloponnesus in southwest Greece. William A. McDonald, professor of classical studies at the University of Minnesota, heads a staff of scientists as MME's director. The project is supported by the university, various foundations, and several private contributors. The scientists spent years making extensive archaeological surveys, a kind of cultural prospecting, in Messenia before the actual digging began at Nichoria in 1969.

Nichoria is about a mile and a half inland from the northwestern coast of the Gulf of Messini. It lies in a position controlling the ancient land route to the east from Epano Englianos, the major ancient city in the Peloponnesus located opposite Nichoria on the same peninsula. The flat-topped ridge is about 1,800 feet long. Erosion has gnawed at the ridge's irregular outline and its shape has surely changed since ancient times. MME's excavations indicate that this ridge once held important settlements. The first was at the beginning of Middle Helladic times (Bronze Age), some centuries before 2000 B.C. A larger settlement existed during the later part of this period, between about 2000 and 1600 B.C., and Nichoria became the center of a local kingdom early in the Late Bronze, or Mycenaean, period (about 1600 to 1400 B.C.).

The evidence also suggests that during the late Mycenaean period, Nichoria became a tributary of Epano Englianos. Just before 1200

B.C., Epano Englianos was destroyed by fire, perhaps during an invasion. No firm evidence has yet been found by MME that Nichoria was burned, though it certainly declined drastically in population and prosperity. But it enjoyed renewed prosperity in the early Iron Age (about 900-700 B.C.), and it appears to have been abandoned from then until medieval times, roughly A.D. 1000-1200.

MME has concentrated on trying to understand how the ordinary people of ancient Nichoria lived. MME scientists are trying to reconstruct the past environment in a town that was not spectacularly wealthy or powerful.

Much of the history and prehistory reported in the past by archaeologists reflects the achievements of only a few people – the monument builders, generals, and kings. We know more about palaces and palace life than we do about how the farmer, potter, or artisan lived. The search for art, gold, and other treasure seems to me still very much a part of classical archaeology. But if the study of Greek prehistory is to describe the ordinary man, archaeologists should add to their ties with art history the scientific thinking found in modern anthropology, anthropological archaeology, and the physical sciences. Archaeology should become more broadly scientific if its reconstruction of any given culture is to be reasonably comprehensive.

At one time or another, scientists from the United States, Greece, and other countries, having various specialties, have taken part in the

Nichoria Ridge, *left,* the site of a village for 3,000 years, is now crowned with olive and fig groves. Aerial survey photographs of the ridge were made from a balloon, *above.*

Nichoria Ridge

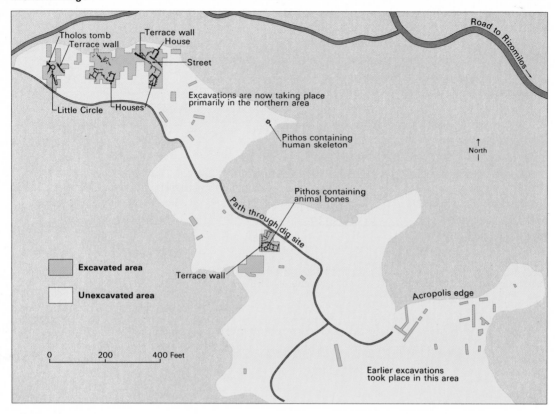

By the autumn of 1972, extensive excavations dotted the crest of Nichoria Ridge. Most of them were based on electrical and magnetic surface measurements.

MME project. This multidisciplinary approach has brought geologists, paleontologists, botanists, photogrammetrists, anthropologists, an agricultural economist, and other specialists to Nichoria.

I knew more about the Greek past than I did of the present when I arrived in Athens, a city alive with the hustle and bustle of every great metropolis. Indeed, much that I saw there–the traffic, noise, crowds, and high prices–differed little from Chicago, New York, or Los Angeles. However, Athens was not at all representative of what I saw elsewhere in Greece during my long, fascinating bus ride through the mountains of the Peloponnesus: Donkeys, narrow roads, poor agricultural communities, and incredibly friendly people were common.

If I felt any sort of cultural shock at being in rural Greece, it was quickly removed when I arrived at the MME dig house. It is a well-built structure of reinforced concrete that has living facilities and space for various project activities. Two large rooms on the upper level serve as dormitories for the roughly 30 staff members. There is also a small alcove area for plant and rock studies and drafting work. The lower level has a large common room for meals, seminars, meetings, and general relaxation. There are also a kitchen and two smaller rooms reserved for the director and visitors. The bathrooms on the

Cameras on big bipods straddle trenches to record the precise location and condition of the finds as the excavators uncover them.

lower level are particularly popular because they alone in the entire dig house have hot water.

Most of the scientific work at the dig house is carried out in the basement. There are darkroom facilities for processing the vast quantity of film shot by the MME scientists. The photo crew can process, develop, and print black-and-white, color, and even infrared-sensitive film on the same day it is shot by excavators in the field. A large part of the basement space is used for pottery analysis. The basement is also used for storing collected material, cataloging small finds, and photographic analysis. There I studied the animal remains.

Greek cooks, under the supervision of the wife of one of the archaeologists, prepared the food at the dig house on a full array of American-style stoves, ovens, and refrigerators. As might be imagined, preparing meals for 25 or more people was a major task. I found that our Americanized version of Greek food was tasty and always fresh.

The bread was always delightfully coarse and substantial, the way bread should taste. Greek food is cooked in, or otherwise prepared with, olive oil, a product used for numerous other purposes as well, such as a cleaning agent, ointment, and medicine. Greek-style coffee takes getting used to. I never quite did.

As many as 30 Greeks were employed by MME, and they did most of the digging. The heat and sun at the site were often intolerable for me, but the Greeks kept right on working. The diggers were paid between 100 and 200 drachmas per day ($3 to $6), depending on their age and experience.

A real Greek house, nearer the site in the tiny village of Karpophora, houses part of the staff and provides additional work space. Although the Karpophora house has no running water, it became so popular that a schedule had to be worked out so everyone could spend time there. The village life made up for the lack of modern conveniences. Even those of us who stayed at the main house spent evenings in the village, although it was a 2-mile hike. We enjoyed the chance to get to know the Greeks better.

The lower level of the Karpophora house originally housed goats, but it had been altered—and cleaned—so that we could use it to store and study the human remains recovered from the site. My own research time was split between the main house, where I studied the animal remains, and the Karpophora house, where I worked on some of the human remains.

When I arrived at the site, it looked much like a battle zone. Trenches dotted the area. There were mounds of excavated earth and rock, the dump areas. Barbed wire fringed some deep trenches—as much to prevent wandering animals and people from falling in as to protect the trench contents from the occasional curious tourist.

Nichoria Ridge has an area about 1,800 feet by an average of over 300 feet and before excavation began, most of it was covered by cultivated olive and fig trees, shrubs, and the characteristic thorny Greek vegetation. The question naturally had arisen about where to begin digging. Early surveys by McDonald and others revealed prehistoric potsherds, or broken pieces of pottery, on the hilltop. Nearby, in the valley to the west of the ridge, the survey team found several man-made mounds that turned out to be part of a Bronze Age cemetery.

The task of assisting in the selection of places on the ridge to excavate had fallen to the project's associate director, geologist George (Rip) Rapp, Jr., of the University of Minnesota, and his electronic probing equipment. In Greece, even if an archaeologist has one of the scarce government dig permits, he is generally allowed to dig only trial trenches before he buys the property. Obviously, it is critical for him to gain as accurate an impression as possible of what is present beneath the earth before he spends his research funds. To get clues as to subsurface features from surface observations, the archaeologist can be helped by two valuable tools of the geophysicist, the proton magnetometer and the electrical resistivity meter. MME's use of these devices marks their first large-scale employment in Greek archaeology at the first stage of an excavation.

The proton magnetometer detects variations in the strength of the earth's magnetic field and helps detect objects such as fired pottery,

The expedition's popular house in Karpophora, *above,* offers the staff a taste of modern Greek culture. The main house, *above right,* at Sikalorache, has American features, but with barracks-life simplicity, *right.*

buried hearths, kilns, and iron objects. The electrical resistivity meter measures the resistance of the ground to an electric current, and is useful in locating large buried features, such as massive stone structures, stone walls, and intact tombs.

In 1969, Rapp and his assistants laid out more than 100 magnetometer grids and about 40 resistivity traverses on Nichoria Ridge. Workers then dug trenches at about 40 places where electrical and magnetic anomalies (abnormalities) were found. The diggers unearthed important archaeological features in 30 of the trenches while 8 contained features of geological significance. One of the two unrewarding anomalies was caused by a discarded 20th-century sardine can.

One important find attributable to geophysical prospecting was a pithos, a pottery jar. It was about 5 feet tall and 3½ feet wide at the center but tapered to a conically pointed bottom. The pithos had been used for a burial about 725 B.C., and it contained a skeleton, bronze ring, decorated vessels, an iron sword, and bronze bowls.

In a preliminary investigation, MME scientists had reconstructed the vegetational history of the region and gained information about

An excavator carefully
probes a burial site
at the Little Circle,
above, a tomb dating
from early Mycenaean
times. Local Greek
workers, *right,* open
a trench. They do most
of the heavy digging
at the Nichoria site.

Archaeologists gingerly clean a corroded bronze pitcher that had just been unearthed in pit 3.

changes in climate. Project palynologists, scientists who study plant spores and pollen, deciphered the past record of vegetation by studying preserved pollen grains. Pollen is produced by flowering plants and is spread by insects and wind. Most of the pollen grains that fall to earth are destroyed in a short time, but those that fall in uniformly wet or dry places, such as bogs, lakes, and desert caves, may be preserved for a very long time. The best pollen record is provided by bog deposits, where pollen sinks beneath the surface and is rapidly covered by a protective layer of fine sediments.

Scientists take samples from successive levels in a bog and separate the pollen grains. Microscopic analysis and counts of the different pollen grains can indicate the type of vegetation that grew in the area while that part of the deposit was accumulating. Problems do exist, however. For example, plants that need insects to disperse their pollen tend to produce smaller quantities of pollen than plants that rely on the wind, and this fact must be accounted for in the analysis.

Palynologist and geologist Herbert E. Wright, Jr., of the University of Minnesota, studied the modern pollen production as well as the ancient pollen record of the southwest Peloponnesus. Knowledge of modern pollen helped him to interpret the ancient record. He had difficulty finding a suitable deposit of ancient pollen anywhere in the entire Peloponnesus, but one was finally discovered in the Osmanaga lagoon, north of Pilos, almost 20 miles from Nichoria. Wright and his assistants analyzed two cores and recognized within them three distinguishable pollen zones. Radiocarbon dating provided a key to the age relationships for various levels in the cores. However, recent research on carbon-14 dating indicates that the dates assigned may have to be shifted in the direction of slightly more ancient ages.

Wright found that pollen zone A, the uppermost and youngest, extends back to at least 250 B.C. and has pine and oak pollen as well as the shrub *Erica*. Human activity rather than natural causes may have caused this pine-woodland environment to be reduced; trees were cut for lumber needed for ship and building construction, for fuel, and to clear the land for crops.

Zone B, underlying zone A, contained an abundance of olive pollen, indicating that large olive orchards existed in the region. The transition from zone B to zone C coincides with a decrease in olive pollen. This transition has been tentatively dated, by radiocarbon methods, at about the same time as the downfall of Mycenaean civilization. It is possible that the change in the abundance of olive pollen marks a shift to growing olives for survival during the Dark Ages of Greece following the demise of the Mycenaeans.

In general, Wright suggests that human activity, not major changes in climate, caused the shifts in vegetation. This challenges the belief of some archaeologists that the Mycenaean civilization declined after a changing climate caused crop failures.

Analysis of the animal remains at the Nichoria site was my responsibility during the 1971 and 1972 excavations. When specimens were found, they were brought to me in cloth bags that held anything from a few slivers of bone to a complex assortment that weighed 2 or 3 pounds. The bags were all identified with pottery, trench, and level numbers, all of which placed the remains that they held within a particular culture and age.

The bones arrived in my work area in the basement of the dig house at the close of each excavation day. There were as many as 30 bags, from all parts of the site, and in no particular sequential order. I had to open every bag and repackage the contents—cloth bags are expensive and must be sent back to the site—and make a quick diagnosis of the contents. For example, one bag contained a fragment of a pig molar, a leg bone fragment from a sheep or goat, some snail shells, and cow tooth fragments. We seldom find animal remains in whole parts, although almost-whole jaws turned up on occasion. Sometimes, the bones are charred by fire and this must be noted.

Unfortunately, there is at this stage little time for detailed analysis or measurement. Data accumulates piecemeal and understanding how it all fits together is difficult. I found time for study whenever possible, but most of the final conclusions regarding the ancient animals must await definitive study. With no assistants, and given the detailed nature of the work, and the oppressive heat, progress often seemed disappointingly slow.

Frequent parties on the rocky beaches, and sailing and sunning provide welcome relief for the staff members from their archaeological work at the dig house. Occasional trips to the nearest large town, Kalámai, allow a relaxing respite from the Greek food prepared at Nichoria.

Fortunately, Greek nights are generally relaxing. The dorms that house the staff at the dig house, though large and airy, can sometimes become hot and uncomfortable. Then, too, living barracks-style for long periods of time is unsettling. Perhaps it is the lack of space all to oneself or just the lack of opportunity to get away from people you are with all day, every day, from the start of the work at 5:30 A.M. until the last lights go out at about 11:00 P.M. In order to get away, to relax a bit, and to sleep comfortably, some of the younger staff members— most of them still graduate students—seek refuge by carrying their foam mattresses up to the roof and sleep beneath a night sky that has to be seen to be believed. The summer nights are generally very clear and stars abound. At full moon, you can almost read. Bats occasionally flutter through the night chasing glow bugs that twinkle on and off, while the nearby Gulf of Messini can be heard lapping gently against the shore. To the west, the lights of the small village of Petalidhi can be clearly seen, while to the east and farther away, the much more numerous lights of Kalámai, the regional "metropolis," are vivid. By dawn, the air becomes crisp and surprisingly cool, and after toast and coffee I was ready for another long session in the basement.

Systematic study of animal remains at archaeological sites is certainly not new. Animal studies form an important part of the research at sites involving very early man or manlike creatures. But too often in the past, information about animals is used as additional information rather than as a basic part of a project's reports. There are important reasons for tracing the geographic shifts of local natural animal populations during this time and determining whether man or the environment was mainly responsible.

My studies at Nichoria indicate that wolves, wild boar, wild cattle, and deer lived in the region at least for a portion of the time that men lived on the ridge. It is important to find out why these species became extinct there. Extensive woodland and grassland probably existed in the past, since those wild animals prefer such environments. And Wright's pollen analysis indicates that the climate has remained essentially the same for the last 4,000 years. Man may be responsible for causing the animals' extinction through hunting or by clearing the land for agricultural purposes.

An interesting discovery in the summer of 1972 was a second large pithos, much like the one dated to about 725 B.C. On some digs, scant attention would be paid to the soil found in a pithos. A washing and screening device developed by Strathmore R. B. Cooke of the University of Minnesota has made possible the large-scale recovery of very small objects from the earth in such jars. The soil from this particular pithos contained an incredibly large assortment of bone and tooth fragments. They were from such animals as snakes, lizards, frogs, toads, rats, shrews, and rabbits. The MME carried these remains back to the University of Minnesota, where Fred Grady, a curator and fossil preparator, and I are now making a detailed analysis of them.

Students sort and catalog collections of small ceramic fragments from the dig site.

The numerous frogs and toads of varying sizes are especially interesting. Amphibians are extremely vulnerable to hot, dry conditions, and their past existence on Nichoria Ridge indicates there must have been standing water there. Without water, these animals could never have survived hot and dry summers. Further study of the amphibians and the other small vertebrates may determine whether the idea of unchanging climatic conditions is accurate. However, we lack a good indication of what the modern small animals are like. This can only be provided by sampling the region.

We found the remains of both wild and domesticated large animals at Nichoria. The domesticated animals include pigs, sheep, goats, donkeys, cattle, and even horses. Only isolated teeth, leg bones, vertebrae, and skull fragments were found, some of them small and poorly preserved. But it would be surprising to find entire animals in midden heaps, the pits where garbage was thrown.

The preliminary findings indicate there are three times as many sheep and goats as pigs, and about twice as many pigs as cattle. Precise numbers are not possible now because the remains are so fragmentary and there is a ton or so of fragments still to be cleaned and analyzed. Closer analysis of this material may allow us to conclude who was eating what at a particular time and part of the site, perhaps revealing changes in economic status.

Bags filled with new finds arrive daily for cleaning and cataloging in basement workroom of the main dig house.

Wild animal teeth from the lower Mycenaean levels show greater wear. Grasses contain a good deal of silica in the form of bits of opal, and this material is extremely abrasive. I hypothesized that grasses were more common in these older times and caused the pronounced wear patterns I could see on the teeth. This hypothesis might have climatic consequences. Wild boar and deer remains in older levels seem to indicate the site had once been near open woodland.

Another feature of the animal remains that troubled me was the absence of shellfish. Only a few fragments of charred oyster or clam shells were turned up. People generally eat these shellfish in quantity, if at all, and the shell debris tends to be sizable. Perhaps the people of that time gathered shellfish only seasonally, and ate them near the shore. If this is true, shell middens should be found near the shoreline of that time, not on a ridge. The same sort of explanation might also account for the scarcity of fish remains at Nichoria Ridge.

The human bones recovered from the site were especially interesting. The remains of 16 or more individuals had at that time been recovered by Thomas Shay of the University of Manitoba, Canada, although the fragmentary nature of some of the remains makes a precise count difficult. At least 10 individuals were found in the remains of a single structure called the Little Circle, a small stone circle a little more than 6 feet across. The Little Circle is a tomb dating from early Mycenaean times.

The condition of the human remains found in the Little Circle sparked a good deal of discussion during the 1971 season. The skeletal remains were in disarray. Although they had not been completely exposed, the intermingling of the remains revealed that a number of individuals had been carelessly stacked in a heap in the center of the circle. The end of the season was rapidly approaching, and in their rush to interpret the available evidence, some of the staff concluded that the individuals had died violently. They cited the disarray of the remains and the fact that there were stones imbedded in three of the skulls. After carefully examining these skulls, I concluded that the stones could not possibly have been the instruments of death; it is far more likely that they had simply pressed against the skulls after death, causing them to fracture. I felt strongly about my interpretation, and several people agreed with me, but others were still attracted by the idea of violent death. Time was running out, however, and the project leaders decided to stop excavation in the Little Circle and complete the work the following season.

In my field report I suggested that a pathologist or coroner experienced in determining if foul play had or had not been involved be asked to join the staff for the 1972 season. The director acted on my suggestion, and the 1972 staff included just such an expert, Dr. Gary Peterson, assistant coroner of Ramsey County, Minnesota. After much work, Peterson concluded that the skeletons he examined showed no evidence of violent death.

I felt that the position I took the previous summer had been justified. I was not an experienced archaeologist, and I feared that my attitude at the time might seem brash. But without free expression and the right to question, there can be no flexibility and advancing scholarship. In fact, there can be no science.

Nichoria is not without its conventional treasure. A nearby tholos, a large, circular stone tomb that originally rose to a dome, was built for an important Mycenaean family. The tomb was approached through a long, unroofed passageway that led into a roofed passage. Although the grave had been robbed thousands of years ago, the expedition, working with the Greek Archaeological Service, recovered jewelry, small personal objects, and bronze pots and weapons from the tholos. However, the main grave, apparently built for Nichoria's local chief or petty king, was empty.

The MME excavations continued undaunted, with scientists from many different backgrounds piecing together the story of what life was like in the many small prehistoric households now being uncovered on Nichoria Ridge. The research patterns and techniques used at Nichoria may help establish a model for future archaeological studies in the Mediterranean region.

For further reading:

McDonald, William A., and Rapp, George R., Jr., editors, *The Minnesota Messenia Expedition,* the University of Minnesota Press, 1972.

New Insights
Into Immunity

By Jacques M. Chiller

**Research revealing that key cells cooperate
to fight infectious agents offers new hope
for victims of cancer and other diseases**

On May 19, 1973, a boy we will call Johnny Carter celebrated his fifth birthday with two large pieces of cake, four glasses of lemonade, an untold number of jellybeans, marshmallows, and peanuts, and a well-earned bellyache. But the day meant much more than the usual birthday to Johnny's parents. For their son, born with a rare defect that normally would have meant early death, had been saved by science when he was 5 months old.

Johnny is one of a few children born every year with immunity defects that render them helpless against bacteria, viruses, and other infectious agents which normal children handle routinely. He and perhaps 10 other children like him are alive today because physicians could inject into their blood streams the critical factor they lacked—the type of white blood cell responsible for immunity. Several recent breakthroughs in our understanding of these cells, called lymphocytes, have made many immunologists optimistic about greater success not only against diseases such as Johnny's, but also against cancer and a host of other maladies.

Immunity comes about as a result of the lymphocytes' jealous guard of the body through their recognition and destruction of all intruders. Experiments in many laboratories throughout the world, including

David Phillip, not yet a year old, was born with an immune deficiency disease that forces him to live in a germ-free plastic chamber. David can be handled only through the long rubber gloves in the chamber's sides.

The author:
Jacques M. Chiller is an immunologist and an associate in the Department of Experimental Pathology at Scripps Clinic and Research Foundation in La Jolla, Calif.

ours at the Scripps Clinic and Research Foundation in La Jolla, Calif., are producing a new picture of how the lymphocytes do this. Much of the most exciting work is uncovering the details of an unexpected division of labor among lymphocytes.

All lymphocytes look alike through a microscope. Because of this, we once assumed that all lymphocytes also function alike. However, there are actually two different types of lymphocytes, each with a different origin and special immunological properties and duties. Both types are descendants of stem cells, which are located in the bone marrow, but there the similarity ends. When the stem cells leave the bone marrow and enter the blood stream, some of them travel to the thymus, a gland located high in the chest cavity, behind the breast-bone. In some way we do not yet understand, these stem cells undergo changes within the thymus to become what we call T (for thymus-derived) lymphocytes, or T cells. They leave the thymus and populate such lymphoid tissues as the blood, spleen, and lymph nodes.

In human beings and many other animals, the rest of the stem cells migrate to these same lymphoid tissues, probably directly from the bone marrow. These lymphocytes are called B (for bone-marrow-de-rived) lymphocytes, or B cells. In birds, the stem cells destined to become B lymphocytes must first journey through the bursa, an organ near the tail. Many immunologists suspect that in other animals there is also a location, which they now refer to as the bursa equivalent, that turns stem cells into B cells. As yet, none has been found.

T and B cells parallel the difference in their origins with a striking difference in the roles that they play in the immune scenario. T cells find and directly attack certain bacteria, viruses, fungi, cancer cells, and whatever else they recognize as foreign. This remarkable ability to search and destroy has earned T cells the nickname "killer cells." Because T cells are particularly effective in ridding the body of cancer

Two Lymphocytes—Two Roles

Stem cells may become T cells, which attack agents they identify as foreign. Or they may become B cells, which form plasma cells that in turn make antibodies to attack intruders. Macrophages clean up after either process. All lymphocytes look alike, *bottom left*, until special techniques mark them clearly as T cells, *center,* or B cells, *right.*

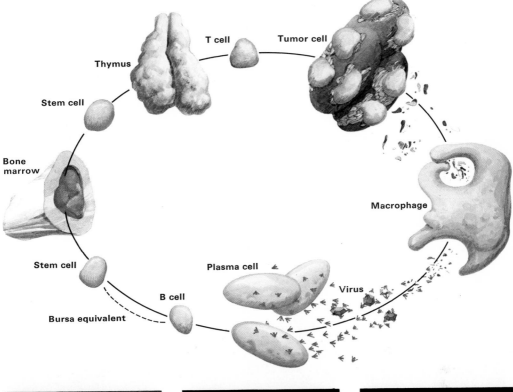

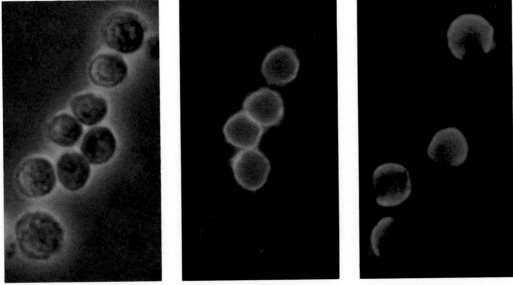

cells, many immunologists suspect that some sort of interference with the T-cell arm of immunity may be a major factor in some cancers.

B cells rid the body of intruders indirectly. They produce and release protein molecules known as antibodies. The cells then begin to differentiate and divide many times, forming plasma cells, and the plasma cells produce and release billions more antibodies. The antibodies travel through the blood stream to all parts of the body. When they encounter the intruder they were formed to combat, great numbers of the antibodies stick to it. Then, the intruder is destroyed with the aid of either another molecule, called complement, or a scavenger cell that is called a macrophage.

Antibody production is set in motion by the intruders themselves. That is, B cells recognize them as intruders and begin to produce antibodies and to divide to form plasma cells. This reaction is extremely specific. Each intruder triggers only one type of B cell. In

A Chemical Triggers Cooperation

Cooperation, in which a T cell somehow helps trigger B-cell activity, may result when a T-cell chemical is released while the receptors of the two cells link with the antigens of an intruder. The B cell then releases antibodies, *opposite page,* left, and produces plasma cells, center, that release more antibodies to attack intruders, right.

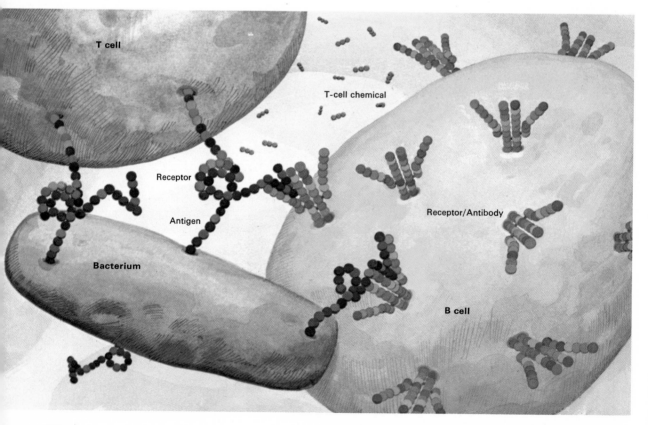

turn, the B cell and the plasma cells it forms produce molecules of only the antibody that will attack the intruder.

The specific intruder against which a B cell produces antibodies is determined by molecules called receptors that protrude from the B cell's surface. These receptors are complementary to—that is, they fit and will join together with—molecules called antigens. These protrude from the surface of all bacteria, viruses, and other intruders, and from normal body cells as well. But, like a key, an antigen fits only one lock, a receptor on the B cell that recognizes it. However, a twist of the antigenic key is not enough to open the B-cell door—that is, to trigger the formation and the release of antibodies. A T cell, which also has receptor molecules that recognize the intruder's antigens, is required.

Until 1966, immunologists did not know that the cells had to work together. We assumed merely that direct contact with the antigen somehow triggered lymphocytes to produce antibodies. Then, experiments performed by immunologist Henry Claman and his colleagues at the University of Colorado Medical Center in Denver showed that antibody production required some sort of cooperation between the two types of lymphocytes.

Claman and his co-workers exposed mice to X rays, which totally destroyed all their lymphocytes and their capacity to produce more.

Antibody

Plasma cell

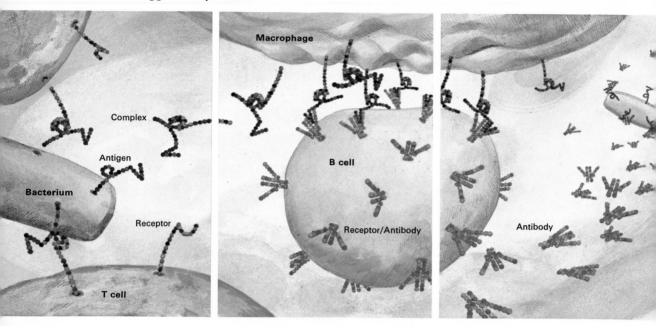

In one theory of cooperation, T-cell receptors link with antigens of an intruder, and the resulting receptor-antigen complexes break free, left, and stick to the surface of a macrophage. Thus, the macrophage surface becomes a concentrated source of antigens. The presence of so many antigens causes the B cell, center, to release antibodies, right.

Next, the scientists injected the animals with lymphocytes from normal mice. Some of the immunologically crippled mice got lymphocytes from the donors' bone marrow (a rich source of B lymphocytes). Others got lymphocytes from the thymus (a rich source of T lymphocytes). Still others got lymphocytes from both sources. Finally, the experimenters injected antigens into the mice and checked their ability to produce antibodies in response. The mice that had been given either thymus cells or bone marrow cells alone did not produce antibodies against the antigens. But the mice that had been given cells from both sources did produce antibodies.

The nature of this T cell-B cell cooperation, or more precisely, how the T cell triggers the B cell, is the puzzle that many immunologists are trying to solve today. A number of them, including Göran Möller of the Karolinska Institute in Stockholm, believe that a B cell would be triggered without a T cell if a large enough number of antigen molecules combined with the B cell's receptors for a long enough time. In the body, however, these requirements are rarely met. Therefore, the T cell somehow either increases the number and duration of antigen-receptor bonds or lowers the number and duration required.

In 1971, immunologist Marc Feldmann and his colleagues Erwin Diener and Tony Basten of the Walter and Eliza Hall Institute of

Self-Betrayal

Some auto-immune diseases occur because an intruder's antigens are similar to self antigens at the point where antibodies stick. As a result of this, the antibodies that are formed to fight the intruders, top, will also attack the body's own cells, bottom.

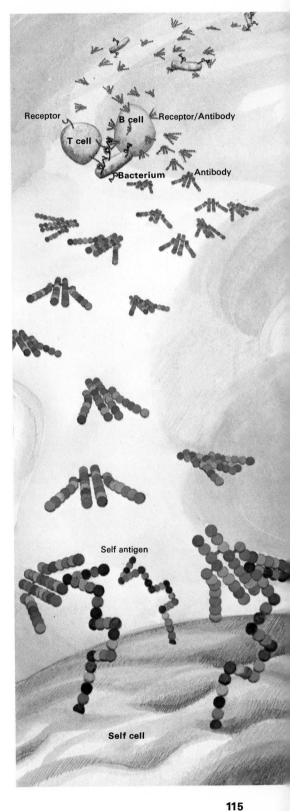

Medical Research in Melbourne, Australia, reported a series of experiments that seem to support this theory. They added two chemical forms of a protein antigen to B cells from a mouse. In one of the forms, the individual antigen molecules were joined together in long chains. In this arrangement, any antigen bound by a receptor would drag the other antigens in its chain within easy reach of other of the B cell's receptors. This guaranteed far more and longer-lasting antigen-receptor bonds than is normal. In the second form of antigen, the molecules retained their normal individuality. The cells that were mixed with the antigen chains formed antibodies despite the lack of T cells. Those mixed with the antigen as individual molecules did not form antibodies until T cells were also added.

Exactly how does the T cell lower the number and the duration of antigen-receptor bonds required to trigger the B cell? Some immunologists have concluded that the T cell releases a chemical that stimulates the B cell, making it more sensitive.

In 1972 and 1973, a number of teams of immunologists in several nations performed a series of similar experiments that indicated that such a T-cell chemical exists. Normally, both T and B cells are needed to trigger antibody formation in a test tube just as they are within the body. The scientists added T cells and antigens to a liquid and then drew off some of the liquid. When they added it to a second tube containing B cells and antigens, the B cells formed antibodies. Therefore, the scientists concluded, the cell-free liquid contained the suspected T-cell chemical.

In 1972, Feldmann reported experiments that led him to another interpretation of how the T cell helps the B cell. In his theory, when the T cell's receptors bind to the antigens of a bacterium, for example, a reaction occurs that releases the receptor-antigen combinations. These combinations then attach themselves, with the antigen out, to the surface of a macrophage. Thus, the macrophage's surface becomes a concentrated source of antigens. All the macrophage need do is present itself to the B cell to assure a sufficient number of joinings between the antigens and the B-cell receptors to trigger the production of antibodies.

To prove this, Feldmann and Basten stretched a membrane across a vessel to form two compartments. They put T cells and antigens in one compartment and B cells, macrophages, and antigens in the other. The pores in the membrane were too small to allow cells to pass from one side to the other, but they were large enough for passage of the presumed T-cell receptor-antigen complex. When all the elements were present, the B cells produced antibodies. However, when any of the participants—antigen, B cells, T cells, or, most important, macrophages—was omitted, no antibodies were produced.

Regardless of which theory explains cooperation, it remains puzzling why nature would choose such a complicated system for her creatures' biological defense. It may be that lymphocyte cooperation provides the balance between two conflicting needs. It is probably imperative that the immune mechanism be sensitive to extremely small quantities of some agents, particularly such organisms as viruses and bacteria that can quickly reproduce and overrun the body. For this reason, it is highly desirable that the T cell triggers the B cell to produce antibodies quicker than it would on its own. But it is equally important that there be some mechanism to keep the immune responses from destroying the body itself. The phenomenon of cooperation is also uniquely suited to fill this need.

The immune system's ability to discriminate self from nonself, tolerating self and destroying nonself, provides a fail-safe mechanism for self-protection without self-destruction. On a cellular and molecular basis, in self-tolerance, those lymphocytes that could recognize the antigens of our own cells and act against them must be eliminated or made inoperative. Because T cells and B cells must cooperate for antibodies to form, lack of response to self antigens by either cell would account for tolerance. In 1971, immunologists Gail Habicht, Bill Weigle, and I performed a series of experiments aimed at determining which cell type is most likely involved. We used antigens altered in a way known to produce tolerance to the unaltered antigens when they are later injected into animals. A variety of concentrations of the altered antigens given under a variety of conditions could make T cells tolerant. On the other hand, B cells very rarely became tolerant.

These findings indicate that in most cases of self-tolerance it is probably the T cells that are absent or fail to function. Therefore, even if the immune system possesses B cells with receptors that recognize self antigens, a lack of T-cell cooperation saves us from self-destruction. And this points up the advantage of lymphocyte cooperation: It enables us to respond to small quantities of nonself molecules and yet tolerate self antigens. Yet, in spite of the seemingly foolproof nature of such a system, there are a number of maladies in which our lympho-

Cancer's Strange Shield

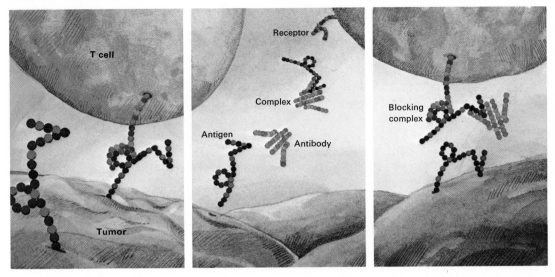

Normal defense against cancer probably begins when a T-cell receptor links to a tumor-cell antigen, left. The tumor may survive by shedding some of its antigens, which form a complex with antibodies, center, and block T-cell receptors, right.

cytes appear to turn against us. These auto-immune diseases include rheumatoid arthritis, rheumatic fever, and several forms of anemia. We do not yet know just how our lymphocytes betray us to cause auto-immune diseases. But our experiments and observations have provided enough information for speculation.

Remember that there are B cells that can form anti-self antibodies, but they do not do so because T cells that can respond to self are either absent or not functioning. One thing that might trigger the B cell would be a change in a self antigen, making it appear to be nonself and thus recognized as such by a T cell. Such a change could occur in several ways. The gene that produces the self antigen could be mutated, or a drug or other chemical might alter it. Viruses might cause cells to produce altered antigens or even new antigens. This could explain why auto-immune diseases occasionally follow severe cases of measles, a virus infection.

In rheumatic fever, there is an unfortunate physical similarity between antigen molecules found on the bacterium that causes the disease and some self antigens found on the heart tissue of the patient. Although B cells recognize the antigens on the heart, they normally produce no antibodies because they get no T-cell cooperation. However, the B cells that recognize the bacterial antigen do get T-cell cooperation, and they produce antibodies that attack not only the bacteria, but also the heart tissue. In this instance, T-cell tolerance to heart tissue is not enough of a safeguard against auto-immunity.

Auto-immunity as a result of even subtle cell changes may be a price we must pay for our lymphocytes' ability to recognize and destroy cancer cells which, although potentially deadly, are merely our own altered cells. The price is very low, considering the infrequency of auto-immunity and the probability that the immune system destroys

117

as many as 100 cancer cells per day that arise by accident in any normal person. Many scientists are now trying to determine exactly how the immune mechanism destroys cancer cells. They hope that by understanding nature's way of rejecting cancer, they may find new methods of treatment. The immune system probably marshals all of its forces to reject cancer cells. But T lymphocytes functioning as killer cells probably do most of the work.

Yet, cancer cells sometimes escape. How? Are the T cells that are capable of killing the cancer cells somehow eliminated or made tolerant? Apparently not, according to immunologists Ingegerd and Karl Erick Hellstrom of the University of Washington in Seattle. In a series of experiments, the Hellstroms and their colleagues have shown that cancer victims have killer T cells that will attack their tumor, but these cells are somehow blocked from doing so by something in the patient's blood. In these experiments, the scientists placed some of a

Mouse from a fast-aging strain, *right,* is old at 4 months of age. At 20 days of age, mouse of same strain, *below,* was injected with T cells of young mice of a normally aging strain. The mouse shows no signs of age even though it is 7 months old.

patient's cancer cells in a test tube with his own isolated blood lymphocytes, and the lymphocytes destroyed the cancer cells. However, when they added the patient's serum—a watery, cell-free portion of blood—the cancer cells were not destroyed. It is probably safe to assume that the killer cells in these experiments are T lymphocytes, but the nature of the blocking factor is still very uncertain. It may be antigens that have escaped from the surface of the cancer cells. Such free antigens could bind to the receptors of the killer lymphocytes and prevent them from recognizing the tumor through the remaining, attached antigens. It is also possible that the blocking factor is antibodies which cover the part of the tumor antigen that the killer cells' receptors would otherwise recognize.

The Hellstroms prefer yet a third possibility. Some of their experiments indicate that the blocking factor may be composed of free tumor antigen that has antibody attached to one portion of it. This antigen-antibody complex could attach to, and block, the killer cells' receptors in the same way that free antigen alone could. But, whatever the blocking factor may be, its discovery provides the faint hope that we can somehow eliminate it, allowing natural immunity to arrest cancer.

That cancer occurs much more often in the aged than in the young may be related to the fact that the immune mechanism, particularly the T cells, decline with age. For example, in humans, the thymus gland grows during childhood, reaching its largest size at puberty. After that, it slowly atrophies with age. As might be expected, the decrease in thymus size is paralleled by a decrease in the number of T cells. This has led some scientists to speculate about the possible role of thymus degeneration and T-cell losses in aging (see LENGTHENING OUR LIVES). In December, 1972, a group of scientists, including Nemo Fabris of the University of Pavia, Italy, and Walter Pierpaoli and Ernst Sorkin of the Swiss Medical Institute in Davos Platz, Switzerland, described dramatic experiments that led them to conclude that the thymus controls aging through its T lymphocyte.

The scientists used a strain of mice that are born with a hormonal disorder that results in several abnormalities including a tiny thymus and greatly accelerated aging. When the scientists provided these animals with either T cells or hormonal treatments that triggered full development of the thymus, they aged normally.

Exactly how T cells might control aging is still a mystery. But a growing number of observations linking immunity to aging will undoubtedly cause a surge of scientific activity in this area. Ultimately, then, the fruits of immunological research may span the life of man—not only save Johnny Carter, but also allow him the pleasure of attending his great-grandson's fifth birthday party.

For further reading:

Greaves, Mel, "Cooperating in Immunity," *New Scientist,* Dec. 9, 1971.
Raff, Martin C., "T and B Lymphocytes and Immune Responses,"
  Nature, March 2, 1973.

Lengthening Our Lives

By Bernard L. Strehler

Current research into the molecular basis of aging may lead to a longer, healthier life span for man

When I was about 5 years old, the kindly old minister of our church died. For the first time, I became aware of a universal biological fact: All living creatures grow old and die. A few years later, when Duke, my little fox terrier, died I began to view aging and its ultimate conclusion, death, as a puzzle rather than an inevitability. This puzzle —why living things start off life vigorously, then slowly decline and eventually die—has fascinated me ever since. It has been a central part of my life as a scientist for nearly 20 years.

Recent scientific developments, some of them in my own laboratories at the National Institutes of Health (NIH) in Bethesda, Md., and later at the University of Southern California, probably signal the beginnings of important new research on aging. These developments have led me to the surprising conclusion that the aging puzzle will be essentially solved by the year 2000—perhaps sooner. We will then have the tools to considerably lengthen man's life span.

The prospects have not always been so bright. The history of *geron-tology* (the study of aging), like that of many other areas of science, is strewn with false starts, false assumptions, and false hopes. Gerontology began in the late 1800s when Russian biologist Élie Metchnikoff recommended removal of the large intestine to combat aging. Metchnikoff believed that the bacteria in this portion of the intestines slowly released poisons that caused people to grow old.

Serge Voronoff, a Russian-born French surgeon, believed that the gradual decline in body function that is part of the aging process resulted from the slow deterioration of various glands, particularly in the sex organs. In his book *The Study of Old Age and My Method of Rejuvenation* (1926), he firmly declared that youthful vigor would be restored if sex organs from young animals were to be transplanted into the bodies of aging human beings.

Among modern biologists, Sir Peter Medawar of England first firmly connected aging with genetics, through which it would later be tied to biochemistry and molecular biology. Elaborating on ideas of British biologist George P. Bidder, Medawar proposed in 1951 that the delayed effects of pleiotropic genes cause aging. Pleiotropic genes have several different effects, some beneficial, some harmful. The pleiotropic genes involved in aging in most land animals and birds are those that also keep them within their proper size range.

There are only two ways that such genes can keep an animal from growing indefinitely. They can either allow the animal to replace all of its parts at exactly the same rate as they are lost, or they can stop the animal from adding any new parts. Many tissues and organs of our bodies, including skin, intestines, and blood cells, replace lost cells right up to the end of our life. Others, including most key tissues and organs, such as muscles, the heart, and the brain, stop adding new cells after a certain stage of life.

The implication is clear: As irreplaceable cells die, the parts they make up deteriorate and perform their life-supporting functions less and less satisfactorily. To put it another way, the parts and, therefore, the living organism to which they belong, age.

The deterioration of irreplaceable cells and their body parts is explained by a universal natural law called the law of entropy increase. Put simply, this law states that all things become increasingly disorganized as time passes, unless energy is used to prevent it. For example, if you tidy up a room, almost anything else you do in the room makes it more disordered. Unless somebody tidies up the room periodically (uses energy), the eventual result of continued activity will be almost complete disorder.

In our body, the molecules within cells, and sometimes even the cells with respect to each other, become increasingly disarranged. When large numbers of the molecules and cells in a vital organ, such as the heart or brain, become so disorganized that they and the organ can no longer function, we die. Biologist Leonard Hayflick of Stanford

The author:
Bernard L. Strehler, a professor of biology at the University of Southern California, has studied aging throughout most of his scientific career.

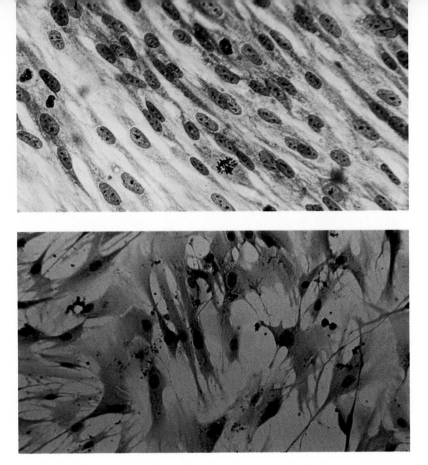

The cells in young human lung tissue, grown outside the body, *above left*, become increasingly disorganized, *left*, as they age and die. This process, called entropy increase, also occurs in many of the vital organs within the body, and is instrumental in aging.

University has shown that many kinds of human cells grown outside the body reach this stage after about 50 doublings by mitosis, a method of cell reproduction. However, there is no evidence that within the body all these cells are limited to 50 doublings.

Even before we understood aging to this extent, we had discovered ways to control it to some degree in experimental animals. For example, as far back as 1917, physiologist Jacques Loeb and biochemist John H. Northrop of the Rockefeller Institute (now Rockefeller University) in New York City discovered that temperature affects aging in the fruit fly, *Drosophila melanogaster*. In their experiments, fruit flies lived for 35 to 50 days at 26°C. (79°F.), for about 100 days at 18°C. (64°F.), and more than 200 days at about 10°C. (50°F.). Thus, within reasonable limits, the life span of this insect roughly doubles every time its body temperature is lowered about 8°C. (14°F.). The same has been found to be true of other cold-blooded animals, including a variety of insects, fish, and tiny pond-dwelling rotifers.

Warm-blooded animals present what at first seems an insurmountable obstacle to such experiments. The body temperature of cold-blooded animals will conform to that of the surrounding environment, but warm-blooded animals use internal mechanisms to keep body temperature at a point that is normal for their species. However, hibernating mammals provide a unique opportunity to compare aging processes in the same warm-blooded animals at different temperatures. Their body temperature is lower during hibernation than when they are active. One such comparison revealed that when hamsters hibernate, their red blood cells live several times longer than the approximately 100 days that they do under normal conditions.

That body temperature should play such a key role is not surprising, because all kinds of chemical reactions, including those involved in aging, slow down as temperature drops. In terms of entropy increase, heat is a great disorganizer. This is easy to see by looking at how heat affects matter. In its coolest state, matter is frozen into a solid, which is its most stable form. For example, water becomes ice. Heat matter, however, and the molecules of which it is composed become more disordered. Ice becomes water. Heat it still more, and the molecules move about even more erratically. The water becomes steam.

If human beings responded the same as fruit flies, we could live for many hundreds of years at reduced body temperatures. There is no evidence that reducing human body temperature by 2 or 3°C. (3.5 or 5.5°F.) would interfere in any important way with body function. If this is so, by reducing body temperature, a human being's life span might be increased by from 25 to 40 years.

The main obstacle to lowering human beings' temperature indefinitely is our automatic temperature-adjusting mechanism. This is already under attack by neurophysiologist Robert D. Myers and his co-workers at Purdue University's Laboratory of Neuropsychology in Lafayette, Ind. Our body thermostat, a key element in the mechanism, is a small section of the hypothalamus, a part of the brain that lies just above the pituitary gland. In experiments reported in 1971 and 1972, the scientists inserted tiny plastic tubes into the hypothalamus of monkeys. The effects of various substances introduced into the hypothalamus through the tubes were then recorded. Small quantities of sodium ions caused the thermostats to raise body temperature, but body temperature dropped by as much as 4°C. (about 7°F.) when small quantities of calcium ions were added.

We may never have to adapt such techniques to human beings, however, because research indicates that we may be able to learn to lower body temperature without artificial stimuli. Australian Aborigines are a good example. They wear very little clothing and live in an environment where warm days give way to nights during which temperatures fall to nearly freezing. The Aborigines sleep in the open, and their body temperature drops several degrees. They do this in part by repressing the reflex that causes shivering, which is a heat-producing reaction to cold. When they awaken, the Aborigines shiver, raise their body temperature, and resume their daily routine.

Experiments by psychologist Neal E. Miller of Rockefeller University indicate that we may all have the Aborigines' skill latent within us. Using a technique called biofeedback, Miller has taught people to control such internal functions as heart rate, blood pressure, and brain

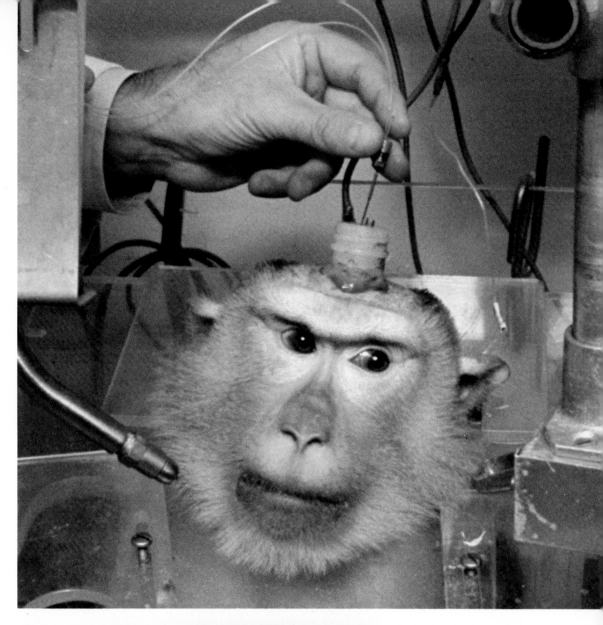

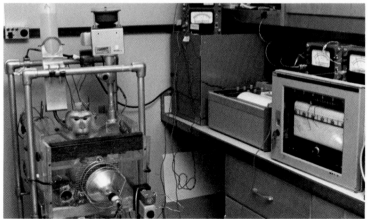

A plastic tube inserted into a monkey's skull allows test solutions to be dripped into the hypothalamus, which contains the thermostat of all warm-blooded animals. The effect on the monkey's body temperature is determined through analysis of records that include how often and for how long he turns on a heat lamp, *left*. Such experiments may lead to ways of slowing aging in human beings by lowering body temperature.

waves. In related experiments, psychologist Elmer Green of the Menninger Foundation in Topeka, Kans., has verified that some persons can deliberately control the surface temperature of their body (see THE INS AND OUTS OF MIND-BODY ENERGY). But it is the regulation of the temperature deep within the body that is critical if we are ever going to be able to "cool it" voluntarily to retard aging. Nobody has yet determined if people can learn this type of temperature control.

Another way to extend our life span might be to restrict our diet. As long ago as 1930, experiments performed by biologist Clive McCay of Cornell University showed that rats fed very limited quantities of food lived almost twice as long as rats that ate normal amounts. In January, 1973, biochemist Charles H. Barrows of the NIH reported experiments in which restricted feeding increased the average life span of laboratory mice from 23.5 months to 28.5 months. Barrows also discovered that there was a 1°F.-decrease in the body temperature of the mice. However, although it probably played some small role, this temperature reduction is too small to account for the increase of almost 25 per cent in life span.

Overeating and being overweight shorten a human being's life span. But there is no evidence that life expectancy can be greatly increased, as that of rats was, by extreme dieting. One study that may ultimately link greater human longevity to restricted diet is that of zoologist David Davies, a member of the Unit of Gerontology at University College, London. Davies studied the large number of aged persons living in Vilcabamba Valley in Loja Province in Ecuador. He found that in this idyllic valley, about 5,000 feet above sea level in the Andes Mountains, ages of 100 or more were not exceptional. Over 7 per cent of the persons living on the farms in the valley are over 80 years old. In all of Ecuador, considerably less than 1 per cent of the population is 80 or over, and in medically sophisticated Western nations, such as the United States and England, less than 2 per cent of the people are that old.

The adults of Vilcabamba consume only about 1,700 calories a day, more than 500 fewer than the average for Ecuador, and roughly half the number consumed by adults in the United States and England. But, other factors may also play a role in the Vilcabamba phenomenon. Some of the valley's inhabitants attribute their long lives to various herb teas they drink. Some physicians who have visited the valley claim the tranquillity and resortlike climate are the major factors.

The gross changes of aging are visible to all of us at a glance. But gerontologists do not think of graying hair and wrinkling skin as aging. We are concerned with tissue and cell changes, and some of these changes can readily be seen with the aid of a microscope. If one compares young and old tissue of the type that does not replenish its cells, such as brain or heart tissue, two differences are immediately apparent. First, cells in the old tissue are arranged in less orderly fashion than those in the young tissue. This is a biological expression of the law

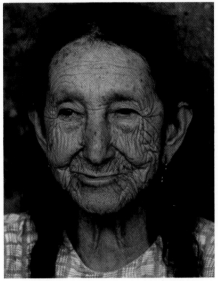

Gabriel Erazo is 125 years old and still works in his garden. Micaela Quezada is 102 years old and still does her own housework. Both live in the valley of Vilcabamba, Ecuador, where the great longevity of many of the inhabitants may ultimately be traced to a low-calorie diet.

of entropy increase. Second, cells in the old tissue contain dark substances called lipofuscin, or age pigments. These pigments were first isolated in 1958 almost simultaneously in several laboratories, including my own. Chemical analysis indicates that age pigments are a sort of cellular garbage that the cells cannot excrete. They are believed to be the breakdown products of a reaction between oxygen and unsaturated fats present in various cell membranes. The reaction, called oxidation, is much like that which causes floor varnish, also containing unsaturated fats, to harden and yellow as it ages.

Scientists have not yet been able to prove whether age pigments damage cells. But it seems highly unlikely that a cell with much of its interior—25 per cent or more in the brain cells of aged rats and human beings—occupied by these nonworking elements could perform as well as it did before they accumulated.

This has led to experiments with compounds called antioxidants, which retard the process that forms lipofuscin. For example, in 1969, physiologist Denham Harmon of the University of Nebraska, Omaha, reported experiments in which he fed mice BHT, a synthetic food additive that is used in small quantities to prevent oxidation of fats in foods such as potato chips. These mice lived about 25 per cent longer than normal. However, vitamin E, another antioxidant, has no such effect, leading some scientists to wonder whether the results with BHT are due to its antioxidant activity or to some indirect reaction such as decreasing the animals' appetite and food intake.

We know that our genes control the rate at which lipofuscin accumulates. The most tragic evidence for this comes from inheritable diseases that cause large quantities of age pigments to accumulate in cells of certain children. Those who inherit the genes for Batten's disease, for example, quickly accumulate vast quantities of lipofuscin in their brain cells. They become blind when they are about 5 years old, and this is followed by gradual mental deterioration and death. Progeria, another disease of children, produces more general premature aging. Children with this rare malady reportedly not only accumulate excessive amounts of lipofuscin in their brain cells, but also become bald and develop wrinkled skin, fragile bones, and hardening of the arteries. By the time they are 10 years old, the unfortunate victims look like wizened old men. Most of these children die of a heart attack at about 12 years of age.

Because genes are at the root of all aging—abnormal or normal—we must understand how and why the genes that cause it work. Fortunately, scientists have made enormous progress in the last two decades in learning what genes are and how they function. For example, we know that certain kinds of genes are "switched on" at just the right time in the life cycle of cells and organs. More important to gerontologists, however, we know that some genes—such as the pleiotropic genes in the cells of nonrenewable tissues and organs—are also "switched off" at specific times. While they were active, these genes produced many of the essential parts of their cells. With the genes turned off, the cells' supplies of the genes' products begin to dwindle as entropy increase deteriorates and destroys them. After a sufficient amount of such deterioration and destruction, the cells can no longer function, and they die. Accumulated damage of this kind in many cells and many organs is what aging really is.

So, the key question in the aging puzzle is: How does a cell selectively switch off some genes? We know that virtually all of the functioning parts of the cells in our body are proteins (including enzymes and structural proteins), which are produced indirectly by our genes. So, we can ask the question more specifically: What makes some genes stop producing their proteins?

All cells, whether they are from bacteria or man, follow a surprisingly simple general pattern when they translate the information hidden within their genes into proteins. A protein is simply a folded linear chain of various amino acids that are attached to one another in a specific order. A gene is a small section of another chainlike molecule, called deoxyribonucleic acid (DNA). The functional units that make up the DNA chain are called genetic bases. The bases form a molecular code in which three bases in a sequence form a code word that determines a particular amino acid.

The first step in producing a protein from its gene is to make a complementary copy of the gene's DNA and, thus, its molecular instructions. This copy is made in the form of another chainlike molecule, called messenger ribonucleic acid (mRNA). Next, the code copy on the mRNA is deciphered by molecules of a second form of RNA, transfer RNA (tRNA). There are 60 or more kinds of tRNA. With the help of an enzyme, each kind can bind a single kind of amino acid to itself and then attach itself to the portion of the mRNA that contains the code word for that amino acid. Then, with the aid of a cell structure known as a ribosome, the amino acids are assembled, one at a time, into the protein for which the original gene was coded.

In my laboratory, my colleagues and I are now trying to find out how this protein-making machinery is switched off during aging. We know the answer to this question in bacteria. Their protein synthesis is shut off at the gene, or DNA, level when a substance called a repressor binds itself to the DNA, blocking the formation of mRNA copies. But, the cells of higher forms of life may have evolved other methods of stopping the production of a protein.

One such method would be to somehow eliminate or interfere with the functioning of one or more kinds of tRNA. Since 1966, my colleagues and I have been searching for evidence that some tRNA molecules in old cells cannot be attached to the amino acids they normally provide in protein synthesis. In our first experiments, we extracted the contents of cells from 1-year-old and 6-year-old rabbits. When we tested the contents for their ability to produce tRNA-amino acid complexes, we could detect no differences.

The cotyledons, or first leaves, of soybeans peek out of the split seed of a young plant, reach maturity after the first top leaves develop, and rapidly age and die as the second top leaves grow. Birth to death takes only two weeks, making cotyledons ideal for aging research.

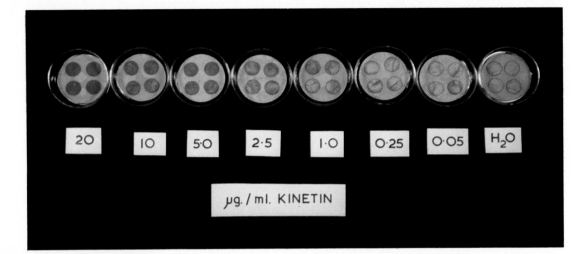

| 20 | 10 | 5·0 | 2·5 | 1·0 | 0·25 | 0·05 | H₂O |

μg. / ml. KINETIN

Disks cut from leaves, then left for four days on paper moistened with different concentrations of the hormone kinetin, yellow and age at different rates. In a lighter mood, the researcher, Daphne Osborne, also painted her initials on a leaf with kinetin and let the leaf age. The experiments indicate that the hormone retards aging in plants.

Although we were discouraged, we reasoned that the genes involved in aging might already be switched off in a 1-year-old rabbit. In other words, unlikely as it may seem, a 1-year-old rabbit might already be fully committed to later aging.

In 1967, because we were having difficulty obtaining research money, we decided we would have to work with a much less expensive experimental organism. We chose the cotyledons, or first leaves, of the soybean plant. Although aging might be different in plants than in human beings and other animals, we reasoned that there are probably many similarities. In addition, plants, and particularly their cotyledons, offer several advantages for our experiments.

Cotyledons lie dormant, folded up within a seed until it sprouts. Then, they unfold, turn green, and supply the new seedling with food until the regular leaves appear and take over. As these leaves develop, the cotyledons turn brown and dry up (age) and fall off the plant (die). All this happens in about three weeks after the seeds have sprouted, making it possible to carry out many experiments on aging in a short time. In addition, the plants require very little space and care.

Neither my colleagues nor I had ever done much work with plants, so we invited biochemist Joe Cherry, on leave from Purdue University, Lafayette, Ind., to join us. Cherry is an expert on the soybean plant, and his help proved invaluable in getting our work underway. One of my students, Michael Bick, and I did many different and revealing experiments with the cotyledons. The most important of these, performed in 1970, was similar to the one we had tried unsuccessfully with rabbit cells. We extracted the contents from cells of young (5-day-old) and old (21-day-old) cotyledons and tested their ability to attach amino acids to tRNA molecules.

To our delight, we found that the extracts from the old cotyledons were very poor at attaching two amino acids, leucine and tyrosine, but the extracts from the young cotyledons had no such difficulty. This is exactly what our original theory had predicted. Further, we found that the older cotyledons contained one or more substances that combine with and inactivate the enzymes that normally attach leucine and tyrosine to tRNA molecules. In other words, cotyledon cells seem to be genetically programmed to make materials that slowly accumulate and ultimately inhibit protein production.

If we find that human cells are also genetically programmed to produce inhibitors that ultimately interfere with the formation of tRNA-amino acid complexes, we may be able to control our aging through a counterattack. The protein-making activity might be turned back on by using antidotes to the inhibitors. Although different combinations of inhibitors may switch off the machinery of different kinds of cells, the total number of inhibitors may be relatively small. If this is the case, we may only need to develop a few dozen specific antidotes and apply them in the right combinations to awaken sleeping genes in any kind of cell.

Antidotes that combat aging already exist in nature. For example, botanist Daphne J. Osborne of the Unit of Developmental Botany at the University of Cambridge, England, has found that a plant hormone called kinetin will keep cotyledons and other leaves green and healthy long after they normally would have died. In 1970, biologist Judith Ilan of Temple University, in Philadelphia, reported on experiments in which she administered a synthetic hormone to adult meal worms. The hormone caused the insects to form a series of new skins, a process that normally occurs only in young, growing meal worms. Ilan's experiments showed clearly that the hormone restored the protein-making machinery of the insects' cells to a younger stage.

In human beings and other animals where aging is tied most closely to cells that have stopped reproducing, the genetic activity in these cells should be particularly revealing. In work we reported in December, 1972, one of my students, Roger Johnson, and I uncovered exciting evidence of a new kind of genetic change that seems to be restricted to older cells of this type.

Johnson and I had set out to disprove the theory that damage to DNA is an important cause of aging. Earlier experiments by many other scientists had already eliminated one form of DNA damage—ordinary mutations—as a cause. However, they did not rule out another possible type of damage, the gradual loss of copies of certain genes that exist by the hundreds, side by side in each cell's DNA. Although the idea seemed improbable to us, we set out to test it by comparing the number of copies of one such gene in brain cells from young and old beagles. The repetitive gene that we chose produces a

Genes and Aging

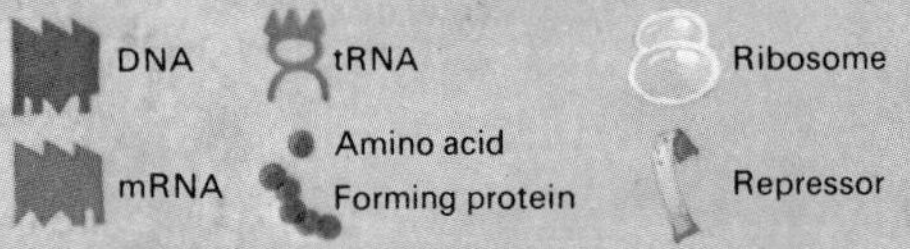

DNA

tRNA

Ribosome

mRNA

Amino acid

Forming protein

Repressor

The genetic activity that produces proteins can be blocked or slowed in several ways to cause aging. Assembly of a messenger-RNA (mRNA) complementary copy of a gene's DNA code, below, can be blocked by a substance called a repressor, bottom. This occurs in bacteria. A number of experiments indicate that in more complex living creatures, normal protein synthesis, right top, is impeded after the mRNA is assembled. In one case, right center, a transfer-RNA (tRNA) molecule without an amino acid to add to the growing chain aborts synthesis, and an incomplete chain of amino acids is released. In another, bottom, too few proteins are being synthesized because ribosomes have become scarce.

Copying the message

Blocking the message

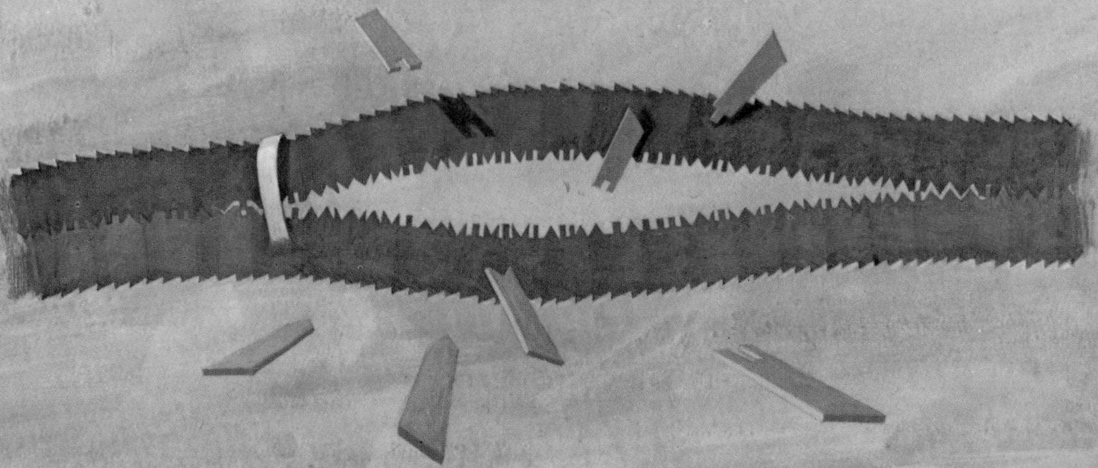

Translating the copy

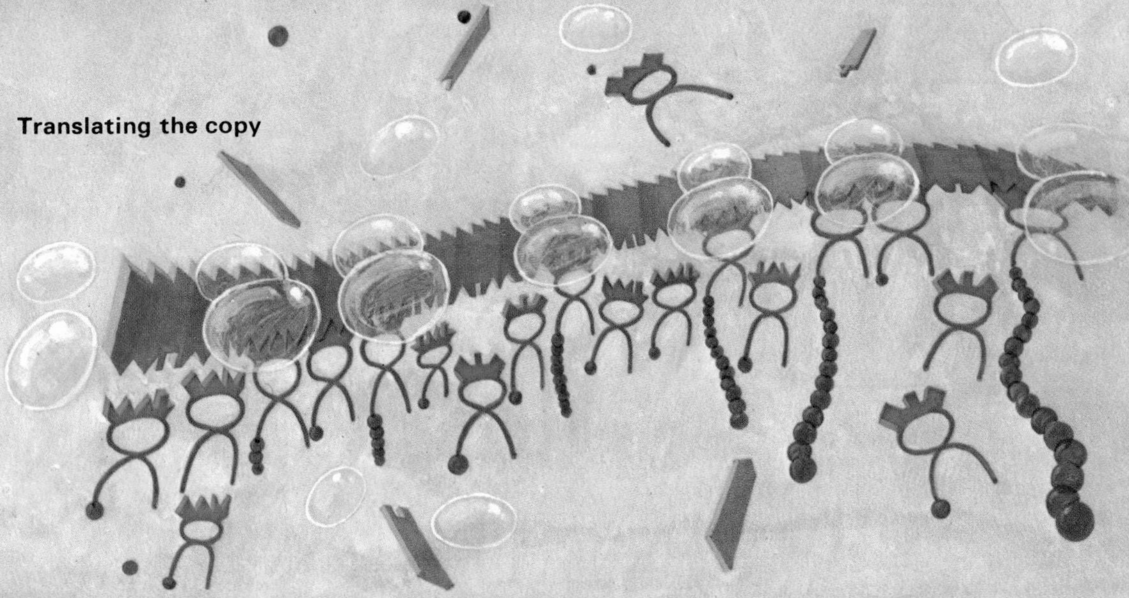

Aborting translation

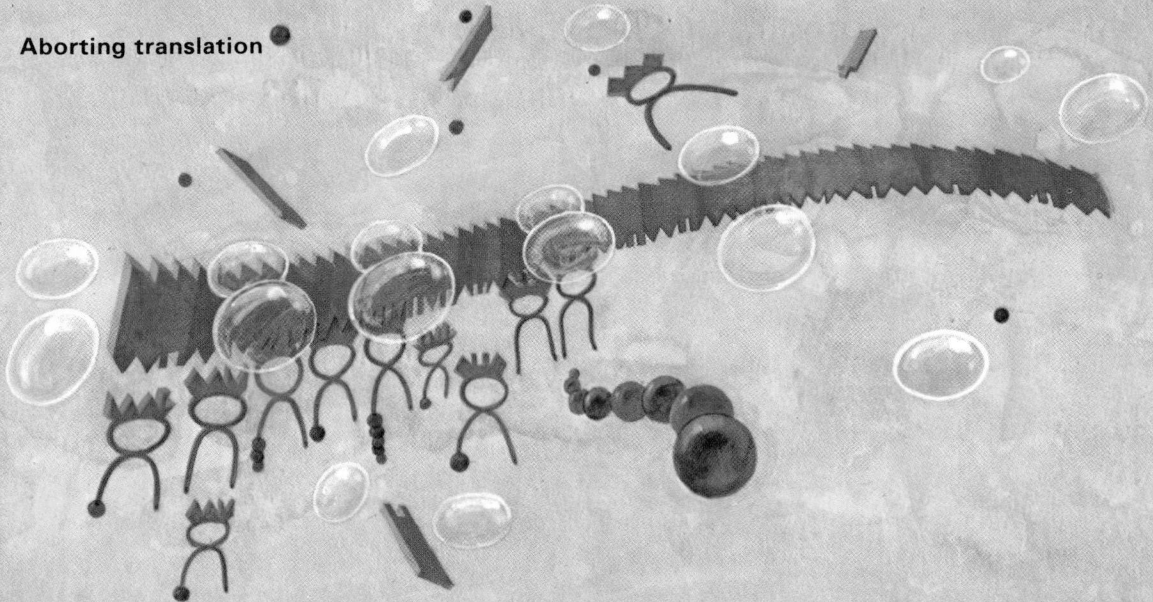

Slowing translation

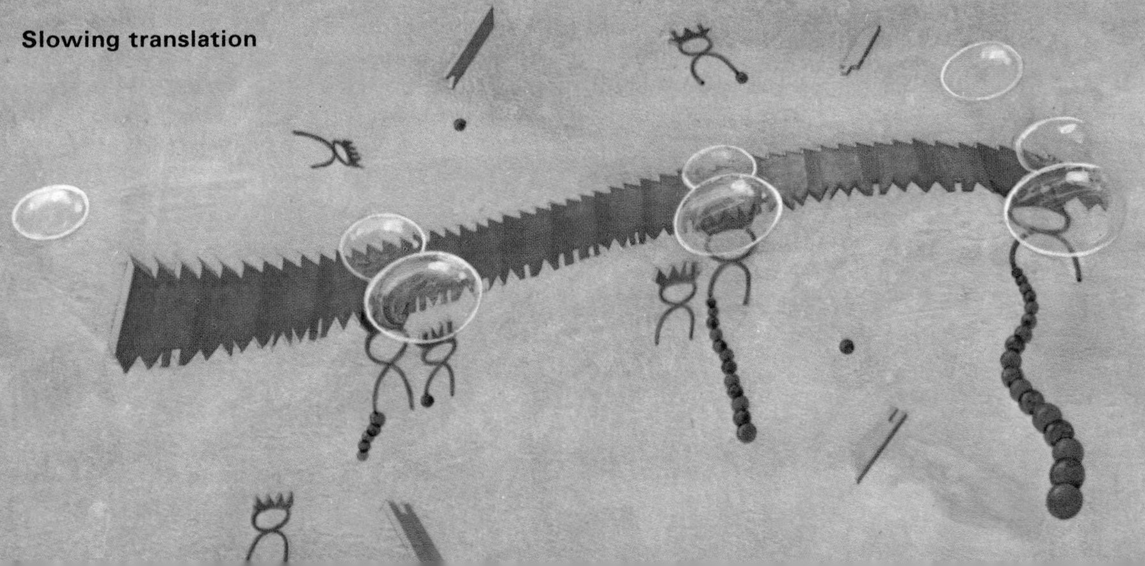

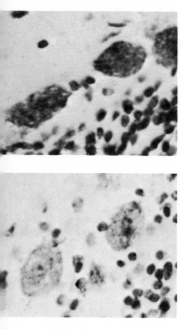

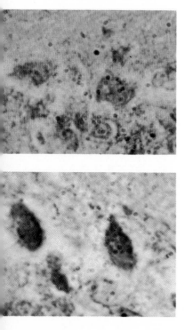

Color intensity shows that a young rat's brain cells, *top*, contain more ribosomes than those of an old rat, *above*. Young rat's brain cells, *below*, also contain far less lipofuscin than those of an old rat, *bottom*.

major constituent of ribosomes, ribosomal RNA (rRNA). This gene is a convenient one to check because the rRNA is assembled directly upon it and can later be used to detect it through a standard laboratory technique called RNA-DNA hybridization.

In our experiments, we first isolated DNA from brain cells of 1-year-old and 10-year-old beagles and trapped it on filters with very fine pores. Next, we immersed the filters in a solution that contained rRNA that had been labeled with a radioactive atom. Because the rRNA matched the DNA genes coded to produce rRNA, it attached selectively to those genes wherever they were in the DNA on the filter.

When we determined the relative number of copies of the gene by measuring the radioactivity on the filters containing the DNA, we found that the brain cells of the old dogs contained about 30 per cent fewer copies of the gene than did those of the young dogs. We also measured the number of copies of the gene in other tissues of these animals, and found that all of those tissues whose cells had stopped reproducing (brain, heart, skeletal muscle) showed similar decreases. However, none of the tissues whose cells were still reproducing (liver, kidney) showed a decrease.

In other words, the protein-making machinery of heart cells, brain cells, muscle cells, hormone-producing cells, sex cells, cells that produce antibodies, and all other nonreproducing cells may stop working effectively for the same reason—too few ribosomes. If further studies should establish this as a universal accompaniment of aging in higher animals, extending the healthy life span might be achieved by increasing the number of ribosomes available in the cells of critical tissues. We know that the genes that produce rRNA increase greatly in number during the later stages of maturation of the sex cells—eggs and sperm—in some animals. This selective gene amplification is under the control of other genes that are switched off before fertilization. Perhaps reactivating these genes in an adult would trigger the production of new rRNA genes and, indirectly, more ribosomes, thereby rejuvenating the individual. If this should prove too difficult, copies of the genes that produce rRNA might be synthesized in large numbers and carried into body cells as parts of specially produced infective viruses. Although many scientists disagree, I believe that either of these methods could be practicable by the year 2000.

How quickly we advance in understanding aging hinges on the quality and magnitude of our research. The NIH spent only about $1-million in 1972 on aging research at the cellular and molecular level. Yet, at the same time, vast sums went to research on cancer, heart disease, and stroke, the three major killers of the elderly today. To me, this seems a tragic oversight, for these and many other diseases can be traced in large part to the gradual deterioration of the aging parts of the body. So, understanding aging may contribute more to the conquest of many diseases than can understanding the diseases themselves. Moreover, the conquest of cancer, heart disease, and stroke will add, at best, only from 7 to 15 years to the average lifetime. This is because the number of deaths due to accidents at age 85 is equal to the number from all causes, including the major killer diseases, at age 70. And, the tendency toward death from accidents is a result of the gradual deterioration of the body, that is, of aging.

Once we control aging, there will be enormous changes in our lives. On the positive side, there will be little or no fear of the pain, impotence, and declining vigor that so many old persons face today. In addition, each of us will have many extra years during which we can work to benefit ourselves and others. For example, you may have the opportunity to pursue several interesting and creative careers—now an impossibility for most of us.

On the other hand, longer lifetimes will create many complex social and philosophical problems. For example, there will probably continue to be a large population increase, causing crowding that could necessitate strict controls on childbearing. Industry, now geared to a 40-year productive lifetime for a worker, will have to adjust. Insurance companies will have to create new risk and premium schedules. The present social security and retirement systems will be outmoded.

Some persons instinctively shrink from new ideas and achievements. They see each advance man has made in understanding nature as a threat to the narrow mode of existence they have learned to deal with. They regard interference with so-called natural laws as distasteful, if not sinful and dangerous—as though all the best things had already happened. Such people believe that manipulating man's life span bodes ill for humanity. But, because I am basically an optimist about the human race and am more impressed than depressed by our past record, I strongly suspect that we will meet the challenges of our extended longevity and make it a blessing.

For further reading:

"Aging: Attacking the Universal 'Disease,'" *Medical World News,* Oct. 22, 1971.

Davies, David, "A Shangri-la in Ecuador," *New Scientist,* Feb. 1, 1973.

Johnson, Roger, and Strehler, Bernard L., "Loss of Genes Coding for Ribosomal RNA in Aging Brain Cells," *Nature,* Dec. 15, 1972.

Myers, R. D., and Yaksh, T. L., "Thermoregulation Around a New 'Set-Point' Established in the Monkey by Altering the Ratio of Sodium to Calcium Ions Within the Hypothalamus," *Journal of Physiology,* vol. 218, pp. 609-633.

Strehler, Bernard L., "New Age for Aging," *Natural History,* February, 1973.

Strehler, Bernard L., "The Understanding and Control of the Aging Process," chapter 7 of *Challenging Biological Problems*, edited by John A. Behnke, Oxford University Press, 1972.

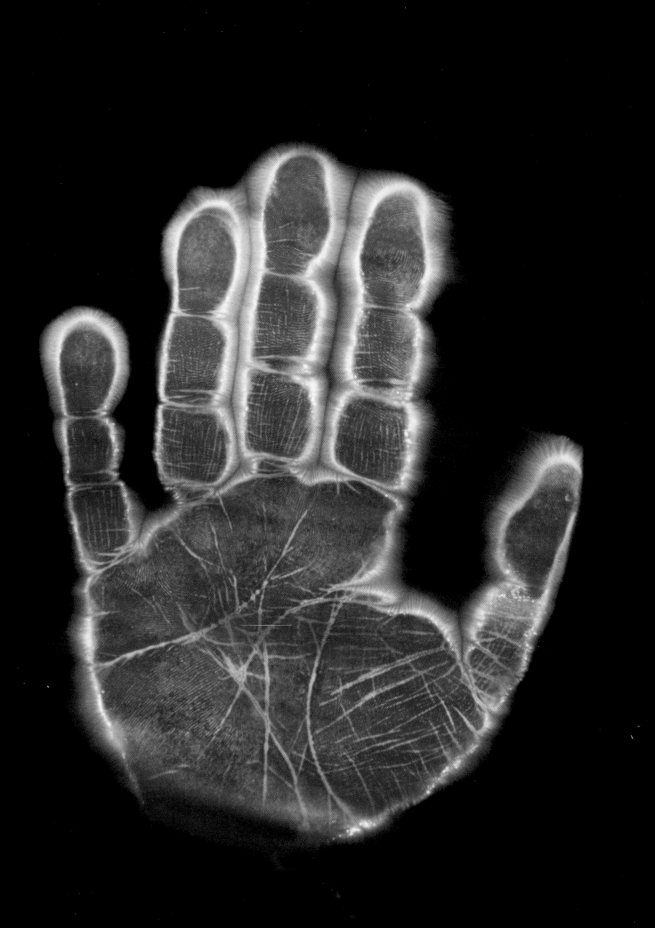

The Ins And Outs of Mind-Body Energy

By Elmer and Alyce Green

A new research frontier is developing in which physics, psychology, parapsychology, and medicine are blending to form a new "science of consciousness"

Seated in a chair in our laboratory at the Menninger Foundation in Topeka, Kans., a 45-year-old Indian yogi named Swami Rama performed an incredible feat. While seven of us watched, the Swami caused a 14-inch aluminum knitting needle, mounted horizontally on a vertical shaft 5 feet away from him, to rotate toward him through 10 degrees of arc. The Swami had been fitted with a plastic mask that covered his nose and mouth. He breathed through a foam-rubber insert which was covered by a plexiglass shield to deflect any "air currents" down to the sides. Even with this, one of the observers was convinced that the Swami had used some method that could be explained by some already known physical law.

We had warned the Swami that even if he succeeded in demonstrating this kind of phenomena not everyone would accept his explanation of how he had done it. He replied, "That's all right. Every man can have his own hypothesis, but he still has to account for the facts."

Energy was recorded emanating from a hand placed on a sheet of film. The film was exposed by an energized copper plate beneath it.

In science, facts have always been more sacred than theories. But a nonconforming fact usually becomes scientifically acceptable only when an enlarged theory is developed that rationally unites the non-conforming fact with the existing scientific data. Yet this does not always hold true, because the emotions of scientists get in the way. Some nonconforming facts are apparently too outrageous to be tolerated and some scientists ridicule them out of existence. They claim that the best explanation for statistically validated parapsychological phenomena is trickery by the experimenters. Others who are intrigued by the nonconforming facts generally remain silent. Heresy can cost them their promotions and reputations. Eugene Condon, former head of the National Bureau of Standards, phrased the threat in this manner: "Flying saucers and astrology are not the only pseudosciences which have a considerable following among us. . . . There continues to be perception, psychokinesis, and a host of others. . . . In my view, publishers who publish or teachers who teach any of the pseudosciences should, on being found guilty, be publicly horsewhipped and forever banned from further activity. . . ."

Nevertheless, some scientists have seriously investigated a host of "unexplainable" phenomena for about a century, and this field of study has grown rapidly in recent years. One of the most interesting and potentially useful areas is control of the autonomic nervous system, through which most psychosomatic (mind-body) diseases are developed. Physicians believe that from 50 to 80 per cent of human diseases are psychosomatic, that is, they result from the body's unconscious reaction to psychological stress. Thus it is possible, in theory, to train patients to control 50 to 80 per cent of their diseases, to handle other psychosomatic problems, and, hopefully, to decrease their dependence on drugs.

We once thought that the autonomic nervous system, which regulates the body's organs, could not be voluntarily, or consciously, controlled to any significant degree. But recent evidence indicates otherwise. Psychologist Neal E. Miller of Rockefeller University has used a system of rewards and punishments to demonstrate that animals can be conditioned to control autonomic processes, such as the flow of blood to various parts of the body. Human beings also can develop voluntary control of the autonomic nervous system—for example, lowering their blood pressure—apparently by learning to control normally unconscious parts of the mind. This kind of learning usually requires visual or audible feedback, such as a light that flashes or a buzzer that buzzes. These cues inform the subject of his success, telling him whether or not he is controlling what is happening in the normally unconscious domain inside the skin.

Although there is a line of separation between the conscious and the unconscious—the voluntary and involuntary nervous systems—this separation apparently can shift back and forth. For example, when we learn to drive a car we focus conscious attention on every detail of

The authors:

Elmer Green heads the Voluntary Controls Program and the psychophysiology laboratory at the Menninger Foundation. A former physicist, he turned to psychology in the late 1950s. Alyce Green is also a psychology researcher with the Voluntary Controls Program.

A volunteer at the Menninger Foundation is wired for simultaneous auditory feedback of hand temperature, muscle tension in the right forearm, and brain waves during attempts to control "involuntary" processes. A laboratory associate examines the polygraph machine, *right,* which shows a correlation between alpha and theta brain-wave states and the physiological changes that are taking place in the subject's body. The data is collected and compiled in a computer located in the master control room, *below.*

muscular behavior and visual feedback. In other words, we manipulate steering wheel, gas pedal, and brakes according to what we see on the road ahead of us. This tells us what we are doing and suggests corrections if, for example, the car heads toward a ditch. Through such feedback we learn conscious control of the striate, or voluntary, muscles. After much experience, driving becomes automatic. We then may drive through a long section of town while thinking about something else and then wonder if we stopped at all the traffic lights.

When this behavior occurs, processes normally controlled by the conscious have temporarily shifted—to the unconscious. When, through feedback, voluntary control is exerted over so-called involuntary processes, such as dilating and contracting the smooth muscles that control blood flow, the shift is to the conscious domain.

In 1964, we began a voluntary-controls research project at the Menninger Foundation to test this conscious control of the unconscious. We set up a laboratory in which we could monitor the physiological variables of our subjects while they practiced autogenic, or self-generated, techniques. Our equipment included an electroencephalograph (EEG) to measure brain waves, an electrocardiograph (EKG) to measure heart rates, galvanic skin response devices (GSP and GSR) to measure electrical potential and resistance of the skin, thermistors to measure skin temperature, and equipment to measure breathing rates and blood flow in the hands. All of these devices were connected to recording equipment in an adjacent room so that we could collect, and later analyze, all the data.

In one series of tests, our subjects—a group of women from the Topeka area—attempted to raise the temperature of one hand by increasing the flow of blood into the hand. Through a technique called passive concentration, some of our subjects were able to raise their hand temperature by several degrees.

Elmer Green adjusts a portable brain-wave machine that provides both auditory and visual feedback. In this experiment, blood flow and hand temperature can be correlated with alpha brain waves.

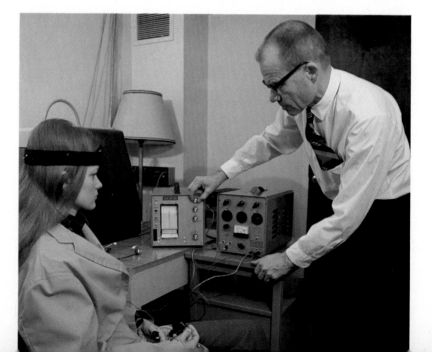

Observing this early work, psychologist Gardner Murphy, then head of the Menninger Foundation Research Department, felt that biofeedback might be useful. This meant connecting the monitoring equipment to visual or audible signaling devices. For example, when a thermistor was connected to a meter or a buzzer, the subject could tell if his attempt to change his skin temperature was succeeding by watching the meter needle or hearing the buzzer. When we combined biofeedback with autogenic training, we found that many people learned to control unconscious physiological functions more quickly than with either one alone. We called this combination of the two systems "autogenic feedback training." Autogenic training supplied a strong, suggestive imagery and biofeedback supplied immediate knowledge of the results. These are powerful factors in gaining voluntary control of involuntary processes, and are of great importance in our continuing research program.

In a few short years, voluntary-controls research throughout the United States has begun to show positive results in alleviating a number of medical complaints. One of these is relief from migraine headaches. Patients have learned to cause their hands to become warmer, an action that relaxes the autonomic nervous system, thereby relieving the migraine pain. Other human malfunctions that can be brought under some degree of self-regulation include erratic heart rate, high blood pressure, Raynaud's disease (which involves deficient blood flow to the extremities), and unconscious muscle tension (responsible for or associated with many unpleasant symptoms).

How does all this take place? Perhaps as follows: According to neuroanatomists, the subcortex of the brain contains a neural network called the limbic system that responds to emotions. Whenever we "have an emotion," the electrical activity of the limbic system changes. This system, however, is linked by many nerve fibers to other sections of the subcortex which contain the neural circuits that control most of the body's involuntary, or autonomic, functions. The exact neural pathways have not yet been traced, but this much seems certain: If we have a thought that is associated with a feeling (and few thoughts are not), the limbic system, through its connections with various control circuits, brings about unconscious changes in some of the body's involuntary functions.

Whatever the exact explanation, the important fact is that if we use a sensitive detector and visual and auditory displays to reveal minute physiological changes, we often can learn to control the sections of the involuntary system that regulate these changes. Theoretically, at least, we should be able to bring under control all our physiological processes with this technique.

This extension of conscious control over involuntary systems has far-reaching implications for psychology and medicine. It suggests that human beings are not biological robots, controlled entirely by genes and the conditioning of life experiences. Migraine, for example, tends

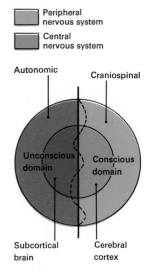

Shifting the States of Mind

■ Peripheral nervous system

■ Central nervous system

Autonomic

Craniospinal

Unconscious domain

Conscious domain

Subcortical brain

Cerebral cortex

The large circle represents the entire nervous system; the small circle only the central nervous system — brain and spinal cord. A solid vertical line divides the system into the conscious domain, which includes all voluntary processes, and the unconscious domain, which includes all involuntary processes. The dotted line suggests that this division is not fixed; there is a continuous undulating separation between the conscious and the unconscious.

to run in families and thus seems to be partly, at least, genetic in origin. When it is brought under voluntary control through autogenic feedback training, the patient is apparently overcoming a genetic predisposition. The freedom gained is not just physiological, however; it has an important psychological component. Many people who learn to control physiological problems find themselves relieved of some emotional and mental symptoms at the same time.

The self-regulation of mind-body energies by consciously controlling normally unconscious functions may, at first glance, seem to be little more than a simple medical advance, but the implications are "theory busting," to say the least. The investigation of voluntary or conscious control of mind-body energies has expanded to include two separate but related areas: Control by the mind of the energy inside the skin (Ins), the domain of psychology, physiology, and medicine; and control by the mind of the energy outside the skin (Outs), the domain again of psychology, but also of physics and parapsychology— the psychic phenomena. Furthermore, Ins and Outs energies are special parts of a general "field of mind" theory, which we will examine later. In a curious blend of Eastern theory and Western technology, a new "science of consciousness" seems to be developing.

Swami Rama, trained in the Himalaya in the discipline of yoga, is contributing to this blend. He came to the United States from India in 1969 and now lives in Palatine, Ill. His guru, or teacher, suggested that he could help bring Eastern and Western science closer together by working with psychologists and medical doctors who are studying mental and physical phenomena. Daniel Ferguson, a psychiatrist at the Veterans Administration Hospital at Fort Snelling, St. Paul, Minn., suggested that our Voluntary-Controls Project might want to study Swami Rama. It would be an opportunity to examine someone with extraordinary control over the autonomic system. In addition, because the Swami appears to have a measure of control over Outs energy as well as Ins, we could also study how the unconscious functions in the relationship between psychology and parapsychology.

Ferguson and the Swami first visited our laboratory in March, 1970. As with our other subjects, we wired up the Swami to record brain waves, heart behavior, respiration, skin resistance and potential, muscle tension, blood flow in hands, and hand temperature. He first made the temperature of the little-finger side of his right palm differ from the temperature of the thumb side by 10°F. He did this apparently by controlling the flow of blood in the large radial and ulnar arteries of his wrist. Without moving or using muscle tension, he "turned on" one of them and "turned off" the other. Later, he demonstrated that he could stop his heart from pumping blood, and could produce specific brain wave patterns on demand.

We asked the Swami how he controlled his heart and blood vessels, and how he consciously produced various kinds of brain waves at will. He explained that these phenomena were possible because, "All of the body is in the mind. But," he added, "not all of the mind is in the body." In other words, each part of the energy structure called the body is literally a part of the energy structure called the mind, although the reverse is not necessarily true.

In the raja yoga school of philosophy, two of the most interesting concepts relating to Ins energy are that every part of the body is represented in the unconscious, and every part of the body also *represents* the unconscious. What potent ideas! They mean that when we extend conscious control over a specific part of the unconscious, as in autogenic feedback training, the associated physiological processes can be brought under voluntary control.

In yogic theory, the mind is not merely a person's perception of involuntary electrochemical changes in the body. On the contrary, the body is only the densest section of a "field of energy" that includes both body and mind. It is interesting to remember that our bodies, like everything else in the universe, are electromagnetic fields with swarms of particles as dense portions. We are almost entirely empty

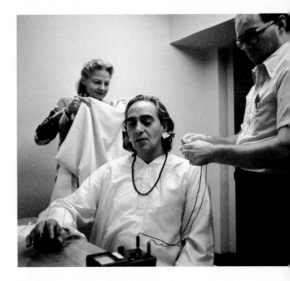

Menninger researchers Alyce Green and Dale Walters prepare Swami Rama for skin temperature and heartbeat experiments.

Controlling the Uncontrollable

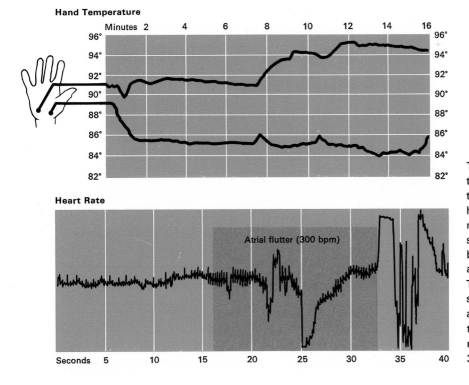

The Swami demonstrated that he can control the temperature of areas of his palm. After 12 minutes he obtained a spread of 10 degrees between the thumb side and little-finger side. The Swami's attempt to stop his heart caused an atrial flutter, raising the heartbeat from a normal reading of 70 to 300 beats per minute.

143

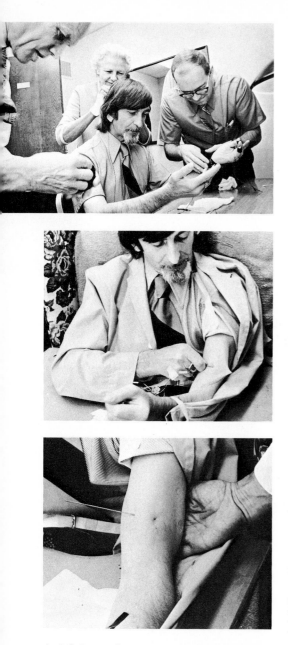

Jack Schwarz demonstrated voluntary control of pain for Menninger scientists. First they wired him to record his brain waves, heart activity, blood flow, hand temperature, respiration rate, and galvanic skin response. Then he pushed an unsterilized darning needle through his biceps. When he removed the needle there was no sign of blood. The monitoring equipment indicated little or no physiological response.

space, although we see ourselves and all nature as solid matter because that is the way we were constructed by evolution to see.

Yogis believe that, without exception, all body processes are mind processes. The mind handles Ins energy because it *is* Ins energy, even though that is not all it is. For mind is an energy structure, and all matter, whether physiological or nonphysiological, is a matrix of energy that is somehow related to mind. In every thought and in every cell, we are part of the general field, but we are normally unaware of this because we are not conscious of our own unconscious.

Swami Rama represents the classical tradition of Eastern philosophy, but Jack Schwarz, a member of our Western culture, demonstrated some of the same types of phenomena as the Swami. Schwarz, who now lives in Selma, Ore., came to the United States from Holland in 1957. Now in his late 40s, he first learned of his ability to control pain when he was a young child.

Schwarz first visited our laboratory in 1971. After we had wired him up in the same way we had the Swami, he produced a 6-inch darning needle that he first rolled in the dirt on the floor and then proceeded to push through the biceps of his left arm. The needle pierced skin, muscle, and a vein. After he pulled it out, the wound bled for almost 15 seconds. Then he said, "Now it stops." Two seconds later, the wound stopped bleeding. In a second demonstration, the wound did not bleed at all. At no time did he appear to be in pain.

The monitoring equipment provided intriguing information. The GSR showed that he was under no unusual stress. His heart rate remained essentially the same. But the thermistors attached to his fingers showed elevated temperature—a sign of relaxation. And his brain-wave patterns showed what we interpret as alert detachment.

In view of what we had learned from autogenic feedback research and from Swami Rama, it was interesting to hear Schwarz explain how he controls his body functions. Control, he says, depends on cooperation from the "subconscious." He does not force the phenomena to take place, but asks his subconscious if it is willing.

When Schwarz was asked to repeat the demonstration, there was a long pause before he said, "Okay." When we asked why he had paused, he

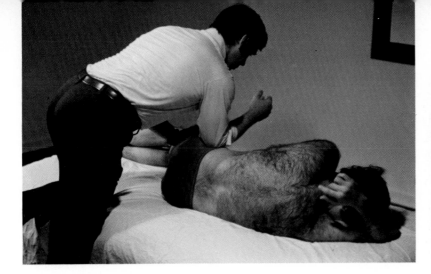

said, "I had to ask the subconscious if it was willing to do it again. When it said yes, then I said 'okay.'" Schwarz also said that part of the delay was due to the fact that his "paraconscious" also considered the situation. He described this level of mind as "wiser" than either the conscious or subconscious. It acts as a kind of intuitional guide.

If we do not have conscious communication with our unconscious (what Schwarz calls the subconscious), it operates as an automaton. That is why a person with a psychosomatic disease cannot control the disease merely by knowing that it is psychosomatic. A certain kind of internal communication is necessary. Indirect control of the unconscious can be temporarily established in human beings by hypnosis, conditioning, or of course, drugs. But the ultimate value of any method of controlling psychosomatic diseases seems to depend on how truly voluntary it is.

So much data on Outs energy has been collected over the last 100 years that many scientists believe the subject can no longer be ignored. The American Association for the Advancement of Science, the American Psychological Association, and the American Psychiatric Association have begun to recognize the need for serious scientific inquiry by sponsoring panels for discussion of parapsychological research. Parapsychology is the study of several kinds of phenomena including *clairvoyance* (seeing without using the eyes), *clairaudience* (hearing without using the ears), *precognition* (knowing of future events with no known source of information), and *psychokinesis* (the movement of physical matter by mind alone).

Psychokinesis, which Swami Rama demonstrated with the knitting needle, is the area that seems most likely to yield scientific facts that are beyond the need for statistical support. One of the earliest scientific investigations was conducted in the early 1900s by Sir William Crookes, chemist, physicist, and a president of the Royal Society of London. After years of research, he announced that he had observed psychokinetic (and other) events under strict laboratory conditions. But he could not account for the facts and would have to learn more before attempting to explain them. Because Crookes could offer no

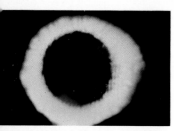

Healer before

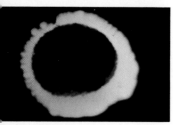

Healer after

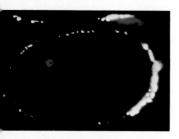

Patient before

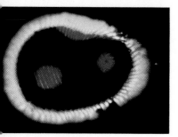

Patient after

Kirlian photographs of finger pads of "healer" and patient indicate an energy transfer during treatment. In this form of photography, film is exposed electrically to record energy patterns that seem to be emitted by plants and animals.

suitable theory, and had no color movies, videotapes, or polygraphs to record information, the evidence he reported convinced only those who observed the events directly. Sir William's less-charitable colleagues thought he had lost his mind.

The next major effort to demonstrate psychokinetic effects took place in the laboratory of Joseph B. and Louisa Rhine at Duke University in the 1930s and 1940s. They first tested whether subjects could influence the roll of the dice by thought. Subsequent experiments by the Rhines and other researchers have amassed statistical evidence showing that the probability that chance alone could account for experimental results is less than one in trillions.

Yet, without personal experience, people remain unconvinced. When presented without a rationale, the idea of Ins and Outs energy seems quite difficult to accept. Swami Rama and Jack Schwarz have a simple theory to explain these phenomena, although it may not be very easy to believe.

According to their theory, not only is all of the body in the mind, but all of nature is a "field of mind." Magnetic, electrostatic, electromagnetic, gravitational, and other fields surround the planet and are special parts of a general planetary field of energy. Human minds are part of this normally unconscious field, and Outs energies can be controlled when we become conscious of the Outs, or extrapersonal, extension of our unconscious. This is a generalization of the theory that explains voluntary control of Ins energy. In other words, we can control both Ins and Outs energies only after we become conscious of our unconscious. Psychokinesis, and all other parapsychological phenomena, as well as control of physiological processes, are included in this field of mind theory.

As far as we know, no one has yet been able to detect the "energy" associated with such psychokinetic phenomena. But it seems only a matter of time until a satisfactory energy detector is built and the field of psychokinesis will be opened for further studies of variables in mind and in matter.

Schwarz demonstrated what could be regarded as parapsychological phenomena—perhaps even psychokinetic—some years ago before physicians of the Los Angeles County medical and hypnosis associations. After the doctors examined his hands, Schwarz put them into a large brazier of burning coals, picked some up, and carried them around the room. Subsequent examination of his hands showed no burns or other signs of heat.

Schwarz's explanation for this is much the same as Swami Rama's. Mind and matter are essentially the same. But we are normally unaware of this because at best we are only slightly aware of our own unconscious and the field of mind of which it is a part. As we become more aware, we can draw on the field of mind for specific powers. In this case, it provided for Schwarz some sort of extremely effective insulation against the hot coals.

Clearly, our capability to regulate our physiological processes has great potential for our well-being. But apart from satisfying our scientific curiosity, why do we bother with parapsychological matters? Partly because there is a potential for misuse of these abilities. As parapsychology becomes scientifically established, we must consider what will happen if human beings can learn to control the minds of other human beings. Evidence of such a possibility is already being seriously discussed by a number of scientists. In *Psychic Discoveries Behind the Iron Curtain,* published in 1970, authors Sheiler Ostrander and Lynn Schroeder discuss the possible use of parapsychological forces for espionage and sabotage. Thus parapsychology confronts us with a number of moral problems.

When the atomic bomb was developed, moral questions were discussed after-the-fact because of the secrecy required by national security. However, psychokinesis is not a secret, and we shall have time to develop rules to guide us in this field of scientific inquiry.

We tend to agree with the existential experts who maintain that the only good "measure and countermeasure" is knowledge and transpersonal awareness. This awareness of the mind operates from a center above or beyond our personal egotism. It transcends the extrapersonal awareness used in developing psychic abilities. According to the field of mind theory, the only dependable guide for the extrapersonal is the transpersonal. Thus it seems that we should support transpersonal research, if only for reasons of safety. If human beings destroy either themselves or their planet it will not be for lack of extrapersonal development, but for lack of transpersonal development.

Whatever else may be said, it seems clear that there are problems in using Outs energy wisely. By default, we have already allowed physical pollution to endanger the planet. There is a disturbing similarity between man's current abuse of nature and his possible exploitation of parapsychological forces for personal gain. Perhaps by serious scientific study of the field of mind, the "science of consciousness," we can avert "psychic" pollution.

For further reading:

Aurobindo, *The Synthesis of Yoga,* Sri Aurobindo Ashram Press, 1955.
 (Available from the California Institute of Asian Studies,
 San Francisco, Calif.)
Barber, T. X., et al. (Eds.), *Biofeedback and Self-Control, 1970: An
 Aldine Annual,* Aldine-Atherton, Inc., 1971.
Green, E., Green, A., and Walters, E., *Journal of Transpersonal
 Psychology,* Volume II, 1970.
Rorvik, David M., "Jack Schwarz Feels No Pain," *Esquire,* December,
 1972.
Schultz, J. H., and Luthe, W., *Autogenic Training: A Psychophysiologic
 Approach in Psychotherapy,* Grune and Stratton, 1959.
Stulman, Julius, *Fields Within Fields . . . Within Fields,* Volume 5,
 Number 1, 1972.

The Bees that Got Away

By Michael Sheldrick

Evil-tempered honeybees have spread through South America and are advancing toward the United States

A new breed of honeybee—one that has brought fear and controversy to much of South America—is moving toward North America. They look like the honeybees with which we are all familiar, but the new bees are far more easily aroused and have a much greater tendency to attack in large numbers. North American beekeepers are watching the approach with a wary eye. They are concerned that accounts of the new bees' ferocity—they have even been called "killer bees"—may stimulate a wave of antibee hysteria. One result could be legislation that, for example, would hinder professional beekeepers by restricting the number or location of hives.

Like many problems, this one began as an effort to improve things. In 1956, the Brazilian government and local beekeepers cooperated to send bee geneticist Warwick E. Kerr of the University of São Paulo to Africa to bring back queen bees of the species *Apis mellifera adansonii*, which inhabits nearly all of Africa south of the Sahara.

An American scientist crouches in a thicket to make a videotape recording of Brazilian bee colonies near the city of Salvador, Brazil.

The author:
Michael Sheldrick is the coordinator of environmental field programs at the Center for the Study of Natural Systems, Washington University, St. Louis. Sheldrick is also an amateur beekeeper.

In Brazil, as in the United States, there are no native honeybees. Races of bees from Europe were introduced into the Western Hemisphere by early colonists. But, while the European bees thrive in the United States, Brazil's tropic and subtropic climate does not agree with them. As a result, they produced disappointingly little honey. The African bees were chosen because they inhabited a similar climate and, although they are extremely aggressive, they produce large quantities of honey. Brazilian scientists planned a breeding program to produce a bee with the relative docility of the European stock and the high productivity of the African.

Kerr traveled widely in Africa, selecting relatively docile queens, already inseminated by African drones, or males. Shipped back to Brazil, 47 of these queens were placed in hives in Kerr's laboratory near São Paulo. Their queen and drone offspring were to provide the African-bee stock for the breeding program. European workers were also placed in the hives to tend the queens. This was perfectly safe for the breeding program, because workers play no role in reproduction.

The queens began laying eggs immediately, and three weeks later, thousands of pure-African honeybees emerged. To keep the queens and drones from escaping, Kerr had fitted the hives with queen traps. Their openings are large enough to allow workers to pass back and forth from the hive, but too small for the larger queens and drones to slip out. A new colony is normally established when the old one becomes too crowded. The queen and about half the workers simply leave the old hive to seek new living quarters after special queen cells have been prepared by the workers. From one of these cells, a new queen will emerge. Keeping the African queens restricted would prevent the uncontrolled establishment of new colonies; keeping the African drones restricted would prevent any uncontrolled breeding between them and European queens from other hives.

But, unfortunately, the hives containing the new African queens were in an isolated location where Kerr and his associates could not keep them under direct observation. By one report, foraging worker bees returning to these hives lost their loads of pollen as they squeezed through the small openings in the queen traps. So, in June, 1957, a visiting beekeeper, not understanding the reason for the traps, removed them. Ten days later, his error was discovered. By then, 26 colonies had swarmed. The African bees and their aggressive genetic heritage had been loosed on the Western Hemisphere.

European bees that had escaped cultivation were basically unsuited to the environment and had not proliferated widely. The escaped bees found what amounted to virgin territory in the Brazilian environment, and they quickly took over the country. Marauding African bees invaded local apiaries, or collections of hives. They killed some bees, drove others out, and crossbred with the rest. The 26 swarms that had escaped soon lost their genetic identity. But the new generations of crossbred bees retained their African ancestors' aggressiveness and their affinity for the tropics. These characteristics were soon spread to virtually all of Brazil's honeybees, making them a new type that scientists now call Brazilian bees.

The Brazilian bee looks like any other, but aggressiveness marks it as different, and makes visiting scientists, *bottom*, thankful for their protective clothing.

The Brazilian bees spread outward from São Paulo in roughly concentric circles, and have been advancing as much as 200 miles a year. Those moving north–toward the equator, the Isthmus of Panama, and North America–advance much faster than those spreading south. This is partly because the Brazilian bees prefer the tropical climate in the north to the cooler climate in the south, but also because the broad, treeless prairies of the south provide relatively few nesting sites.

The new bees are now well established in Argentina, Bolivia, Paraguay, Uruguay, and parts of Peru. There appears to be nothing to prevent them from entering Central America, Mexico, and the United States. Kerr has estimated that, unhindered, they could reach Panama in 4 to 6 years and the United States 8 to 10 years later.

The Brazilian bees probably could not survive the winter in most of the United States. European bees survive by clustering–forming a ball –and consuming honey they have stored. Constant generation of heat within the ball keeps all the bees warm. The Brazilian bees do not cluster, but that does not exempt the United States from the threatened invasion. Commercial beekeepers in the north commonly restock their hives each summer with bees that are shipped north from southern California and the Gulf Coast states, warm areas where the aggressive Brazilian bees could easily survive.

With continuing reports of the Brazilians' ferocity and their unchecked spread north, the anxiety of U.S. beekeepers grew. Since the

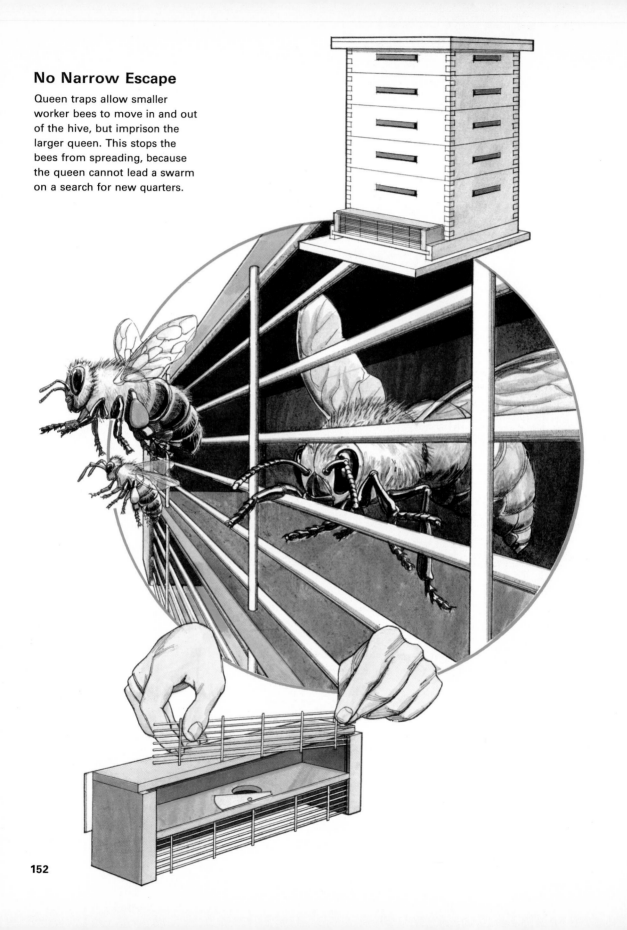

No Narrow Escape

Queen traps allow smaller
worker bees to move in and out
of the hive, but imprison the
larger queen. This stops the
bees from spreading, because
the queen cannot lead a swarm
on a search for new quarters.

U.S. Department of Agriculture (USDA), through its Entomology Research Division, supports most U.S. bee research, many of the beekeepers looked to the USDA for help. The department commissioned a study of the problem by the National Research Council (NRC) of the National Academy of Sciences. The NRC chose a committee of entomologists, ecologists, and geneticists to evaluate the situation. The committee visited Brazil in December, 1971, and made its formal, final report in June, 1972.

Agreeing with Kerr's earlier evaluation, the NRC committee reported, ". . . there is no known geographic or climatic barrier . . ." to keep the Brazilian bee from North America. Even the absence of normal nesting sites would hinder their passage very little. The committee discovered that the insects were often indiscriminate in choosing such sites. They reported finding them in ". . . old armadillo burrows in the ground, hollows in termite nests on the ground or in trees, empty boxes, holes or cracks in cliffs, and in cracks in a building. In warm climates, Brazilian bees also make perennial colonies with exposed combs hanging from rocks or tree branches."

Although their feverish stinging behavior has apparently been overdramatized, the Brazilians' ferocity poses the biggest problem. Because most people know little about bees and are naturally afraid of being stung, the potential exists for near-hysteria. Something approaching that may prevail in Brazil, where hundreds of unsubstantiated reports of deaths from bee attacks have been published.

A single sting from a Brazilian bee is no more harmful than that of any other bee, but it is much more likely to be followed by a frenzied mass attack. The NRC committee says Brazilian colonies are extremely sensitive to disturbances. They communicate alarm within and between colonies and launch massive attacks on intruders much faster than do other bees. "The slightest disturbance at or near the hive entrance can set off a chain reaction that explodes within seconds," the committee reported. "Whole apiaries may go out of control. Hundreds of bees become airborne and pursue and sting any animals or people within 100 meters [more than 300 feet] of the hive." The vigor and duration of such attacks in northern Brazil greatly impressed committee members. In properly reserved language, they reported that "some committee members were stung severely, even though they anticipated the problem by wearing protective gear, and by massively smoking hives."

One of Kerr's colleagues devised a test of the bees' aggressiveness, later used as a model by the NRC committee. A worker in protective clothing dangled a small, black leather ball in front of several hives. European bees began to sting the ball in an average of 19.3 seconds; the African bees waited only 2.9 seconds. While European bees left an average of 26.1 stings in the ball, the African bees left an average of 63.7 in the ball and an additional 39.3 in the tester's glove. The committee scientists, in their tests of hives throughout Brazil, used a piece

A leather ball at a hive entrance helps measure bees' aggressiveness. The number of stings in the ball is an index of the bees' irritability.

of black suede about an inch square. A worker dangled it about 4 inches in front of the hive until the stinging began, but for no more than 3 minutes. He then moved the leather directly to the hive entrance and jiggled it for 30 seconds more. In 66 such tests, the bees left an average of 35.2 stings in the suede.

The most ferocious colonies were those closest to the equator. In several of the tests conducted there, the number of stings ran well above 100. But the scientists were unable to make as many tests near the equator as they would have liked, because many apiaries in this northern region were located near humans and animals. "Indeed," the committee said, "some of the beekeepers are already in debt for damage done to neighbors' animals."

Even in Rio Grande do Sul, Brazil's southernmost province, where bees were supposedly less aggressive, some were very easily angered. In one test there, the committee's worker accidentally jarred the hive "slightly." The bees poured out and the first sting came almost instantly. The worker—in full protective garb—fled immediately, but the frenzied bees chased him two-thirds of a mile. The piece of suede was stung 92 times. No one counted the stings in the worker's clothing.

The Brazilian bees also tend to move about rather frequently, and this adds to the problem, because it leads to more contacts with human beings. European bees will remain in their hives even if food and water are unavailable, but, under similar pressures, Brazilians will abandon the hive to find another nesting site. Because they are more prolific than other bees, they also tend to swarm more often, up to seven times a year as opposed to once a year for European bees.

Clouds of swarming or migrating bees have often been seen cruising about 100 feet above Brazil's landscape. Many of the serious, unprovoked attacks on humans were apparently by groups of bees that swooped down unexpectedly from such heights.

The public health problem that might result from an invasion by Brazilian bees is difficult to evaluate. In the United States, 20 to 30 persons, nearly all of them highly allergic to bee venom, die each year from bee stings. Fortunately, only a small fraction of 1 per cent of the U.S. population is probably so hypersensitive. Nevertheless, a population of aggressive Brazilian bees in the United States would probably make more attacks and inflict more serious or fatal stings. And, because they build hives in almost any sheltered location, they might settle readily in urban or suburban areas.

A more serious impact from a Brazilian invasion could be on crops. More than a billion dollars worth of crops each year depend on pollination by bees. It is not unusual for beekeepers to follow the crops in a manner similar to that of migratory farmworkers, renting out their bees to growers. The beekeepers set up their hives, perhaps one per acre, as flowering begins, and move on when it stops. The hives are sometimes moved as many as seven times in a single season, and beekeepers doubt that the touchy Brazilian bees would tolerate such

Members of a committee of the National Research Council inspect Brazilian beehives in the northern tropics where the bees were particularly vicious.

treatment. Brazilian beekeepers report that, unless the queen is put in a special case before a hive is moved, the other bees will kill her. In an orchard, Brazilian bees could also cause serious problems for anyone working there, particularly if near the hives. Even agricultural machinery passing some distance from the hives has enraged Brazilians.

Still, the invaders have their strong points. They are hard workers and prodigious honey makers, up and working in the fields early and still foraging after dusk. Even on cold or rainy days, they turn out to work rather than remain in the hive, as do the European bees. And they carry larger loads of nectar on each trip. The result is that their production is usually higher than that of European bees. In a comparison Kerr made, they produced up to four times as much honey. Increases of 50 to 100 per cent were more common, however.

In spite of the Brazilian bees' productivity, the NRC experts suspect an invasion might not increase U.S. honey production. The Brazilians might quickly multiply and establish wild colonies in areas where nectar is abundant, and commercial honey production could be reduced by competition for the nectar.

The same circumstances could lead to robbing, a characteristic of all bees, but one that seems especially common among the Brazilians. Instead of collecting nectar and evaporating the water to form honey, bees enter others' hives, remove the honey, and bring it back to their own hive. An epidemic of robbing can bring an apiary's honey pro-

In quaint English, a
Brazilian beekeeper pays
eloquent tribute to the
ferocity of the new bees
in this letter published
in *The Speedy Bee,*
a beekeeping journal.

++++++

Dear Sir:

I have now about 50 colonies mastered by Italian queens plus 100 colonies with African blood.

The door-yard is forbidden for my children. When they come from the school, in a hurry, speeding up the last 300 feet, shouting "open the door! Open the door!. . .", a swarm of African bees makes a comet's tail behind them and stops only against the slap of the wire mesh frame of the door.

No dogs can stay in the yard. They must dwindle in the bush soon at the morning. The chickens, those still alive, do have spoiled eyes.

No guest is sufficiently brave to come during the day.

A sweating salesman, embracing a handful of Holy Bibles, struck desperately the ring. When my wife came from her kitchen, the man was sprinting off, killing bees like a windmill, while the Bibles lay scattered along the way.

A taxi-driver was gentle enough to carry my bag just to inside the house. The Africans, disliking the fumes and noise, encircled the car and kept the driver prisoner in the house for half an hour. Then I told him: "The bee venom is good for rheumatism. Take one more dollar tip to increase your courage and face the bees." So he do. One jump to the car, the slap of the car's door, the furious maneuvering of the windshields, a fight, the Sign of the Cross, a violent acceleration, the whirling of the tires and the brute faded away rapidly.

The African bees are fierce creatures. They'll arrive to the U. S. Be sure of that, sooner or later.

Geraldo Peixoto de Freitas
Caixa Postal 6
35175 Acesita, MG - Brasil

+ + + + + +

duction to a halt because the robbery victims may respond by robbing another hive. One case leads to another until, eventually, hundreds of colonies have ceased gathering nectar and making honey and simply rob—and sometimes even kill—each other.

If Brazilian bees reach the United States, apiaries near cities and towns will probably have to be abandoned, further reducing production. Even the number of hobby beekeepers—those with only from 1 to 10 hives—would probably fall drastically, since few would be willing to deal with the vicious new breed.

These gloomy forecasts—a greater number of serious or fatal bee stings, serious cuts in pollination, and smaller honey crops—are rejected by some bee experts. In its northward passage through countries where there is a substantial beekeeping industry, they say, the invader may well be genetically pacified through interbreeding with local bees that are better established than the Europeans were in Brazil. Mexico, with its relatively large beekeeping industry, would be the likely location for such pacification. Even if the Brazilian bees arrived unchanged, some of these observers argue, they would not have much impact. Such optimists may be correct. Other insect pests that are spreading through parts of the United States, such as the fire ant with its painful sting or the brown recluse spider with its potentially fatal bite, actually do little total harm. Beekeepers might adjust their practices to the Brazilians' pugnacity and, in the end, even come to admire them for their hard work. The NRC committee found that this is just what happened in Brazil. Indeed, some beekeepers were supplied with mild, European queens in hopes of modifying the aggressive strain. But "beekeepers ultimately killed such queens ... because they [the colonies] produced less honey" than colonies of Brazilian bees, the committee said.

Even though there is disagreement over the effects the Brazilian bees might have, once they arrive, most scientists agree that we should try to prevent their entry into the United States. The Isthmus of Panama is the logical place to stop them. The NRC committee recommended several possible ways and said, "... all approaches should be initiated promptly ..." since time was short and the most successful approach could not be identified in advance. Among the proposals: Lure swarms and migrating colonies to trap hives; spray poison on nests; set up poison bait stations; and erect a "genetic barrier."

The NRC scientists who studied the new bees prepared a map showing how quickly they have spread.

Artificial insemination of honeybees could help create a strain of barrier bees that might intercept and genetically pacify the Brazilians before they can spread to the United States.

Of all the suggestions, the genetic barrier to the passage of Brazilian bees seems to be the only method that may be both feasible and acceptable to the Central American countries in the bees' path. Such a barrier would be made up of a dense population of bees that could compete in every way with the Brazilians. Most important, the barrier bees would have to be less vicious than the invaders, able to crossbreed with them, and able to produce less vicious offspring that would be less likely to migrate and less prone to robbery.

Creating the genetic Maginot Line would involve establishing the barrier bees in hives throughout a zone wide enough—perhaps 50 to 100 miles wide—to ensure that no Brazilian could reach the northern edge. No such bees now exist. Tailor-making a successful barrier bee would require careful selection of strains of tropical bees. European bees would not do, as the Brazilians' current success demonstrates.

Once a barrier bee candidate was developed, its ability to compete against and pacify the Brazilians would have to be tested in some isolated area, such as an island. Both bees would be released, and their interaction carefully monitored. Ultimately, scientists could measure the effectiveness of the test bee by checking hives in much the same way that the NRC scientists did.

Chief supporter of the barrier is Walter Rothenbuhler, a behavioral geneticist at Ohio State University who has worked with Kerr. Rothenbuhler has proposed that, once a suitable barrier bee is found, it be established in large numbers of hives in the barrier zone. Next, efforts would be made to increase the population of barrier drones, since the barrier's success would hinge largely on those drones, rather than

Brazilian drones, breeding with Brazilian queens. Beekeepers already know that drone production is increased when a beehive becomes crowded. Thus, they might use undersized hives with large colonies in the barrier zone. They might also stock the hives with what they call drone combs. Queen bees normally use the larger cells of such combs for eggs that will hatch into drones, which are appreciably larger than worker bees. Under such a program, when Brazilian bees arrived at the barrier zone, their drones would be far outnumbered. The likelihood that Brazilian drones would inseminate Brazilian queens—or even barrier queens—would be greatly reduced.

Of course, some Brazilian drones would successfully mate with Brazilian queens and some with barrier queens. But resulting Brazilians or Brazilian-barrier hybrids that took up wild nesting sites would also be destroyed through a careful program of seeking out and destroying all wild nests. Even though some wild Brazilian colonies might survive, they would not be overwhelmingly successful. The genetic character of the barrier bee and the width of the zone, together, would ensure that any colony that passed through would be genetically changed during the journey. The aggressiveness that now characterizes the Brazilian bees would have been bred out. Some scientists are much less optimistic about the possibility of creating a genetic barrier. Samuel E. McGregor, apiculturist at the USDA bee laboratory at Tucson, Ariz., remarks, "I believe that the proposal to use the barrier bee is impractical. . . . To develop such a bee, with all the resources the USDA and state bee specialists could hope to muster, within time to stop the [Brazilian] bee in the Panama Canal Zone area, would be extremely remote . . . we have been unable to develop really disease-resistant bees, . . . pesticide-resistant bees, cranberry-pollinating bees, or even superior honey-producing bees. Why then should we expect to find and develop the 'barrier' bee when we have no idea what characteristics it should have."

Essentially, then, man saw a need to manipulate nature by using his understanding of genetics—and the attempt backfired. Instead of answering the need, genetics apparently has given us a new and larger problem. In responding to it, we may well have to resort to the use of genetics again. And even if this attempt succeeds, we will not have found a solution. The Brazilian bees, evil-tempered product of our fumbling earlier efforts, will still infest much of South America.

For further reading:

Elton, Charles S., *The Ecology of Invasions by Plants and Animals,* Wiley and Sons, 1958.

Final Report of the Committee on the African Honey Bee, National Academy of Sciences, 1972.

Nogueira-Neto, Paulo, "The Spread of a Fierce African Bee in Brazil," *Bee World,* Autumn, 1964.

Portugal Araujo, Virgilio de, "The Central African Bee in South America," *Bee World,* Autumn, 1971.

It's A Dog's Life

By Michael W. Fox

Man has made the dog in his own image, and produced a frustrated species that can ill cope with its modern environment

A veterinarian from Chicago called me recently about a patient–a large 5-year-old Weimaraner–that had suddenly attacked its owner. With the owner also on the phone, we proceeded to figure out the dog's motive. Apparently the pet became insecure and jealous after the owner bought a second dog, a young female puppy. The older dog was now extremely possessive of its toys, gulped down its food in seconds, and threatened its master whenever disciplined. The attack occurred when the owner reprimanded the dog for trying to steal a beef roast from the kitchen table.

I suggested that the owner make the dog more secure by indulging him–petting and praising him constantly–and virtually ignoring the younger one

for awhile. This is not unlike advice a pediatrician might give a mother and father whose older child is jealous of the new baby.

I get increasing numbers of calls and letters from veterinarians, trainers, and owners about such behavior problems. As a veterinarian who has studied the development of normal and disturbed dogs, I try to help people understand their pets. With the number of dogs increasing so rapidly and city life becoming ever-more hostile to their natural habits, we must expect them to have more and more behavior problems. The closer we can come to understanding our dogs' needs, the better we can care for them. Love is not enough.

But for me, love was the beginning. After I graduated from the Royal Veterinary College, London, in 1962, my interest in the inborn brain disorders of dogs led me along an unexpected path. A serious omission in veterinary training, I soon realized, was the study of how animals behave and develop. Before I could hope to determine whether a dog was abnormal, I needed to know how the brain and behavior of puppies normally develop. Thus, my first research was to find how their reflexes, such as the blink and startle reflex, and their inborn behavior, such as chasing moving objects and scent marking, develop while the puppies are growing up.

During five years of this research at Jackson Laboratory in Bar Harbor, Me., and at the Galesburg (Ill.) State Research Hospital, I was increasingly impressed by the similarity in canine and human development. Dogs developed disorders that were quite like those psychiatrists describe in troubled children—hysterical paralysis, depression, diarrhea, hives, and asthma. At the time, I felt that some of these disorders might be related to the fact that dogs have had a close relationship to human beings for thousands of generations. Perhaps the process of domestication, by which men slowly changed the inherited behavior and appearance of dogs, had some bad side effects leading to problems in dogs today. Would animals that had never been bred as pets show similar disorders?

To find out what domestication might have done to the behavior of dogs, I began to study their undomesticated relatives. I had an easier job finding these doglike species, or canids, that were undomesticated than anthropologists and transcultural psychiatrists have in finding "uncivilized" human beings. I was by then at Washington University in St. Louis, and several zoos in the area provided me with pups of wolves; coyotes; coyote-dog and wolf-dog hybrids; jackals; and red, gray, swift, kit, and Arctic foxes.

I bottle-raised most of the pups at home, beginning before their eyes opened so they would be accustomed to people and would behave naturally in my presence. As I observed them growing up, I studied how they responded to sudden noises, to strangers, and to new objects in their environment. Compared with pet dogs, these tame but undomesticated canids were, for instance, easily disturbed by sudden, unexpected stimuli, but also were more inquisitive and alert.

The author:
Michael W. Fox is a veterinarian who has a Ph.D. degree in psychology from London University. He teaches courses on animal behavior at Washington University, St. Louis.

Because dogs behave more like wolves than coyotes or jackals, writers and naturalists in the past concluded that the domestic dog, *Canis familiaris*, developed from the wolf. Most zoologists and paleontologists also believe the dog descended from the wolf, possibly from the Asiatic race *C. lupus pallipes*.

My comparisons of the behavior of dogs with that of undomesticated canids lead me to believe, however, that the dog is not simply a domesticated wolf. Rather, the wolf and the dog seem to be close cousins and evolved from a common ancestor before man started domesticating animals. The first dog, the prototype for *C. familiaris,* was at one time widespread in Europe and Asia. Today its relatively pure form is found only in Malaysia (for example, the New Guinea singing dog), and in Australia (the dingo).

When the dog was first domesticated is also a mystery. The oldest archaeological evidence found with human remains indicates that by 10,000 B.C. men kept dogs as close companions. Both the dog and primitive man were hunters and scavengers. Dug out of their dens as pups and raised by human foster parents, dogs without fear of man would make valuable allies to track, flush, drive, dig out, retrieve,

Tugging at leashes held by professional dog walkers, city pets lead the way to a nearby park. More and more owners hire others to exercise pets because they lack the time.

Dogs confined to a pen or car tend to bark more for attention, and, when encountering strangers, to overprotect their territory. A dog on a leash is often much more aggressive than one running free because it may see its owner as territory it must defend.

ambush, and corner game. Those dogs that did not work well, were shy of people, or disloyal and disobedient were probably killed and eaten; there would be no point in feeding noncontributors. Over thousands of generations, this selection process must have fixed the behavior men desired in dogs—and this long trial-and-error process is termed domestication. Primitive men must have also crossbred their dogs with wild dogs, wolves, jackals, and coyotes. The early Eskimos had dog-wolf hybrids and the Plains Indians had dog-coyote hybrids. Highly selective breeding to fit a dog to the terrain and kind of game man hunted resulted in regional breeds. Later, the crossing of these breeds, which had been genetically and geographically isolated from each other for generations, could account for much of the incredible variety in our domestic dogs today.

When man began to farm and settle down to live in one place for longer periods of time, he developed other types—guard dogs, herders, and war dogs—to protect him and his property. As he became more affluent, pet dogs, whose only duties were to entertain and please their masters, found a place.

The great differences between these dogs, however, are primarily physical—contrast the size and appearance of a Great Dane with that of a Chihuahua. Behavioral differences between breeds are less pronounced. Although certain breeds have been specialized in such traits as guarding, pointing, tracking, retrieving, and herding, all these traits are present in every dog.

From very early times, one trait man must have emphasized in breeding dogs is dependency; the more dependent a dog, the more trainable it was, and the more willing to work for its master. In addition to keeping only those dogs that had desirable traits, primitive man also bred the dog to reduce undesirable "wild" traits, such as shyness and fear of sudden noises or movements. He had to accept some compromise, however. In selective breeding, eliminating an undesirable trait or intensifying a desirable trait often interferes with other behavior. For example, the German shepherd, used as a war dog to locate snipers or buried mines, had to be extremely alert and sensitive. But the breeding programs developed supersensitivity in many of the animals. This made them much too frightened and shy to be useful to soldiers who needed them in combat zones.

Another trait man had to breed out of dogs was fear of strangers, which can be seen today in wild canids. The wild canids seem to have a limited ability to associate with strange animals; wariness and even fear of strangers begins increasing when the canid is about 4 to 5 months of age. In the wolf, this natural tendency keeps packs apart and ensures that pack members will stay together in their own territory. In guard dogs, the wariness of strangers may have been preserved. But it is a highly undesirable trait in pet dogs. Breeders must have tried to eradicate it, yet this fear is common in pets that are referred to me because of their behavior problems.

Today, the ideal pet dog shows neither fear nor aggression toward strangers, except perhaps on its own territory. Some dogs, however, are friendly only to those persons who took care of them when they were between 4 and 12 weeks old. They seem to have partly regressed to the undomesticated temperament. When a stranger approaches, the dog typically flees or furtively tries to avoid contact. But when it cannot escape because it is cornered or on a leash, it may suddenly attack, "imagining" that it is being threatened. Virtually impossible to coax or reason with, such a "fear-biter" is notoriously difficult to handle and rehabilitate.

This behavior tends to be inherited in certain strains of such breeds as German shepherds and American cocker spaniels, supporting the theory that genes rather than experience cause fear-biting. This trait often appears when a particular breed becomes popular and the dog is mass-produced to meet the demand. Breeding standards are relaxed, especially for purebred dogs, and some dogs are born with wild traits.

Three important aspects of interrelated wild behavior—territorial instincts, sexual behavior, and aggression—have also been changed by domestication. These, and other genetic "fossil" needs, such as status rivalry, marking behavior, roaming, and hunting, are still being modified in adapting dogs to modern life.

How fitted is the dog to the relatively new phenomenon of urban life? And how fitted is man, for that matter? Both were once wandering hunters, but their lives now are very different. City life limits space for people and animals on the one hand, and on the other, eliminates the once natural environment of woods and field. What needs are now unsatisfied or frustrated for man and dog alike? Leash laws and fences separate and alienate dogs from each other, preventing them from behaving naturally. Man, too, is experiencing the changes of a new social life—a family that includes only the father, mother, and children, who seldom see their grandparents, aunts, uncles, and cousins. Alienation, anonymity, and identity crises are the growing pains of man the hunter and gatherer adapting to a new life style.

Crowding and frequent encounters with strangers lead both man and dog to overdefend their territory. Home base becomes excessively important as one's refuge and identity. Consequently, as insecurity mounts, both dog and its master are more likely to attack an intruder rather than simply threaten him. Next time you walk through a suburban area, note how each dog responds as you near its territory and how the owner reacts if you step on his lawn. Similarly, a dog on a leash is more aggressive than one not on a leash because the owner may be a territorylike reference point from which the dog directs defensive aggression. Also, next time you are parking your car, note how another car owner reacts if you accidentally touch his car—his territory—with your bumper or car door.

Primitive man's actions in keeping certain dogs must also have changed the dog's sexual behavior to increase its fertility. Wild canids have only one heat period a year, and the males produce sperm only during this short breeding season. These wild animals reach sexual maturity between 1 and 2 years of age and most of them have only one mate for life. Domesticated dogs, however, are sexually mature when they are about 6 to 8 months old, males constantly produce sperm, and males and females are sexually promiscuous. Puppies, too, show more sex play than wolf or coyote pups. Man made the dog hypersexual, then took away its sex life. Few urban dogs are ever satisfied sexually. A male, aroused by the urine odor of a female in heat, must, indeed, be frustrated, and so must the female, who is usually confined during her heat period.

Despite these frustrations, there are still enormous numbers of unwanted puppies born each year. Unfortunately, three out of four dogs given to animal shelters are put to death because homes cannot be found for them, while strays are easy prey for disease and automobiles. I believe a massive program of castration and spaying would, in fact, be a kindness to urban dogs. Spaying was once done only to prevent the problems associated with a female in heat, and the high price for the operation in most veterinary hospitals is still based on its being a convenience or luxury. But there now are urgent social and ecological reasons for encouraging the operation.

Few realize the magnitude of the dog and cat population. There are no accurate records of how many dogs are born each year, or how many are destroyed by their owners or killed in traffic accidents. Each year, animal shelters in the United States kill about 13.3 million dogs and cats—more than 12 per cent of their total population—at an annual cost of nearly $100 million. Three-fourths of this sum goes to kill unwanted or stray animals, most of which are healthy.

Another barrier to the operation is the owner's attitude. I know dog owners who practice birth control themselves, even to the point of having no children. Yet they ask: "Why put this on our pets, too?" Perhaps they gain something emotionally by having a fertile, productive pet. Many would never consider having their dog or cat sterilized, and more than one childless couple would like their female pet to live naturally and have at least one litter. Parents, too, want their children to have the experience of seeing a litter born and raised by its mother. But as the human birth rate drops, the pet population—more child substitutes for substitute parents—increases. There are, in fact, no drawbacks to castration or spaying, either psychologically or physiologically, except perhaps for the owner.

A sterilized animal makes a better pet for some people because it is often less aggressive. Normally, a dog reaches maturity in terms of emotions and temperament when it is about 1 or 2 years old. Then it begins to test its human "family" and other dogs to see how much social freedom and status it can get. Similarly, a young wolf of this age will test its pack mates. An aggressive or overindulged pet, however, often finds it can become the ruling member of the household pack.

Clothed and curbed, the urban dog is devoid of opportunities to follow his instincts.

Packs of stray dogs pose problems for the public, including traffic hazards. Scientists in St. Louis are studying these strays to find out why dogs form packs.

This is the genesis of the canine delinquent. Like a spoiled child, the dog shows no respect for its elders, does not respond to discipline, and may even bite when reprimanded. I know of more than one canine delinquent that has taken over the bedroom and will let no one come in. At other times, it may allow the wife to enter, but the subordinate husband must sleep on the sofa.

In the wolf pack, the status hierarchy is strict and the pups soon find their places. This makes the pack more stable and reduces fighting over rank. In the human family, a well-adjusted dog is one whose master has dominated it psychologically, and occasionally physically, early in life. A leader wolf puts an unruly subordinate in its place by glaring, growling, and even seizing it by the muzzle and forcibly pinning it to the ground. Similarly, a person should discipline a pup, not by striking it with a rolled newspaper or a stick, but by using the "language" dogs understand. Give it a direct stare, a growled command ("No, Fido!" or "Down, Fido!"), and if necessary, grab the dog by the scruff of the neck with one hand, the tail with the other, and shake him; or grab his muzzle and hold him down until he submits by whining or yelping and ceasing to struggle.

Leader and parent wolves do not always browbeat the pups though. They play with them (wrestle, chase, and stalk them), lick and groom them, and share their food. Affection, humor, and discipline are subtly balanced. Dog owners could learn much from wolf parents. Many owners, however, simply do not wish to accept responsibility for their pet's well-being; they merely want an object for their affection and indulgence. Puppies raised in this way rarely make satisfactory pets.

When the owner fails to discipline a pet, gives in to its every whim, and smothers it with affection, it will probably grow up to be a dependent "perpetual puppy." Such excessive dependency is aggravated in many breeds, such as the poodle and Yorkshire terrier, because they are basically extremely dependent.

An overdependent dog is more likely to develop a variety of emotional and physical disorders when its relationship with its owners is threatened by a sibling dog or child rival or when its relationship is changed by time spent in a kennel. Jealousy, aggression, depression, refusal to eat, intestinal disorders, asthma, convulsions, and hysterical paralysis of one or more legs are but a few of the dog's possible reactions. So it would seem that as people have fewer children and raise dogs as child substitutes, more and more veterinarians will face the problems that are now familiar to child psychiatrists.

Despite thousands of years of domestication, dogs still have not lost a number of basic ancestral urges. They need to hunt and chase moving things. Cars and children on bicycles may become substitute hunting prey. Dogs also like to mark scent posts. Urinating on trees and hydrants is like leaving a calling card. In the wild, canids obtain important information as to who came by and how recently at these

Dogs like to hunt and chase moving things. This bicycle rider may have become substitute prey for a German shepherd trying to satisfy its instinctive needs in a city park.

Living by the same principles as wild animals, stray dogs in cities find shelter in abandoned cars and make their "kills" at garbage cans. But natural instincts cannot always cope with all the facets of modern urban life, and the animals fall victim to automobiles and disease. As a result, few stray dogs ever live to an old age.

scent posts. And dogs seem instinctively to want to roll in foul-smelling material. Perhaps they have an aesthetic sense of smell beyond ours. They may "wear" a strong odor for much the same reason that human beings, a more visually aesthetic species, wear bright ties or blouses. Dogs need to roam and to scavenge for garbage and carrion, even though there is ample food at home. Many dogs get sick after such feasts but still prefer what they find while roaming to the canned delicacies their owners provide at home.

These basic urges are difficult to eliminate because they are so deeply ingrained; without them, an animal in the wild would have little chance of survival. But what value can these urges have for urban dogs today, which are always leashed or fenced in and rarely, if ever, meet each other? In sympathy with some of these needs, many owners still allow their pets to run free, even against local leash laws.

But even free-roaming dogs have problems in meeting their basic needs in the city. Alan M. Beck, an urban ecologist now working with me at Washington University, studied dogs in Baltimore from 1970 to 1972. He learned that there are two kinds of stray dogs: Those that have no owners and fend for themselves independently, and those that have a human home that they leave and return to at will. Both, however, have reverted to the wild and live in urban areas by the same principles as animals in the wild.

These strays quickly outwit two dogcatchers in Baltimore. They can apparently recognize their "predators" in time to escape from them.

One of the more interesting aspects of their behavior is how the natural inclination to scavenge for food affects their relationships with other dogs. In the wild, the sociability of wolves and coyotes is influenced by the size, distribution, and abundance of prey. Where food is scarce, packs are small and family life is less stable. Likewise in the city —where food is scattered in small amounts—dogs tend to stay apart. They scavenge alone or occasionally in pairs to avoid conflict and competition. If three or more do forage together, they undoubtedly will fight over what little food they find in a garbage can. When food is abundant, however, the dogs form permanent packs that scavenge together. Such packs have been found living off city dumps.

Urban dogs occasionally form temporary packs that may be dangerous. They may frighten and even attack children, delivery men, and especially older people. It is not clear what social factors give rise to this packing phenomenon. A female dog in heat, a temporary abundance of food (picnickers in a park, a street festival), an early morning ritual "get-together" at a particular time and place (market square or park) are all possible situations that trigger the forming of packs.

Free-roaming dogs can cause serious social, ecological, and medical problems in the cities. As a social problem, for example, these dogs frighten people, and occasionally bite them. Recent surveys show a marked increase in the number of dog bites in the last 10 years. This is due in part to the rapid increase—estimated to be well over 1 million—in the number of unwanted stray animals. In Brooklyn, N.Y., for example, 13,322 bites by stray dogs were reported in 1971, and probably an equal number were not reported.

Fear of dogs is a major barrier to communication between people and dogs. A person who expects all dogs to be unfriendly may be more likely to be bitten. And once bitten, his expectation is reinforced and he develops a permanent fear. Beck and I hope to break this vicious cycle by freeing people from such catastrophic expectations so they can predict a dog's intentions. We are trying to discover the canid signals that certain people perceive and misperceive and how accurately they can interpret a dog's language.

A dog communicates with other dogs and with human beings mainly through its body language. A number of subtle facial expressions and tail and body postures and movements show a dog's emotions and intentions. It usually combines these with eye signals, for example, when it stares in threat or looks away when submitting. It flicks its tongue in submitting and greeting, and it may whine a greeting, growl threateningly, or yelp submissively. A dog combines these signals to increase, decrease, or maintain distance from other dogs and human beings and to show that it is aggressive, submissive, or wants to play or be cared for.

Beck and I plan to make videotape recordings of interactions between a docile dog and three groups of children: Those who are not afraid of dogs; those who are afraid of dogs because they have been bitten; and those who are afraid even though they have never been bitten. These will help us to identify what makes a dog more likely to attack or to behave defensively. By analyzing the cues a person gives to a dog—tone of voice, facial expressions, eye contact or lack of it, tense or relaxed shoulders, clenched fists, jerky or smooth slow movements, shifting of weight, and body posture—we hope to determine how a dog decides whether a person is confident, friendly, hostile, frightened, or frightened but bluffing. We will put our conclusions to the test by having an experimenter, protected by a padded suit, check out these body signals on selected dogs. Ultimately, we hope to be able to eliminate fear of dogs by showing people how to behave in order to avert an attack. A basic course for schoolchildren might help reduce the high incidence of dog bites.

Besides frightening people, free-roaming dogs probably contribute to a large number of automobile accidents when drivers swerve to avoid hitting them. Beck and I are also surveying this possibility by examining police and insurance reports of traffic accidents.

Beck's Baltimore study showed that the marking behavior of urban dogs has a significant ecological impact. Repeated soaking in urine kills many young trees that dogs use as scent marking posts.

Allowing one's dog to roam freely during the day could be a biological health hazard to the members of the household. Dogs very often

Scientist Alan Beck, *above*, posed a stuffed mongrel on the Washington University campus to see how dogs waiting for their student owners would react. Two dogs approached it as they would a stranger. One watched, circled, barked excitedly, then expressed anxiety, and tried to get it to play. The other dog, less excited, sniffed inquisitively, but seemed to prefer to watch the first dog's actions.

carry fleas, whose bites can be severe in children who are allergic to them, and ticks that may transmit Rocky Mountain spotted fever. Dog bites may cause rabies and a deadly, jaundicelike disease called leptospirosis.

Fecal contamination of sidewalks not only is a public nuisance but also a very real health hazard. It is the primary source of *visceral larva migrans*, a rare disease in man. Children who play in or eat dirt contaminated from feces containing the eggs of dog worms are especially vulnerable. They may get fever, coughing spells, abdominal pain, an enlarged spleen, and become quite weak. If migrating larvae reach an eye, the lesions could cause blindness. In order to control this disease, it is imperative that all puppies be wormed regularly and their stools examined for parasite eggs, that all urban dogs be licensed and given regular fecal examinations, and that owners pick up dog droppings and either incinerate or bury them.

All dogs are banned from the city of Reykjavík, Iceland, because of an epidemic of a more dangerous disease, echinococcosis. When accidentally ingested from dog feces, the eggs of *Echinococcus,* a tapeworm, can lead to the formation of cysts in various internal organs, including the liver and brain. I do not see the need for such extreme measures in other parts of the world as yet. But if local leash laws are not enforced more strictly, dogs may someday be barred in many urban areas.

Allowing dogs to roam free in rural areas can also have devastating effects. Many of these dogs are strays, others have been abandoned by owners who got a puppy for their child and did not realize that raising a puppy takes time and responsibility. Not having the heart to take it to an animal shelter for humane destruction, they let "fate" take care of it and let it fend for itself in the country.

Such dogs then revert to the wild state and form temporary or permanent packs. These "feral" dogs are having a significant impact in some areas of the United States, killing livestock and competing for carrion and prey with natural predators such as foxes and coyotes. Feral dogs are also breeding with coyotes in many areas, notably in Arkansas and Missouri. Many hybrid coyote-dogs are quite large and are usually mistaken for wolves.

Future consequences could be severe if man allows dogs to spread from rural to wilderness areas. A Colorado man called me recently, outraged that a wildlife management official had shot his two malamutes because they were chasing deer. He thought that his dogs had a right to be out in the woods, following their natural instincts. What he did not realize was that dogs are no longer able to coexist with other creatures in the wilderness because their presence is now causing ecological imbalance.

Man is not dog's best friend because he has made the dog in his own image: He has projected his needs for freedom, virility, strength, and courage onto his dog, and finds support in it for other needs—companionship, status, affection, and a child substitute. But dogs have needs, too. Only when man learns to see the dog for what it is—and himself for what he is—can he free his dog from some of the frustrations of living in the modern world.

For further reading:

Beck, Alan M., *The Ecology of Stray Dogs: A Study of Free-Ranging Urban Animals,* York Press, 1973.

Beck, Alan M., "The Life and Times of SHAG, a Feral Dog in Baltimore," *Natural History,* October, 1971.

Djerassi, Carl; Israel, Andrew; and Jochle, Wolfgang; "Planned Parenthood for Pets?," *Bulletin of the Atomic Scientists,* January, 1973.

Fox, Michael W., *Behaviour of Wolves, Dogs and Related Canids,* Harper & Row, 1971.

Fox, Michael W., "Neurosis in Dogs," *Saturday Review,* Oct. 28, 1972.

Fox, Michael W., *Understanding Your Dog,* Coward, McCann & Geoghegan, Inc., 1972.

How Canids Communicate

Michael Fox interviews a kit fox, one of the recording stars in his vast tape library of canid sounds.

Wolves, foxes, coyotes, and dogs communicate with a variety of whines, howls, barks, and growls. This vocal language complements a body language. The phonograph records in **Part I,** "The Music of the Canids," present a fascinating collection of the sounds canids use to regulate the distance between themselves and other species and to express their emotions—whether they are aggressive or submissive, fearful or friendly, need attention, or are in pain. **Part II,** "Decoding a Dog's Body Language," explains why it is important to pay close attention to a dog's facial expressions, body postures, and movements. It shows how the animal reveals its feelings by the positions of its ears, head, rump, tail, and legs. It explains how to "read" the whole dog: How to know when it is safe to approach and make friends, or what to do to avoid or resist an attack.

Coyote

New Guinea Singing Dog

Husky

Red Foxes

Chihuahua

Gray Fox

Timber Wolf mother and pup

Dingo

Wire-Haired Dachshund

Arctic Fox

Coyote-Beagle hybrids

Decoding a Dog's Body Language

You can tell how aroused a dog is by "reading" its body language. Dogs display subtle facial expressions and tail and body postures and movements. They combine these with sounds, eye signals, and tongue movements. But to interpret a dog's intentions accurately, you must also understand its concepts of *territory* and *distance*.

A dog will be more aggressive toward a stranger on its home territory than on neutral territory—in the park, for example. A dog also has other, more subtle, concepts of space. Surrounding every dog are two invisible auras. Its reactions when you intrude into them depend on whether the animal is timid or bold and if it knows you or not. A dog may challenge you on its own territory, but will tolerate you within its *social distance* if "introduced" or given the okay by its master. Once acquainted, the dog may allow you to enter its *intimate distance* and pet it. However, if you misinterpret a dog's signals or act too quickly, the dog may attack.

Will Fred Make a Friend...

A new subscriber on Fred's paper route has a dog. As Fred approaches, he sees the dog become alert, **1,** when its ears stand up. He knows he has entered the dog's *territory* when it stands up and barks, **2.** Realizing the dog is just proclaiming its territory, Fred proceeds slowly; he neither wants to frighten nor encourage the dog. When it greets him, **3,** with an open-mouth "smile" and panting, its ears relaxed and tail waving high, Fred thinks the dog will let him enter its *social distance.*

Territory

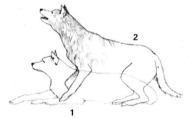

Fred moves slowly and steadily, talking in a gentle voice. To move too fast or stare at the dog would make it tense. It watches him closely, **1,** listening with its ears forward. Still uncertain, the dog circles and avoids Fred, **2,** keeping him outside its social distance. Fred knows that the dog is somewhat anxious, because it licks its lips. When it begins to relax, **3,** then Fred does, too.

Social distance

2

3

To make friends, Fred kneels, talks, and smiles. Then he slowly extends his hand into the dog's social distance, and waits for it to let him into its intimate distance. The dog bows, lowers its tail, licks its lips, and pulls its ears back in submission. When it licks his hand, Fred has a new friend.

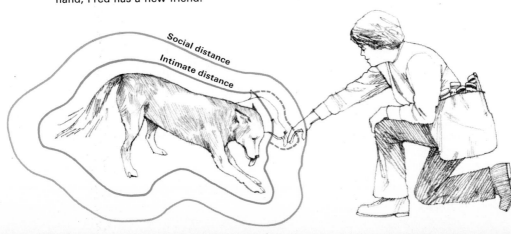

Social distance

Intimate distance

Territory

1

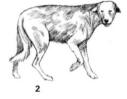

2

3

. . . Or Contend with a Foe?

But what if the dog is feeling especially bold the day Fred arrives? In this case, when he enters the dog's territory, **1,** it tests him with threatening lip-curled snarls and growls, ears forward and alert, hackles up, and the tip of its tail wagging stiffly. To see if the dog is bluffing, Fred walks on slowly and confidently, talking to the dog, but not staring. When its bluff fails, the dog submits, **2.** It crouches, tail between its legs, hackles lowered, and ears and lips pulled back as it moves away. This indicates it is safe for Fred to proceed. But then the dog tests him again, **3:** It makes a stand, staring and threatening him with a snarl, ears forward, muzzle wrinkled, hackles completely up, and legs and tail stiff. Fred is cautious, yet must show confidence; he stands still, smiling and talking.

182

If Fred's challenger had shown some fear while making a previous stand — held its ears slightly back and its tail between its legs, even though it was growling, standing stiffly, and had its hackles up — he might try this dangerous bluff. If Fred succeeds, **1,** the dog shifts its weight, lowers its hackles, and pulls its ears and lips way back. Tail between its legs, **2,** the dog slinks away, beaten, **3.** *Caution: Trying to bluff a challenging dog with stares, shouts, or brandishing a "weapon" may cause the dog to attack.*

The best strategy when the dog makes a stand is passive indifference, which might confuse the dog. With one arm ready for action, Fred watches the dog out of the corner of his eye. A wrong move, even a step back, could bring an attack. At first, **1,** the dog growls with a piercing stare, ears forward, and hackles high. If Fred does not react, it may get bored and go away. But if it attacks, **2,** he thrusts his paper bag forward against the dog's jaws, then knees the dog — hard — to knock out its wind. He must then slowly back away.

MORGANTOWN

Getting Around The Rush Hour

By Samy E. G. Elias

**A small town's experiment with personal rapid
transit may provide a partial answer to the
mass transportation problems of major cities**

Anyone who has spent any time in Morgantown, W.Va., knows
that the city has a traffic problem. I learned about it one summer day
in 1965, when I drove my family into Morgantown, ready to start my
new job as associate professor of industrial engineering at West Virginia University. It took us more than an hour to drive 1 mile. Cars
and buses clogged Monongahela Boulevard, one of two hilly routes
that connect the university's three campuses. The nightmare worsened
in the downtown area as Monongahela Boulevard's four lanes fun-
neled into two-lane Beechurst Avenue. I soon discovered that Morgantown's traffic is that bad every day.

Something needed to be done. So my colleagues and I in the university's industrial engineering department began looking for a solution.
Our search was to lead to a major breakthrough in urban transportation. Seven years later, on Oct. 24, 1972, we dedicated and began
testing a new method of moving people within urban areas – personal
rapid transit (PRT).

In a PRT system, the emphasis is on personalized service, and this is
what makes it different from older, more familiar rapid transit systems, such as subways. Although several types of PRT systems have
been designed, with different kinds of vehicles and rights of way, they
all work somewhat like an automatic elevator. When you press a call
button in a PRT station, a vehicle is sent automatically to pick you up
and take you nonstop to your destination.

Our PRT project was born between 1965 and 1967 during conversations among university staff members about the causes of Morgan-

PRT Routes in Morgantown

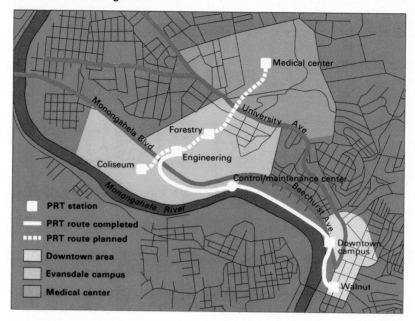

The hilly terrain around Morgantown, W. Va., has hindered development of a good, modern road system. Morgantown's only through streets are two-lane University Avenue and four-lane Monongahela Boulevard, which funnels into two-lane Beechurst. The solution seems to be a PRT system with an exclusive right of way.

The author:
Samy E. G. Elias is the chairman of the industrial engineering department at West Virginia University.

town's traffic congestion. Steep hills rise along the banks of the Monongahela River, which passes by the city. These hills, we knew, had hindered the development of a good road system. In fact, only two roads pass all the way through the city. Houses are built on the sides of the hills, close to the roads. So, the city could neither build a superhighway nor widen existing routes without incurring enormous expense and moving or destroying many houses. Because of this situation, we began to suspect that the solution to the city's traffic problem lay in some form of rapid transit with its own elevated right of way.

However, we also realized that attacking the problem was not simply a matter of applying rapid-transit technology. Engineers are also members of the larger community. For this project, we knew we would need the cooperation and approval of many agencies, but above all, we needed the support of the public. So in 1967, my colleagues and I began lecturing to civic groups about Morgantown's traffic problem and how we might solve it. We enlisted the aid of city and state political leaders, and within two years had enough community support to form a project-coordinating committee, made up of university staff members, city and county officials, and private citizens. We then presented our idea to the U.S. secretary of transportation. In July, 1969, we received federal financing for a study to determine if it would be feasible to build a rapid transit system in Morgantown.

As the basis for our study, we used the systems approach. In this method of finding workable solutions, the engineer assumes that a complex system is composed of smaller subsystems. Each subsystem has direct or indirect effects on all the other subsystems.

186

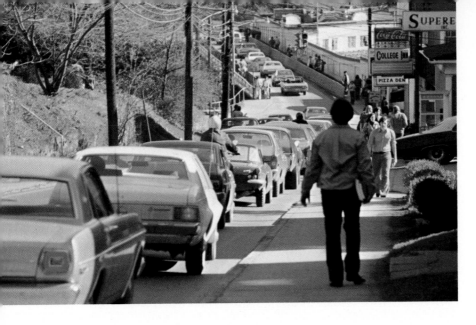

Because traffic clogs the narrow streets when students change classes, a 1½-mile drive between the campuses of West Virginia University can take up to an hour.

The systems approach to a problem involves several logical steps. The main ones are: Define the problem and determine the various factors that make up the problem; make a list of possible solutions and test each one to determine which is most suitable; and, finally, select a solution and further test it to assure that it will not upset any of the subsystems. The subsystems involved in Morgantown's traffic problem included climate, terrain, townspeople and student population, and operating costs for a public transit system.

We began defining the problem by reviewing the pattern of Morgantown's recent growth and concluded that its transportation problem was directly linked to the tremendous growth of West Virginia University. Enrollment had increased from 5,238 students in 1950 to 17,341 in 1969, and was still growing. Since the university's main campus is located in the densely built-up downtown area, the university could not expand there. In 1961, two other campuses, Evansdale and the Medical Center, were built in sparsely settled areas to the north and northeast of the city. The university also built a coliseum at the edge of the Evansdale campus.

Although the maximum distance between any two of the three campuses is only 1½ miles, the steep terrain makes walking or bicycling between campuses difficult. So the university started a bus service for students attending classes on the outlying campuses. Since 1965, the university has increased its fleet of buses from 2 to 18, trying to keep pace with the growing student population. But the bus service itself created a whole new set of problems.

Our investigators made surveys of student travels and found that they were making about 50,000 bus trips between the three campuses each week. However, we also estimated that because of the inadequate bus service, as many as 10,000 students and staff members were using private automobiles. The traffic congestion that results when the students change classes makes it difficult to get the buses to the right place at the right time. The buses crawl up the hilly roads, slowing other traffic, which compounds the congestion, and the buses fall behind schedule. Students frequently miss classes.

To solve this problem, the university first extended the interval between classes from 10 to 20 minutes. Even then, the buses could not move the students quickly enough, so the university went back to a 10-minute break between classes and discouraged students from scheduling consecutive classes on different campuses. This forced many students to schedule classes on the basis of the bus schedule rather than their academic interest.

While the bus service was disrupting students' careers, we found that the movement of students between campuses was disrupting the traffic pattern throughout the entire community. Near the main campus, where four-lane Monongahela meets two-lane Beechurst, university traffic merges with that of downtown Morgantown. Students, shoppers, and workers compete for the same limited roadway and parking space. Cars inch into the main campus area at about 5 miles per hour, slowed by students crossing University Avenue to change classes and by other cars searching for places to park.

We had now defined Morgantown's traffic problem and confirmed our suspicion: More roads, cars, and buses were not the answer. The next step was to examine new concepts in mass transportation to see which ones were suitable for our situation.

The mass transit troubles of other cities taught us that if our solution was to succeed financially, it must operate with a minimum of labor. Almost all public transit systems in the United States are in financial trouble. According to the American Transit Association, ridership has declined by at least 85 per cent since the early 1950s, forcing transit companies to cut service or increase fares, or both. This, in turn, discouraged more people from using public transportation. Meanwhile, operating costs skyrocketed. Employee payrolls consume about

Each fiberglass PRT car has room for up to 21 passengers, 8 seated and 13 standing. The vehicles run on their own concrete guideway.

How the PRT Car Operates

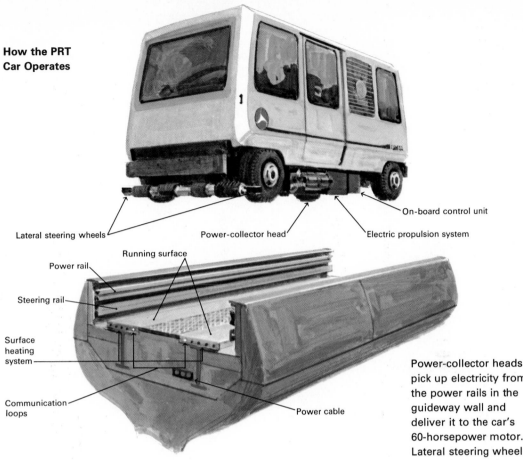

Lateral steering wheels

Power-collector head

On-board control unit

Electric propulsion system

Running surface

Power rail

Steering rail

Surface heating system

Communication loops

Power cable

Power-collector heads pick up electricity from the power rails in the guideway wall and deliver it to the car's 60-horsepower motor. Lateral steering wheels run against steering rails. Communication loops transmit radio signals between the car and the computer system. A heating system melts ice and snow on the guideway running surface.

80 per cent of all public transit revenue. If it was to avoid this pitfall, our system would have to be operated by as few people as possible.

Our researchers came up with a list of more than 130 new concepts in mass transportation, ranging from trains powered by the force of gravity to vehicles that floated on a cushion of air. To sort out all these different concepts and find the system with the highest chance for success in Morgantown, we went through a process of elimination.

In addition to being economical to operate and efficient enough to lure people out of their cars, the system we chose would have to operate over hilly terrain, through the ice and snow of cold winters. It would also have to be nonpolluting and be installed with a minimum of disruption to the community. Eventually, we narrowed our list to three systems, and engineering studies showed that all three could do the job. So, in the last phase of our feasibility study, our evaluation team carefully went over each system again, and finally recommended a PRT system that was later designed by the Boeing Company of Seattle, the Bendix Corporation of Ann Arbor, Mich., and Frederic R. Harris, consulting engineers in Stamford, Conn.

The system proposed for Morgantown was computer controlled and electric powered. Small, rubber-tired vehicles would carry passengers

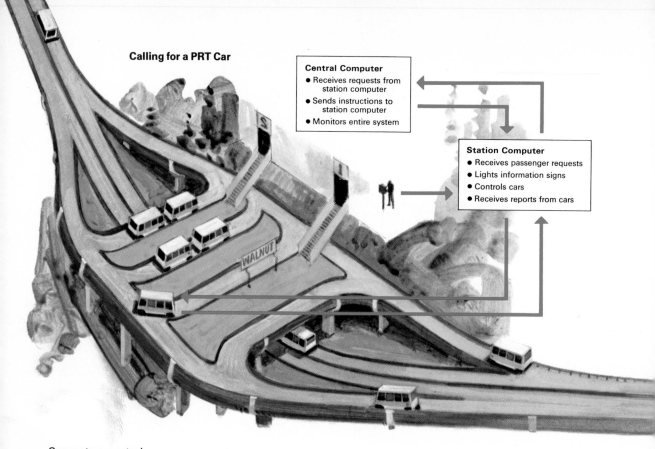

Calling for a PRT Car

Central Computer
- Receives requests from station computer
- Sends instructions to station computer
- Monitors entire system

Station Computer
- Receives passenger requests
- Lights information signs
- Controls cars
- Receives reports from cars

WALNUT

Computers control Morgantown's PRT system automatically. A passenger calls for a car by pressing one of the buttons on the destination-selection unit, *below*. Computers select a car to take him to his destination and light signs guiding him to the right car.

along 3.6 miles of two-way elevated guideway. There would be a total of six stations—one at the Medical Center, two in the downtown area, and three at the Evansdale campus. Automobile parking lots would be built near three PRT stations. A control and communications system, consisting of a central computer and 6 station computers, would control a fleet of up to 256 vehicles. The PRT system would be capable of moving 6,000 persons an hour with 80 cars running. Because it was so highly automated, only about 35 people would be needed to operate this entire system.

We released the final report on our feasibility study in August, 1970. By that time, the project had the solid support of the city, the county, the university, and West Virginia's congressional leaders. It had also attracted the interest of the U.S. Department of Transportation's Urban Mass Transit Administration, and in April, 1971, the federal government awarded Boeing, the overall system manager, $43.8 million in contracts to develop the system design.

By October, construction had started. Laying out the PRT route caused almost no disruption in the community. We had to move only two houses and a junkyard. The first completed part of the network, connecting the downtown area with the engineering building on the Evansdale campus, is scheduled to begin carrying both townspeople and students early in 1975.

The Morgantown PRT system will provide two types of service: Scheduled service for the peak traffic periods, and personalized service at all other times. For scheduled service, the computer programs vehicles ahead of time to provide nonstop service between any two stations. During rush hours, the computer constantly makes surveys of the passenger load. This enables it to schedule an adequate number of cars for each day and hour.

For the personalized service, individual passengers request nonstop trips between stations. The cars do not operate unless there is someone to ride on them. This results in a saving of electric power. Power accounts for about 9 per cent of the system's total operating cost.

One of our considerations in choosing an electric-powered system for Morgantown was the fact that it is pollution-free. Some people might say that all we did was remove the source of pollution from the individual cars to the power plant that supplies the transit system. Yet if this is the only achievement, it is of great importance, since the center of cities need immediate relief, while power plants are usually constructed away from congested areas. Also, it is easier and less expensive to control pollution at one source, the power plant, than at thousands of sources, the private cars.

To better understand the workings of the Morgantown PRT system, suppose you are a student at West Virginia University. You are about to take a ride. It is an off-hour, so the system is operating for personalized service.

From the time you enter the station, the only part of the system aware of your presence is the computer. Inside the station, lighted signs guide you to the automatic fare-collection gate. Either you have prepaid your fare and have been issued a monthly or yearly pass, or you buy a single-trip ticket from an automatic ticket machine in the station. You insert your pass or ticket into a fare-collection unit, which

In the control center, operators monitor the central computer, which watches over all traffic on the guideway and in the stations. Ramps lead off the main guideway to the PRT station, *below left,* elevated 14.5 feet above street level. Inside the station, *below,* a car stops at a platform to unload passengers.

Laying out the PRT right of way caused minimum disruption of the town. The guideway by-passes residential areas, *above*. Workers assemble the guideway wall on an elevated section that runs between buildings, *right*. A short section of the two-way PRT guideway, *below,* is built below ground level to pass underneath a road.

opens the entry gate and activates the destination-selection unit. The selection unit is a small box on the station platform on which there is a button for each station. You select your destination by pressing a push button, which then lights up, just as on an elevator.

The central computer receives your request and begins a sequence of searches. First, the computer looks for an empty vehicle in the station and directs it to take you to your destination. Or, if a PRT car is already in the station loading passengers for the same destination as yours, the computer will instruct it to wait for you. Otherwise, the nearest available vehicle will be sent to pick you up. You should not have to wait longer than five minutes.

Within the three main stations—Downtown and two Evansdale stations—there are two platforms for passenger loading and unloading and four channels that bring the cars to the platforms. There is room for four cars to stop on each channel. The other stations have one passenger-loading platform. Because the stations are built off the main guideway, through PRT cars can by-pass them without stopping or even slowing down.

The computer lights up display signs directing you to the car that has been reserved for you. For example, if you are going to the Walnut Street station, you would look for a sign reading "Walnut Street" lit up next to a waiting vehicle. You find and board the car assigned to you, the door automatically closes, and you are on your way.

You will find that the PRT car is clean, quiet, and comfortable. In designing the vehicles, we gave major consideration to passenger comfort. The cars—15½ feet long, 6 feet wide, and 9 feet high—carry up to 21 passengers, 8 seated and 13 standing. Each vehicle is powered by a 60-horsepower motor.

The exterior and interior ceilings and walls are reinforced fiberglass. The fiberglass seats are colorful, durable, and easy to clean. Each car is equipped with a temperature-control unit that provides air conditioning, heating, and ventilation. Urethane foam panels in the walls, ceiling, and seat supports provide thermal insulation and reduce noise.

Pneumatically controlled lateral "steering" wheels at the front of the vehicle extend to run against rails on either the left or right wall of the U-shaped guideway. One wheel is always touching the guideway wall to control steering. These steering wheels also permit on-board switching. For example, when the computer tells a car to bear right, its right lateral wheel is extended to contact the right side of the guideway, which it follows around the turn. Unlike a conventional railroad track, no part of the guideway itself moves to switch cars. Power-collector heads on each side of the cars pick up electric power. Either the car's right or left power-collector head is extended to make sliding contact with the guideway power rails. The rails are housed in the sides of the guideway and carry 575 volts. Buried in the guideway's concrete surface are communication lines, safety sensors, and heating pipes for melting ice and snow.

Other potential PRT systems include Rohr Industries' Monocab, with a car that hangs from a monorail, and an air-cushion vehicle with its own guideway, made by Transportation Technology, Inc., *bottom.*

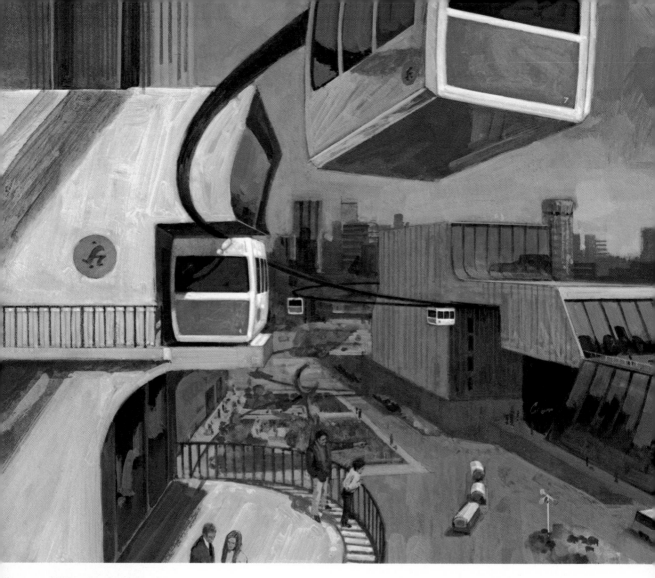

PRT systems may soon be used in combination with or as a link between buses, trains, and other forms of transit that bring people into the city.

Through the big picture windows you can see the Monongahela River as the PRT car travels quietly along its guideway. Various sections of the guideway are built at ground level, above ground level, and for a short distance, underground through a tunnel. At the PRT stations, the guideway is elevated 14½ feet to avoid interfering with normal street traffic.

The PRT cars travel along the guideway at speeds up to 30 miles per hour (mph). A minimum of 15 seconds could separate your car from one in front of or behind you. In this almost bumper-to-bumper condition, you can readily understand that the most critical element of the PRT system is the control and communication system. It consists of a central computer, computers at all six stations, and communication links between the central computer and the station computers and all PRT vehicles. There is also an independent collision-avoidance system controlled by sensors buried in the guideway.

The central computer, housed approximately midway in the guideway network, supervises overall transit operations. It also receives destination requests from the station computers and sends commands back to the stations. Signals between the central computer and the station computers travel along a cable buried in the guideway.

The station computers receive and relay passenger requests from the destination-selection units in the stations and control the passenger-information signs. Through communication loops, antennas buried in the guideway surface, the station computers transmit speed and stop commands and steering signals to the vehicles. Each station computer controls all vehicles in its zone on the main guideway and in the station. Through special antennas and electronic units on board, each PRT car receives instructions from the local station computer and transmits back information about its condition, such as motor temperature and tire pressure.

The tasks of each computer are best understood in terms of how the cars are controlled on the ramps and the main guideway. Before deciding to send a vehicle onto the guideway, the central computer must not only know where each car is at the time, but also—to avoid collisions—know where each car will be minutes or seconds in the future.

Picture the guideway with imaginary slots traveling over it. Each slot could conceivably be occupied by a PRT vehicle. The spacing between slots represents the minimum allowable time and distance between cars. These slots exist only in the "imagination" of the computer. Because it has an image of these slots traveling around the PRT network, it can locate any slot on the guideway at any point in time and determine whether it is empty or occupied.

Each car that is about to move is assigned a slot by the central computer, which tells the station computer when to release the car. The car then merges with its assigned slot on the main guideway and stays in this slot all the way to its destination.

Because you are traveling at 30 mph in a driverless car, you may ask how safe is this system and what would happen if something should go wrong. In designing the Morgantown PRT system, we heavily stressed safety requirements, including backup systems for any components that might break down. Every station has a backup computer on stand-by in case the first should fail. Each vehicle has two separate, independent braking systems. The computer and control system constantly tests itself to be sure all control functions are working properly. If it detects a malfunction, the computer operator will be promptly alerted. At any time, the operator on duty can override the computer and control the system manually. In the event of an emergency, the computer system will move vehicles to an area where passengers can get out. If this is impossible, it will automatically cut off current to the power rail so that the passengers can walk along the guideway.

The vehicles are made of fire-retardant, energy-absorbent material, reducing the danger of passenger injury in a collision. In case of power failure, each car has its own emergency power supply to apply the emergency brakes, open the doors, and provide ventilation and communication with the control system. The cars are also equipped with chemical fire extinguishers and emergency two-way radios.

The doors will not open until the car stops, and the doors must be fully closed before the car can start. If the doors should open while the car is moving, it will automatically brake to a stop. Each car has two emergency exits that can be opened from either inside or outside. An overload sensing device will prevent a car from leaving the station if there are too many passengers aboard. And, if the car goes too fast, electronic controls on board will slow it down. These are just a few examples of the safety features we built into the system.

Your first ride on a PRT is coming to an end. The station computer at your destination has been instructed by the central computer to stop the car. It pulls off the main guideway, slows down on a ramp similar to the exit ramps on an expressway, and stops at the station platform. The door opens automatically and you step out. This trip through Morgantown, which could take an hour by bus or automobile, has taken you only five minutes.

Although Morgantown, with a population of about 30,000, is far smaller than New York City or Los Angeles, its traffic problem is a miniature version of those in larger cities. Traffic jams became commonplace in all urban areas during the 1950s, with the rapid growth of suburbs and the rise in popularity of the private automobile. Millions of cars poured into the cities each day, their exhausts fouling the air. Construction of highways and parking lots could not keep pace with the increasing number of commuters. Simultaneously, public transit service sharply declined. And so our urban transportation crisis began, and each year the problem gets worse.

The PRT concept as a possible solution is beginning to receive serious attention throughout the world. At Transpo 72, an international transportation exposition held in Washington, D.C., from May 27 to June 4, 1972, four PRT prototypes were on display: Rohr Industries' Monocab, which hangs from a monorail; an air-cushion vehicle by Transportation Technology, Incorporated; the Ford Motor Company's rubber-tired car that runs on an epoxy-coated, aluminum guideway; and the Bendix Dashaveyor, another rubber-tired vehicle with a concrete guideway.

The Japanese are building an experimental PRT system, and studies are also underway in England, France, Germany, and other European countries. Denver is considering a PRT system, and forms of PRT are used to transport passengers from parking lots to terminals at the Tampa, Fla., and Dallas–Fort Worth airports.

However, in spite of all our expectations for PRT, we are aware of its limitations. It is not and will never be the solution to all the world's transportation problems. I do not believe there will ever be just one solution. Each area has a unique set of conditions that created its transportation problem. The solutions will depend largely on the local environments and will probably be composed of more than one type of transit system. It could be a combination of a subway system, a PRT system, private cars, buses, or commuter trains.

PRT is only one term in the mass transit equation. The potential for other new forms of mass transportation is exciting and the opportunities are unlimited—smaller vehicles separated by less than a second; PRT systems that carry autos "piggyback" style; private cars that can be switched to automatic control. What technology can deliver is limited only to what our imaginations can conceive.

For further reading:

Elias, Samy E. G., "The Computer as the Heart of Personal Rapid Transit Systems," *Computers and Automation,* June, 1972.

"Attacking the Mass Transit Mess," *Business Week,* June 3, 1972.

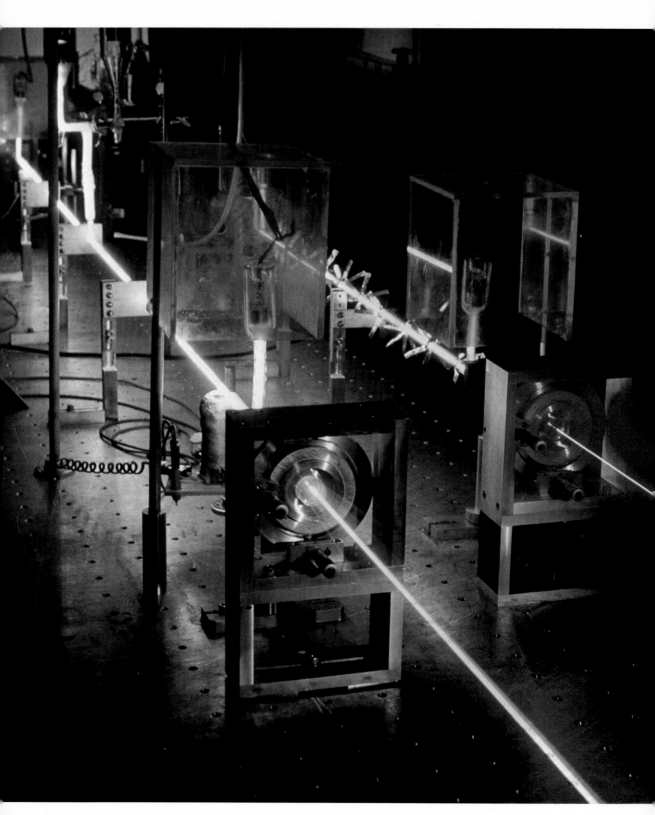

Lasers Today: Still Not In Full Focus

By James L. Tuck

These unique light resources are finding many uses, but not yet the ones our fantasies had prophesied

The first pulse of laser light flickered for a millionth of a second in 1960. It came from a small rod of pink synthetic ruby. The next year, a steady laser beam was obtained from a low pressure mixture of the gases helium and neon. Now, 12 years later, laser beams from many substances are used in countless ways, with many more yet to come.

One familiar misconception, however, is found in science fiction—the motion-picture hero in danger of being sliced apart by the villain's searing laser beam. And on a television show, the hero's laser gun knocks down attacking capsules from outer space. Both of these remarkable lasers must wait for the future; lasers today cannot produce the concentrated energy needed to perform such feats. For example, the $10-million laser at the Lebedev Physics Institute in Moscow produces the most energetic pulse of light in the world—600 joules, or 600 watt-seconds, of energy. Yet, this is less energy than your car battery puts out when you step on the starter for one second.

Probably the most powerful steady running laser in the world is a gas dynamic laser at AVCO Everett Laboratory in Everett, Mass. It requires about 70,000 horsepower, or 50,000 kilowatts (kw), to operate. Experts say it produces a power of 60 kw as an infrared beam. But 60 kw is not much power either—a good-sized electric arc welder puts out that much power, and so does the motor of a compact car climbing a hill. Nevertheless, since the laser's power is concentrated, it can drill through a concrete wall 2 inches thick in two minutes.

Similarly, at the Hughes Aircraft Company in Malibu Beach, Calif., the beam from a high-powered $50,000 ruby laser can pierce a $1/16$-inch steel plate with a single pulse. That pulse may have only a couple of joules of energy–about twice what a pocket flashlight uses every second. But the ruby laser can unload its couple of joules in as little as one ten-billionth of a second. Such a pulse or packet of light is only 3 centimeters (cm)–about an inch–long. But the power, or energy per second, of this laser is absolutely stupendous: 2 joules divided by 10^{-10} second equals 20 billion watts. The power in that flash of light is $\frac{1}{10}$ that of all the electric power produced in the United States.

How can lasers produce such power? In the early 1960s, a variety of lasers were developed and a variety of uses proposed. These proposals were based not only on the ability of lasers to concentrate energy, but also the unique form of the energy. To understand how lasers provide these advantages, we need to review how they work. The word itself partly explains this. Laser is an acronym for Light Amplification by Stimulated Emission of Radiation. The laser uses atoms or molecules to produce its light. They may exist as a stationary gas or as a gas flowing in some kind of duct. They may be embedded in a crystal, or even as molecules of a dye dissolved in a liquid. The region containing these atoms or molecules is called the laser medium.

Atoms have different energy levels, or states, that can be likened to the steps of a ladder, going upward–1,2,3,4. The atoms in an object at room temperature are mostly on the lowest step, or ground state. They can absorb energy and be excited to higher levels.

Atoms emit light by dropping down the steps of the ladder. They usually do this spontaneously, in about, say, 10 billionths of a second. But they can also be stimulated to fall down by light having exactly the right wave length. This stimulated process used in lasers has an extraordinary property. The light an atom emits falls precisely into step with the stimulating wave, strengthening the wave and making the light brighter. That stimulated emission had to occur and be in step with the stimulating wave was deduced by the German physicist Albert Einstein in 1917, long before it was known experimentally.

One more detail: An excited atom in, for example, state 4 may quickly drop to 3 and from 3 to 2, but it may not drop spontaneously to 1. Many lasers depend on atoms in such metastable states.

The atoms can be excited by heating the medium, by illuminating it with an intense flash of light, and by causing an electric discharge within it. When light of the right wave length passes through these excited atoms, it stimulates them to drop from, say, state 3 to 2 and emit light. But atoms in state 2 absorb the light and rise to state 3. Overall, the wave will weaken from absorption by the atoms in state 2, because in an ordinary temperature distribution there are more atoms in state 2 than in any of the states above it. The same weakening applies to all the other states. If growth is to exceed loss, we must have more 3s than 2s or more 4s than 1s, and so on. This is an inverted temperature distribution, which has to be created if a laser is to work.

An enormous flash of light gives an inversion by exciting all the atoms to near their highest energy levels. But the atoms fall back to

The author:
James L. Tuck is associate physics division leader at the Los Alamos (N. Mex.) Scientific Laboratory. He is a leading expert in the attempt to produce controlled thermonuclear fusion.

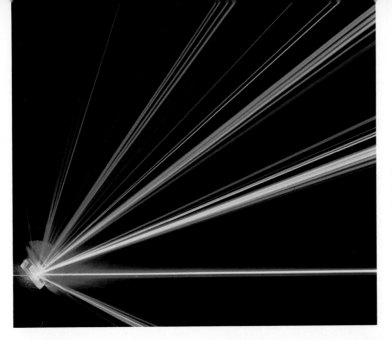

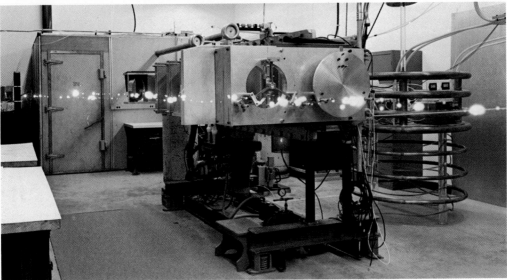

A diffraction grating separates the pure colors in the beam of this low-powered helium-selenium laser, *left.* A giant pulsed laser, *below,* uses carbon dioxide to emit an infrared beam with energy so concentrated it causes sparks in its invisible path in air.

state 1 very rapidly, and it is difficult to pump them up faster than they fall. This is where a metastable level, say state 3, helps. Atoms get trapped there, and you can get laser action by stimulated emission from 3 to 2. This is essentially what the first laser did in 1960. The medium in that laser was a synthetic ruby, a crystal of aluminum oxide, and the laser atoms were the chromium "impurities" in the crystal. Powerful xenon flash lamps did the pumping.

Another way to get inversion is to heat gas to a high temperature and then suddenly chill it, before those atoms in the upper states can fall down to the ground state. It is possible to cool gas rapidly by expanding it as it flows at near the velocity of sound through a nozzle (like the gas flowing out of the nozzle of a jet engine).

Laser Action

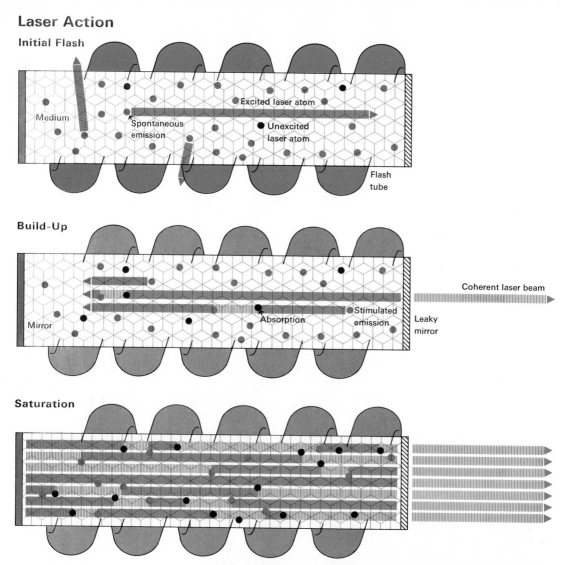

Initial Flash

Medium

Excited laser atom

Spontaneous
emission

Unexcited
laser atom

Flash
tube

Build-Up

Coherent laser beam

Mirror

Absorption

Stimulated
emission

Leaky
mirror

Saturation

A laser transforms some of the flash lamp energy, top, into a coherent beam of light, bottom. Only spontaneous emission from an excited laser atom that heads perpendicular to the mirror can pass back and forth through the medium and stimulate other emissions (and absorptions). The light quickly grows to saturation, producing a very strong beam through the leaky mirror. The ruby laser, *right,* thus emits a beam both highly directional and with all of its light waves in step.

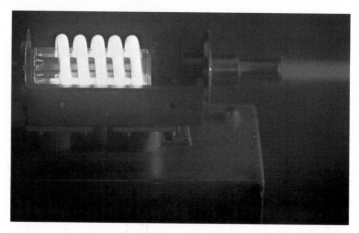

If, for example, state 4 is inverted with respect to 3 for an instant, a laser working from 4 to 3 is possible. Gas dynamic lasers work on this process. These high-powered devices look quite different from gas-discharge lasers and use heavy machinery, such as jet engines.

The inversion drives the laser. But even though a state is metastable, some spontaneous emission will always occur. So there will always be enough light of the right wave length to trigger the lasing. The light wave grows as it passes through the excited medium, feeding on the inversion; the number of atoms in the metastable state falls and that in the state below it rises. When they become equal, lasing stops.

The rate at which the wave grows depends on how many metastable atoms there are. The growth of the wave also depends on how strong the stimulating wave is. A weak wave might grow slowly, and it would take many passages through the medium before the inversion is gone. The rate of growth shows up as the distance a newly born wave has to travel before it reaches full saturation, or "avalanche," strength, capable of eating up all the inversion in one pass.

This distance–the saturation distance–is extremely important in making the medium lase in a desired direction. In a laser, the saturation distance is made, say, 50 times longer than the laser medium so that the light wave will grow very little in one pass. Mirrors placed at both ends of the medium reflect the light wave back and forth. In 50 reflections, only the light that hits the mirror just right will have grown to saturation strength. The beam is directed very accurately in a straight line between the mirrors. Most lasers use mirrors that reflect about 90 per cent of the light and transmit 10 per cent. If only 10 per cent escapes at each reflection, the light continues to build up as it crisscrosses the laser medium. But after about 50 reflections, virtually all of the light pulse has been drained away to the outside.

A laser beam is a unique form of light. How we can utilize it depends on its properties. Light is a wave motion, and the wave length is quite short for visible light, approximately 10^{-5} centimeter. But a light wave ordinarily has an irregular series of breaks, or changes of phase, up to a million times each second. This is because the atoms in all ordinary light sources emit independently. Each atom emits a train of waves, up to billions of waves long, and the overall light is a random mixture of huge numbers of such trains. Such independently emitted light is called incompletely coherent.

But not laser light. The organizing effect of stimulated emission makes a laser beam almost perfectly coherent. All the wave trains are in step. Perfect coherence gives laser light a unique and useful property–the beam is truly parallel. In a searchlight, the mirror reflects the light from different points on the hot carbon of the electric arc into parallel beams, which fan out. A searchlight beam gets bigger and bigger and weaker and weaker the farther it goes. A laser beam, however, grows up parallel. It spreads only by virtue of the finite wave length of light–by diffraction–and therefore much less.

A good application of this parallel beam is as a lecturer's optical pointer, which looks like a flashlight but has a power socket cord on one end. It uses a low-powered helium-neon laser. When the lecturer presses a button, a fine beam makes a bright spot on a blackboard or projection screen perhaps 50 feet away.

The "laser ranger," a very precise surveying instrument, is another application. It is really laser radar and is sometimes called lidar. This device sends out a light pulse that reflects off a distant object. The time it takes the beam, traveling at the speed of light, to make the round trip, provides a precise measurement of the distance. No great power is involved here, only a truly parallel beam.

A good reflector makes it easy to get a strong signal back from a distant object, and the corner reflector possesses remarkable properties for this. Cheap versions of it are used in those small nighttime reflecting disks on highways. The reflector is a block of glass with three faces in the back end like the corner of a cube. The three faces must be precisely at 90° angles to one another, so that a beam of light entering

A curb, the blind man's nemesis, is foiled by the Laser Cane, *below.* Radarlike coherent beams of infrared radiation are aimed above, ahead, and down to detect obstacles and trigger warning tones. Another medical use, based on the sharp focus of the beam, is in experiments to attack skin cancer, *below right.*

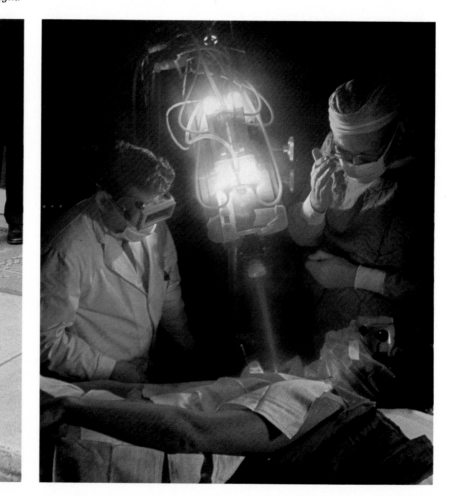

Laser-scan photography reproduces a drawing of an airport runway on a TV screen. Working in complete darkness, the helium-neon laser beam rapidly sweeps an area and the reflected light is electronically formed into an image.

the reflector makes three reflections inside and leaves on a path exactly parallel to its path on the way in. With an ordinary mirror at a distant reflection point, it might take days to get the angle exactly right, so that the reflected beam goes back to the starting point. But the corner reflector need only be roughly pointed at the lidar unit.

Laser radar was used to measure the distance to the moon in 1972. Very accurate measurement can detect such things as wobbles in the moon's orbit and influences of such other objects as the sun, which in turn provide important information about the solar system.

Because the moon is 238,000 miles from the earth, on the average, even the slow spreading of a laser beam over such a distance is serious. The high-powered ruby laser pulse was estimated to be a mile wide when it hit the moon. The reflecting system, set up on the moon by the Apollo 17 astronauts, consisted of a metal plate that had hundreds of corner reflectors set into it.

The beam coming back from the moon was about 9 miles wide when it reached the earth. Only very delicate photomultipliers on the big Lick Observatory telescope in California could detect the return pulse: It was far too weak to be seen or photographed. Still, the round trip times of many pulses measured the distance between the earth and the moon with unprecedented accuracy—to within 6 inches.

Laser light is also a valuable tool where focusing to a fine point is important. If you put a convex lens in the beam of a searchlight and try to focus it down to a point, the best you can get is a little image of the white-hot carbon. In the same way, the best and hottest focus of the sun you can get is a little image of the sun's disk—it never goes down to a pinpoint. The atoms in the searchlight arc and on the sun radiate independently, and their light does not give a parallel beam.

A coherent laser beam can be focused down to the smallest size that the wave length of the light will allow, an area of about a ten-billionth of a square cm. You can hold your hand in the beam of a low-powered laser of, say, 1 watt, and barely feel it. Yet, when the beam is focused by a lens, the light intensity becomes stupendous. One watt falling on 10^{-10} square cm is equivalent to 10^{10} watts per square cm. (The Hoover Dam hydroelectric plant puts out about 3×10^9 watts.) Such a

Laser light excites a jet of gas, *below,* in a new type of chemical analysis. The specific light frequency excites the molecules, and a mass spectrometer then analyzes products of the reaction. Laser beams are also used to find art forgeries. Beam causes smoke to rise from painting, *right.* The vapor can then be analyzed to determine the paint's age.

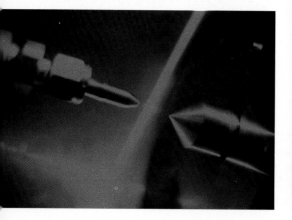

focused beam instantly vaporizes any material it falls on. This explains why we have to be so careful to shield our eyes from all but the lowest-powered laser beams. The lens of the eye can focus the beam to a point at the retina, burning a tiny area. The first medical use of the laser – retinal surgery – used this phenomenon.

A laser of medium power, say a carbon dioxide (CO_2) laser, can emit a steady infrared beam carrying 100 watts. It can be focused down to make a very good knife. In cardboard carton manufacture, a metal knife was used to cut corrugated paper, but it got dull quickly and the machines had to be stopped while it was sharpened. Now, a laser beam burns through miles of paper and never gets dull.

In some clothing factories, cloth cutting also was done by knife. But the knife got dull and the cloth crumpled, causing much waste. The focused beam from a medium-powered laser is now being used for this kind of cutting – and it is never blunted.

The high energy intensity of the focused beam can drill as well as cut. Tungsten is an extremely hard metal that is used for the filaments of electric lamps because of its high melting point. It is so hard that diamond, the hardest material known, is the only material that can be used in the dies that draw down tungsten wire. But how do you make the die; how do you drill a fine hole in a diamond? Until now, the only way was to grind it out using a spinning needle coated with diamond dust, a slow and expensive method.

But now, a medium-powered laser, say, 1 joule per pulse, focused down to the desired wire diameter, does the job quickly. With each pulse, the laser beam evaporates a small quantity of diamond in a puff of smoke. You can guide the beam's powerful focus down through the diamond like a drill.

A much higher-powered laser should be able to do more, however, than drill larger holes. Theoretically, its focused beam should be powerful enough to trigger a nuclear fusion reaction. This would be a giant step toward solving our energy crisis.

Fusion power is more desirable than the fission process used in the nuclear power plants today. It should be safer because there are no dangerous fission products to bury and no massive chain reaction to keep under control. Nuclear fusion uses light elements—deuterium (D), tritium (T), and lithium—instead of the heavy elements, such as uranium and thorium, used in nuclear fission.

For many years, scientists have devoted a great deal of research effort to try to get fusion to work. But they cannot heat the D-T fuel hot enough to cause its atomic nuclei to fuse and then maintain the process. Most of the fusion research today involves containing the hot D and T in a magnetic bottle, the only bottle that would not be burned up. But so far, the researchers have not been able to keep the fusion reaction going long enough to repay the energy expended in the

The vivid reflection of a laser's pulse aids surveyors in measuring distances with great accuracy. They check the time it takes for the pulse to travel to a distant mirror and back.

Focused laser beams controlled by computers are now being used on industrial production lines to cut fabrics.

heating-up process. The problem is very difficult; the hot D-T plasma is very quick to wriggle out of the confining magnetic bottle.

Meanwhile, scientists have begun to wonder if a focused laser beam could be made strong enough to blast a pellet of fuel to the extremely high temperature that is necessary almost instantly. The density of frozen solid hydrogen at 4° Kelvin is nearly a million times higher than the gas density in the magnetic bottle devices. The denser the hot fuel is, the faster it burns. So for hot D-T fuel at the density of the solid D-T pellet, the burning might be enough to repay the initial heating, before the pellet has exploded. If this should be so, there would be no need for a magnetic bottle or any other confinement.

You can imagine how such a laser fusion reactor might function: You drop the D-T pellet into a strong boiler. The laser pulses, and the pellet explodes, generating neutrons and heat. Tritium, which is about half the fuel used, is much too expensive to consume as a source of energy—it costs $3 million a pound. But it is possible to recover and even breed it. This would be done by coating the inside of the boiler with a 1-meter-thick layer of circulating molten lithium metal. The heat and the neutrons are absorbed there, and the lithium nuclei break apart in absorbing neutrons, regenerating the expensive tritium. The hot lithium would be circulated through the steam boiler of a normal thermal power plant to generate the electricity.

It is possible to increase laser power by combining laser mediums. Such a system is called a giant pulsed laser. A laser directs its beam into another pumped-up laser medium that has no mirrors, empties the inversion there in one pass, sends it to a third medium, etc. These laser media are called laser amplifiers. Their job is not to lase on their own, but to work on the pulse from the master laser, or oscillator. If you prevent the light beam from leaving the medium in the oscillator, and then let only one strong pulse out through an electronic shutter and amplify that, you can get all the energy stored in the laser amplifiers in one giant pulse. This pulse could be focused on the pellet.

The world's leader in this attempt is the laser at the Lebedev Institute in Moscow, which has nine parallel amplifier stages. The laser medium is neodymium atoms in solid glass rods about 2 cm thick and 100 cm long. The rods are all pumped at the same time by xenon flash tubes clustered around each rod. The primary laser is one neodymium-glass rod, and the shutter lets out only one pulse at a time. Using prisms and mirrors, this pulse is divided into three parts, which are passed to three amplifier rods. The amplified pulse is subdivided again, and passed on to nine rods. These rods put out a total of 600 joules, and the shortest pulse is an amazing 2×10^{-10} second.

But power is not the only important factor in trying to make fusion power. The efficiency of the laser also becomes of paramount importance. Neodymium-glass lasers are very inefficient—at best, 1 per cent. That is, for the 600 joules put out in the beam, the flash lamps have to be fed 100 to 1,000 times as much. This is an experimental device,

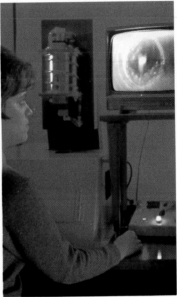

A 10,000-watt CO_2 laser, *above left,* is used to weld, cut, and heat-treat materials, such as stainless steel, *left.* In a more delicate feat, the operator watches a TV monitor, *above,* as a laser beam drills a hole a ten-thousandth of an inch in diameter through a diamond die.

Lebedev's Laser-Fusion Experiment

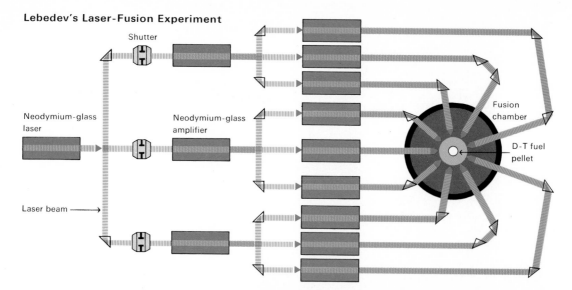

Shutter

Neodymium-glass
laser

Neodymium-glass
amplifier

Fusion
chamber

D-T fuel
pellet

Laser beam →

The world's most powerful laser, at the
Lebedev Physics Institute in Moscow,
brings together laser pulses from nine
parallel neodymium-glass channels in an
attempt to achieve nuclear fusion. The
multiplied beam would fuse and explode,
releasing energy. Experiments in the
United States, such as at the University
of Rochester, *right,* are just beginning.

hardly one expected to generate power. The experiment has an element of financial and personal risk, too, which must add greatly to the
excitement. When neodymium glass (or any other solid medium) is
pumped to the limit, as these rods are, a step higher in power usually
cracks the glass. Researchers have to be extremely careful to prevent
any light from bouncing back into the rods from the target, or Moscow
could have some expensive glass breaking—as much as $100,000 worth
of glass shattered in a millionth of a second!

The efficiency of most lasers—ruby, helium-neon, helium-cadmium,
and gas dynamic—is low, between 0.1 per cent and 1 per cent. This
makes them useless for power production. The CO_2 laser, which has
reached 22 per cent, is the popular candidate for laser fusion.

Getting fusion to be competitive is going to be tough by any
method. I am an optimist, and I estimate that fusion by the magnetic

bottle process will produce energy more economically than fission by the year 2000. The laser-powered method has been under investigation for no more than two years, and problems are appearing. For example, in the explosion, one pellet might generate only from 1 to 5 cents worth of electricity. How do you prevent the explosion (and the molten lithium) from spewing from the holes in the boiler and smashing the laser and focusing equipment? Windows of any known material will not work; they will not stand up to the laser beam.

Does that mean that laser fusion research should not be strongly encouraged? Not at all. Laser pellet explosions would seem to be ideal for space propulsion. Imagine simply tossing the pellet into a rocket nozzle. Zap!, and the push of the reaction products out of the back provides the rocket thrust. As of now, laser space propulsion would be too expensive because half the essential fuel of a pellet fission reactor is costly tritium. And there would be no place on a spacecraft for a molten lithium recovery and circulation plant.

However, the price of tritium will drop when commercial fusion power becomes a reality, because fusion power reactors will make more tritium than they consume. So, perhaps the price will drop low enough by the year 2010 to allow laser fusion space travel.

The pinpoint focusability of the laser beam that made it unique for laser-pellet fusion also made it unique for tacking down detached retinas, in which the eye's own lens provided the focusing. Lasers have other medical applications—for example, they can be used for burning off warts or knocking out accessible tumors. However, the laser is not the only tool that can be used this way. Any other concentrated heat source, such as an arc light, can and has been used. The wonder and promise of the laser has led some to speculate—by analogy with X rays—that there might be cancer-curative properties in the laser beam. This is probably expecting too much. There does not seem to be any reason why laser light should have any specific effect on cancer cells. Furthermore, X rays penetrate easily through intervening tissue and bone, and laser light does not.

There are also studies throughout the world on applying lasers to communications. A laser-light beam, modulated in the manner of a radio wave, could carry millions of messages simultaneously. Similarly, researchers are studying laser holography as a means of storing information. Such message storing is needed for tomorrow's electronic computers, and perhaps even for the libraries of the future.

The laser is, indeed, a wonderful device, and nobody yet knows all the things it will do for us.

For further reading:

Lubin, Moshe J., and Fraas, Arthur P., "Fusion by Laser," *Scientific American*, June, 1971.

Silfvast, William T., "Metal-Vapor Lasers," *Scientific American*, February, 1973.

Engineering The Environment

By Michael Reed

**A mountain-carving project in Australia
has rerouted rivers to produce electrical
power and bring new fertility to the land**

With the ceremonial pull of a switch on Oct. 21, 1972, Australian Governor General Sir Paul M. C. Hasluck officially marked the completion of the $900-million Snowy Mountains Scheme to bring power and water to the people of southeastern Australia. For 23 years, engineers, technicians, and laborers had been blasting through mountain granite to reroute the abundant water that collects high in the Snowy Mountains, part of Australia's Great Dividing Range. To do this, they built 16 large dams, laid 50 miles of aqueducts, and bored more than 90 miles of tunnels through the mountains.

When nature had her way, most of the water from the melting snow in these mountains formed the headwaters of the Snowy River. During the summer, the swollen Snowy plunged down the eastern slopes, ran through the wet coastal land, and spilled into the Tasman Sea.

Meanwhile, farmers in the dry but fertile Murray and Murrumbidgee valleys west of the mountains gambled with drought. Although the Murray and Murrumbidgee rivers also rise in the Snowy Mountains, they receive only a fraction of the run-off that goes to the Snowy River. In good years, these rivers supplied farmers with all the irrigation water they needed. But a dry year meant hardship or disaster.

However, irrigation water was not the only reason behind the Snowy Mountains Scheme. Australia, a growing industrial nation, needed power to run its factories and light its cities. So a major part of the project is a series of power stations that the water passes through as it falls 2,500 feet down the western slopes. This generates pollution-free power for the states of New South Wales and Victoria.

Men and machines battled
the rugged terrain to
build the spillway for
Geehi Dam and reservoir,
above. Steel arch forms
are set in place, *left,*
in the Snowy-Eucumbene
tunnel that connects
Tantangara Reservoir
to Lake Eucumbene.

The author:
Michael Reed is
Managing Editor
of *Science Year.*

The heart of the Snowy scheme is its main reservoir, man-made Lake Eucumbene, which lies about 4,000 feet above sea level on the eastern side of the mountains. With a surface area of 56 square miles, it can store almost 4 million acre-feet of water (enough to cover 4 million acres with 1 foot of water). The Tantangara Reservoir, with a capacity of 206,000 acre-feet, lies 10 miles north of Lake Eucumbene. Fifteen miles south is Lake Jindabyne with a 558,000-acre-foot capacity.

The Snowy Mountains Scheme is divided into two sections: The Snowy-Tumut Development diverts water into the Tumut River, a tributary of the Murrumbidgee; and the Snowy-Murray Development feeds the Murray. There are four hydroelectric power stations in the Snowy-Tumut Development and three in the Snowy-Murray section.

Work continued on the Snowy project 24 hours a day in all kinds of weather. Despite the ice and snow of winter, workers make progress on the Island Bend Dam, *left.* A night-shift construction team pours concrete on the floodlit Blowering Dam site, *below left,* while sparks from a welder's torch, *below,* illuminate a pressure tunnel at the Murray 1 Power Station.

Snowy Mountains Scheme

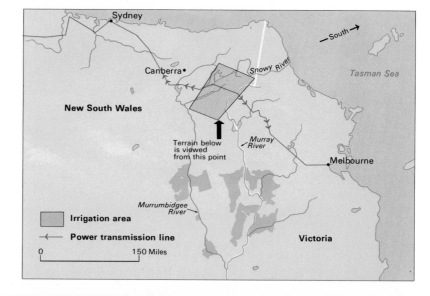

The Snowy Mountains Scheme covers about 2,000 square miles of rugged mountainous terrain in southeastern Australia. A network of dams, reservoirs, and tunnels divert water from the east slopes to the west slopes of the Great Dividing Range.

Sydney

Canberra

New South Wales

Snowy River

— South →

Tasman Sea

Terrain below is viewed from this point

Murray River

Melbourne

Murrumbidgee River

☐ **Irrigation area**

←— **Power transmission line**

0   150 Miles

Victoria

Murrumbidgee River

Lake Eucumbene

Tantangara Reservoir

Great Dividing Range

Tumut Pond Reservoir

Tooma Reservoir

Tumut 1 Power Station

Tumut 2 Power Station

Talbingo Reservoir

Tumut 3 Power Station

Blowering Reservoir

Blowering Power Station

Tumut River

Water from the three large reservoirs is transported to the western side of the Great Dividing Range by huge tunnels drilled through the mountains. If Lake Eucumbene is the project's heart, these tunnels are its veins and arteries. In the Snowy-Tumut section, the Murrumbidgee-Eucumbene Tunnel carries water 10.3 miles from Tantangara Reservoir to Lake Eucumbene. From there, the Eucumbene-Tumut Tunnel transports the water 13.8 miles through the mountains to Tumut Pond Reservoir, where it begins its westward descent through reservoirs and the power stations. The 14.6-mile Eucumbene-Snowy, the 9-mile Snowy-Geehi, and the 6.1-mile Jindabyne-Island Bend tunnels transport water from lakes Eucumbene and Jindabyne to other Snowy-Murray reservoirs and power stations.

Sixteen huge dams were built to form the reservoirs. The tallest is Talbingo, a 530-foot-high structure on the Tumut River. The second highest is 381-foot Eucumbene Dam, which forms Lake Eucumbene.

When the Snowy Mountains Scheme began in 1949, it was apparent that Australia alone could not supply the labor needed to build

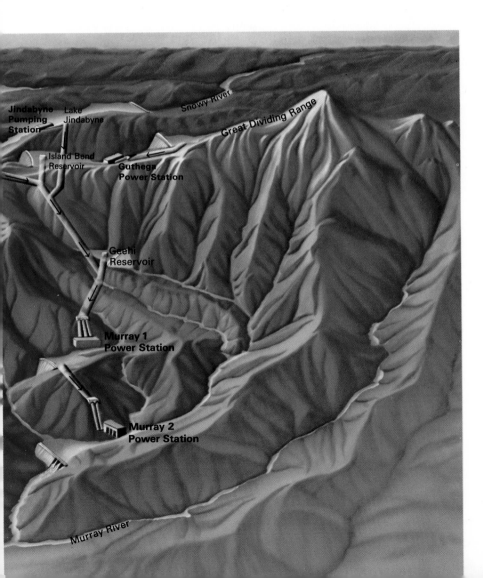

Geehi Dam traps water for the Murray 1 Power Station, one of the seven power stations in the Snowy scheme.

this gigantic network. So contractors, engineers, and workers were recruited from 30 other nations. In 1959, at the height of construction activity, 7,300 workers were employed on the project.

Seven towns and 100 camps were set up to accommodate this great influx of people. More than 1,000 miles of roads and trails were built to move men and machines. Pickup trucks in summer and snow-cats in winter brought workers from the newly built towns to the tunnels and dam sites where they worked 3 shifts, 24 hours a day, 6, and sometimes 7, days a week. Eucumbene Dam was completed in two years; the schedule had allowed for four. Tunnels were also finished well ahead of schedule. As a result of these great efforts, the Snowy project was virtually completed in 1972 – three years ahead of schedule.

The completed project made all this labor worthwhile. Farmers in the Murray and Murrumbidgee valleys can now depend on almost 6-million acre-feet of irrigation water a year, 50 per cent more water than was available before the Snowy Mountains Scheme was built. This has not only made established farms more reliable and productive, but has also opened up an additional 1,000 square miles to farming. And from the Snowy project's power stations high in the mountains, transmission lines stretch out, ready to carry a peak 3.74-million kilowatts of electricity to homes and industries throughout southeastern Australia—from Canberra to Melbourne to Sydney. Once again, man has used his technology to change an environment. But this time, people are cheering the results.

Water redistributed by the Snowy Mountains Scheme irrigates areas in the Murrumbidgee Valley, *above.* Power lines, *right,* carry electricity from the Snowy power stations to cities throughout southeastern Australia.

The New Rainmakers

By Charles L. Hosler

Until they know more about the life of a cloud, meteorologists can never know if their attempts to make rain or to suppress hail will be successful

In the summer, when there is little rain over southern Florida, meteorologists in Miami carefully study the cumulus clouds. If they decide that the prospects are good, a two-engine, propeller-driven airplane is ordered aloft to drop brightly burning silver iodide flares into the clouds. The meteorologists hope that the smoke particles pouring from the flares will trigger rainfall on the drought-stricken area to the north, around Lake Okeechobee. Also in the summer, when hailstorms have been known to damage crops near Boulder, Colo., clouds sweeping off the Rocky Mountains are seeded by small airplanes to increase the number of ice crystals and thus prevent the growth of large hailstones. These are experimental efforts, but cloud seeding is used each year in many other parts of the world in efforts to increase rainfall, prevent hail, or dissipate fog.

Yet, after 25 years of experimental and operational weather modification, we still know comparatively little about how to make it rain

when and where we want. It is still difficult to determine, on a given day or with a given group of clouds, whether man's intervention has a positive, negative, or any effect whatsoever on the rainfall.

For example, a casual look at the long-term data from places where seeding has been tried, such as the Sierra Nevada mountains in California and the upper Colorado River Basin, reveals nothing startlingly different from rain or snowfall patterns observed before the seeding began. Only after a careful statistical analysis that measures target and control areas with historical data does it become clear that—in some cases—there was a 10 or 15 per cent increase in precipitation, which probably resulted from cloud seeding. But in Missouri, a similar analysis shows just as clearly that a significant decrease in precipitation resulted after cloud seeding.

If we are on the verge of making significant breakthroughs in our understanding of weather modification, as was suggested in the National Advisory Committee on Oceans and Atmosphere 1972 report to the President and Congress, it will happen only if we expand our studies to learn a great deal more about what is happening within the clouds themselves. We probably now have sufficient technology to do this, but we would need many more researchers and more equipment to carry on the work. And, of course, this would mean the allocation of considerably more federal funds.

Cumulus clouds, a type often used in cloud-seeding experiments, are in such a delicate balance with their environment that minor changes in the manner in which water condenses in them have major effects on how they develop and on how much rain they produce. For instance, even a slight change in the updraft or an increase in the amount of atmospheric dust in a small cloud can help to produce a thunderstorm. The situation resembles one in which floodwaters build up behind an earth-filled dam, then overflow and suddenly demolish it, releasing a tremendous wall of water that rushes downstream. Energy stored in a cloud can be released just as suddenly, but inadvertently, by human activities. For example, smoke particles from city stacks can affect rainfall. The science of cloud modification involves finding those times and places when nature's energy dams can be breached by small, but intentional, actions of man.

There are hundreds of factors, however, that determine how clouds develop. Every condition of the earth's surface affects the air that blows over it. The air is modified by the surface temperature, wetness, slope, and roughness, and by the particulate matter that may leave the surface to enter the atmosphere. Because the earth's surface contains infinite variations, it becomes a tremendously complex job to determine where and how a cloud will develop.

Clouds form as air ascends. The volume of air being lifted expands and cools. As the air cools, some of the moisture in it condenses into tiny water droplets and ice crystals. Actually, a cloud contains little water in liquid form. If all the water droplets in a stratus cloud 2 miles

The author:
Charles L. Hosler
is dean of the
College of Earth
and Mineral Sciences
at Pennsylvania
State University.

thick were brought to the ground, it would produce less than 1 millimeter of rainfall. To produce heavy rain, clouds must grow rapidly, continuously drawing more moisture up from lower levels of air to replace that which falls.

Subtle variations in the process by which water in a cloud is transformed from gas to liquid or solid can determine whether raindrops or snowflakes form, and whether rain, snow, or hail will fall to the ground. The total amount of water that condenses out of a cubic centimeter of the cloud may be constant, but whether the water is distributed as thousands of tiny droplets or a few hundred larger ones will make a large difference in how much precipitation, if any, actually falls to the ground.

Water vapor in the atmosphere will condense only if there are particles, such as bits of dust or other air pollution, for it to condense around. If the air in which a cloud forms has only a few suitable particles, large droplets of water will coalesce around them. In this case, rainfall is likely, since larger droplets will collide more easily with other droplets to produce large raindrops that will not completely evaporate on the way down. On the other hand, if there are a great number of small particles, water will condense on them as many small droplets, and the likelihood of rainfall reaching the earth is lessened.

You can perform a simple experiment to illustrate the effect of particles on the formation of droplets. Fit a one-hole stopper and a few inches of glass tubing into a large glass container, such as a gallon jug. Put enough water in the jug to cover the bottom, and attach a foot of

Tiny particles of ice break off and are electrically repelled, *left*, as water vapor condenses and freezes on the cold, round surface. Air bubbles cover a freezing water drop, *above*, before bursting and thus scattering electrically charged microdroplets. Such studies help explain rapid changes in a cloud.

rubber tubing to the end of the glass tube. In a darkened room, shine a flashlight beam through the container. Suck as hard as you can on the rubber tubing to reduce the air pressure in the container, and you will see a cloud of water droplets form. The reduced air pressure cooled the air and some of the water vapor condensed on dust particles in the air.

Suck on the tubing again to lower the pressure in the container and pinch the tubing end shut to maintain this pressure in the vessel. Then, hold a burning match near the end of the tubing and let the tubing open slowly above the tip of the flame so that the soot particles are sucked into the container. When you suck on the tubing again, you will see a much denser cloud form inside the jug. The same amount of water condensed, but now there are many more particles to form droplets on, and the many droplets formed scatter more light.

Astounding as it may seem, in a cloud composed entirely of tiny water droplets, from 1 million to 10 million individual droplets must collide and combine to produce a single large drop of rain. In many such water clouds, therefore, no rain reaches the ground. We have all observed such layers of clouds and fog that persist for many hours, and yet yield no precipitation.

On the other hand, some cumulus clouds form and produce a downpour within 30 minutes. In these clouds, some droplets must be much larger than others so they can fall and bump into the smaller ones. Through laboratory studies in vertical wind tunnels, scientists have verified theoretical calculations of the efficiency with which one size of water droplet captures another. And, while flying through clouds in specially equipped planes, scientists have captured droplets on a film and counted and measured them. In this way, they have established a relationship between the size of the droplets and the probability and speed of rain formation. But the mixture of various types and sizes of ice crystals and other nuclei with various-sized water droplets creates a system so highly complex that much more laboratory work and field observation are required to understand raindrop growth and the movement of particles in a cloud.

In fact, scientists still cannot always explain why ice forms in clouds at all. Thus, many scientific mysteries must be solved before we can predict how much ice will occur in a given cloud. Experiments that were conducted by physicists Irving Langmuir, Vincent Schaefer, and Bernard Vonnegut in 1946, and were repeated by hundreds of others since then, have shown that the difficulty in changing very cold water droplets to ice crystals can be overcome.

Most cloud droplets freeze only when the temperature reaches a low −39°C. (−38°F.). In fact, left to its own devices, the condensed water in clouds at temperatures well below freezing (0°C.) is almost entirely liquid. There is an energy "barrier" that prevents ice from forming in the "supercooled" droplets. Sea salt, particles of soil, and other forms of dust in the air, produce many droplets, but only a few ice crystals per liter of air at temperatures down to −20°C. (−4°F.). Adding ice

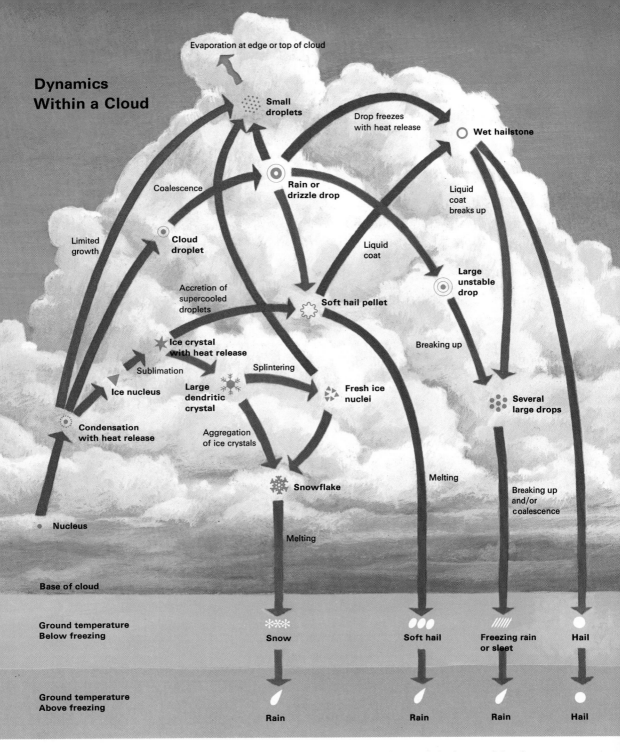

Dynamics Within a Cloud

Evaporation at edge or top of cloud

Small droplets

Drop freezes with heat release

Wet hailstone

Coalescence

Rain or drizzle drop

Limited growth

Cloud droplet

Liquid coat breaks up

Liquid coat

Accretion of supercooled droplets

Large unstable drop

Soft hail pellet

Ice crystal with heat release

Sublimation

Splintering

Fresh ice nuclei

Breaking up

Ice nucleus

Large dendritic crystal

Aggregation of ice crystals

Several large drops

Condensation with heat release

Snowflake

Melting

Nucleus

Melting

Breaking up and/or coalescence

Base of cloud

Ground temperature Below freezing

Snow — Soft hail — Freezing rain or sleet — Hail

Ground temperature Above freezing

Rain — Rain — Rain — Hail

Many processes can cause a growing cumulus cloud to precipitate water as rain, snow, or hail. One or more of these activities may take place, depending on the cloud's temperature and size, the strength of its updraft, number of nuclei, and amount of moisture in it. The stronger the updraft, the larger will be the precipitation particles that hit the ground, though often the precipitation evaporates before it reaches the ground. The rates of droplet and ice formation are also important influences on the process. In addition, the rates of heat release and water accumulation have an effect on cloud growth.

A silver iodide flare, burning brightly under a cloud-seeding plane over Colorado, adds seed particles to a hail cloud, increasing the number of tiny ice particles, and reducing the number of large and destructive hailstones.

crystals to a cloud causes the original ice particles to grow very rapidly. Water molecules leave the water drops in greater numbers than leave the ice crystals at the same temperature. Thus, the ice crystals grow at the expense of surrounding water droplets. The large ice particles also grow by colliding with many small water drops.

By introducing something very cold into a cloud containing supercooled water, a region can be temporarily cooled below −39°C. Billions of tiny ice crystals are generated, and these will grow rapidly. If you have a chest-type freezer or cold box where the temperature can be held at −18°C. (0°F.) or lower, you can demonstrate this effect. Shine a bright light into the box and exhale into it. All you will see is a gray or white cloud of tiny water droplets. In all probability you will see no ice crystals, or at most only an occasional one. Now hold a small piece of dry ice over the cold box and scratch it with a stick or nail so that tiny flakes of dry ice fall into the box. When you exhale into it again, in about 15 seconds you will see a dramatic change. The very cold particles, at −79°C. (−109°F.), cause some ice crystals to form and then grow, using up the water that evaporates from the liquid droplets produced by breathing into the chamber. This simple experiment is the basis for most cloud seeding.

Physicists have found that water condensing upon particles of some other solid materials will form ice at higher temperatures. Silver iodide is the most celebrated of these substances because it produces some ice at temperatures as high as −4°C. (+25°F.). Fogs at belowfreezing temperatures are routinely dissipated by this type of seeding at a few airports in various parts of the world.

The results of seeding clouds remain in doubt, however, because of our inability to monitor the natural variations in the multitude of processes that occur in clouds. In some clouds, for instance, no ice crystals or water drops ever grow large enough to bump into smaller ones, and no snow or rain will form regardless of how much seeding is done. In others, nature produces enough ice crystals to result in snow or rain, regardless of man's intervention. In this case, man may actually reduce the amount of precipitation by adding extra ice crystals, because the production of more condensation nuclei will result in more but smaller raindrops or snowflakes. Most of the resulting precipitation then evaporates before it reaches the ground.

Experiments have shown that cloud seeding sometimes causes cumulus clouds to expand vigorously upward so that additional air and its moisture become part of the cloud. This increases the chance that precipitation will reach the ground. When the meteorologists in Miami seed clouds, they usually attempt to induce this kind of growth. They also sometimes can get the seeded cloud to merge with several other clouds as it expands. This greatly increases the chances of the occurrence of a significant rainfall.

The growth of a cumulus cloud depends on the relationship between the temperature in the cloud and that of the surrounding air.

The cloud also must retain a slightly higher temperature and buoyancy than the air surrounding it. Cumulus clouds develop and grow large only when the temperature of the surrounding atmosphere decreases with altitude slightly faster than does the temperature of the cloud. A number of complex factors determine the vertical temperature and moisture distribution in the surrounding atmosphere. But the horizontal distribution will be reasonably constant over large areas.

Temperature changes in a cumulus cloud that is growing vertically upward are determined by natural laws and can be calculated accurately. But the air outside always mixes with the cloud, and this confuses the picture. The heat needed to maintain the cloud's buoyancy is released when water vapor condenses to form liquid drops, or when water freezes to form ice. For each gram of water condensed into liquid, roughly 600 calories of heat are released; for each gram of water frozen, about 80 calories are released. If the cloud is large and the mixing area at the edge is far from its center, continued heat release guarantees continued buoyancy. However, mixing at the edges of a small cloud or a tall and thin cloud brings in air that is relatively dry as well as cold, and water must evaporate from droplets or ice crystals to saturate the incoming air. This is a cooling process and it is likely to decrease the temperature in the cloud so much that the cloud will not stay warmer than the surrounding air. Then the cloud stops ascending.

Since the water in a cloud usually remains liquid even at temperatures much below freezing, a cloud has a reserve supply of heat from ice formation. Nature may never tap the supply, or only when the cloud reaches great heights where the temperature is quite low. In spite of the heat release by condensation, the development of many cumulus clouds is restrained by a layer of air above the cloud in which the temperature decreases so slowly with height that the rising cloud becomes the same temperature as the environment and loses its buoyancy. However, if the cloud droplets can suddenly be changed to ice by seeding, the additional heat released may heat up the cloud so that it can break through this layer of warmer air. The result may be spectacular growth of the cloud, leading to rainfall.

Cloud buoyancy also depends on the amount of water or ice particles it carries, and on the level in

A seeded cumulus cloud quickly grows and produces rain. The sequence, by scientists at Florida's National Oceanic and Atmospheric Administration (NOAA) Laboratory, shows the cloud, from top to bottom, at the time of seeding, 9, 19, and 38 minutes after seeding when the cloud had become fully developed.

the cloud where this accumulates. The release of heat through seeding changes both the upward velocity of the cloud and the particles' sizes and how fast they fall. Thus, predicting the course of events in a cloud is a computational problem that taxes the fastest and biggest computers now available.

For cloud-seeding attempts to succeed regularly, we must be able to assess the physical state of the cloud and the surrounding air–the vertical distribution of pressure, temperature, and humidity–so that we can predict what the natural course of events will be. In addition, we must be able to measure the liquid-water content of the cloud, the range of droplet sizes in various portions of the cloud, and its ice-forming potential. This involves measuring the vertical velocity in various parts of the cloud and the rate with which the cloud is mixing with its environment. A dominant factor in determining whether a cloud will blossom into a large cumulus rain cloud or simply become a small, dissipating cumulus cloud is the rate at which the cloud mixes with the air around it.

When these things are known, mathematical models of the cloud can be used to calculate ahead in time to find the natural course of events in the cloud. This assumes that we can project how, for example, the measured spectrum of drop sizes in the cloud will change as a result of additional condensation, evaporation, and interaction between the particles in the cloud. But we do not yet know precisely how

Many types of scientific equipment are needed to monitor the wide variety of changes that seeding may cause.

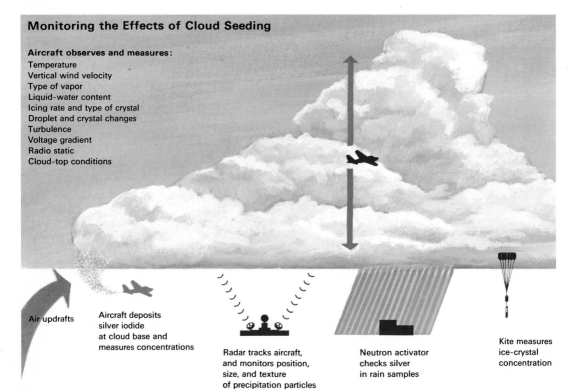

Monitoring the Effects of Cloud Seeding

Aircraft observes and measures:
Temperature
Vertical wind velocity
Type of vapor
Liquid-water content
Icing rate and type of crystal
Droplet and crystal changes
Turbulence
Voltage gradient
Radio static
Cloud-top conditions

Air updrafts

Aircraft deposits silver iodide at cloud base and measures concentrations

Radar tracks aircraft, and monitors position, size, and texture of precipitation particles

Neutron activator checks silver in rain samples

Kite measures ice-crystal concentration

Scientists launch a 12x20-foot research kite at the National Center for Atmospheric Research in Boulder, Colo. The kite carries instruments to record imprints of snow crystals on strips of film. The equipment helps determine the effectiveness of seeding clouds with silver iodide particles.

the spectrum of droplet sizes will change in a cloud, or how many droplets will grow into raindrops. We know even less about the effect of the droplets' surface properties and the effect that electric fields in the atmosphere have in promoting or hindering coalescence. We do not even know at what temperature the droplets will turn into ice.

A principal weather-modification goal of current cloud-seeding experiments is to be able to regulate that temperature. Until we know if nature will turn the cloud droplets into ice in a given cloud, we cannot be certain that cloud seeding will have any effect. Many cloud-chamber experiments have shown that the temperature at which water changes to ice depends on the types of nuclei in the air. These nuclei, which may be dust or smoke particles, serve as centers upon which water and then ice forms as the air cools. But in a cloud, successive condensation, freezing, and evaporation of water often alter the behavior of the particles and hence change their ability to have water or ice condense on them. Particles collected by airplanes that fly through clouds are extremely varied, and the number of special ice-forming particles may change by a factor of a thousand within a few minutes. It is obvious that the cloud chambers in which we create clouds on earth to count the ice crystals do a poor job of simulating real clouds.

We are now designing a cloud chamber for use in the space shuttle that is scheduled for launch into space in the late 1970s. While the shuttle is in orbit under near-zero gravity, a cloud will remain in the small cloud chamber without dissipating or hitting the walls. Cloud particles can then be grown, frozen, melted, and electrified in the chamber on the same time scale and temperature, pressure, and humidity conditions that occur in a real cloud.

In some cases, even the spectacular growth of a cloud may not result in rainfall reaching the ground. If seeding causes a much taller, bigger cloud that processes more water and the water falls out the bottom of the cloud, the seeding produces rain. But suppose the seeding increases the vertical velocity in the cloud so much that the condensed water moves to the top of the cloud and high-speed winds shear off the top and carry all the water away. Then, seeding decreases the amount of rainfall in the region.

In order to get the information necessary to successfully and consistently modify the weather, we will have to go much further than we already have in developing remote sensing techniques, using radar, infrared-sensing devices, and unmanned aircraft. We must use the recently developed Doppler radars and multiple-frequency radars to learn about the water distribution and the motion of water within clouds. The use of several radar wave lengths will detect the sizes of water droplets, and how they are distributed and moving in the cloud. Only a few Doppler radars are now available, but meteorologists can use them to measure the motion of precipitation particles in both the horizontal and vertical directions. They measure the apparent change in frequency of radar waves caused by the motion of the particles. Infrared-sensing devices can be used to determine a cloud's temperature at various heights.

Unmanned aircraft, or drones, can penetrate clouds in areas too dangerous for manned aircraft. They can continuously monitor cloud particles, temperature, and the speed of air and particle movements in a cloud. Drones can be almost any size, and can relay a variety of data

Air inversion patterns over a model of downtown Denver are created by dry ice in the weather laboratory at Colorado State University, Fort Collins.

that ground stations need. Weather-satellite measurements already promise to ease the job of collecting much of the data that meteorologists need. They are particularly helpful in tracing the vertical structure of the atmosphere. By measuring the emission of electromagnetic waves in a restricted portion of the infrared spectrum, we can obtain the vertical temperature distribution, which is vital in determining a cloud's modification potential.

We may be able to measure the liquid-water content of clouds by measuring the absorption of microwaves by water in clouds. This is possible because the oceans emit microwaves. These radiations can be monitored by satellites. The amount of liquid water in the atmosphere determines to a large degree the intensity of the radiation reaching the satellite. Where there are clouds between the seawater and the satellite, there will be less radiation reaching the satellite because of the absorption. Thus, cloud water and precipitation potential can be assessed. Visual and infrared images should enable us to identify the cloud fields that are most susceptible to modification. Photographic versions of this data are already being used in weather prediction.

There is little doubt in my mind that many cumulus clouds and cloud systems can be modified for the benefit of mankind without adversely affecting large numbers of people. There will always be a conflict—between the farmer or rancher and the operator of a golf course or amusement park, for example—over what kind of weather they would like on any given day. When crops and food supply are threatened by lack of rain, however, there is an overriding public interest to be served in producing rain if the atmosphere provides the circumstances man might alter.

None of the experiments that need to be done, and none of the improvements in our observation systems that need to be made will place anyone in jeopardy. Until we know the true potential of cloud modification, we are in no position to decide whether it is economically, legally, or morally feasible. Consequently, this research should be vigorously pursued.

For further reading:

Battan, Louis J., *Cloud Physics and Cloud Seeding,* Doubleday, 1962.

Mason, Basil J., *Clouds, Rain and Rainmaking,* Cambridge University Press, 1962.

Mason, Basil J., *Physics of Clouds,* Oxford University Press, 1971.

National Academy of Sciences—National Research Council, *Weather and Climate Modification, Problems and Prospects.* Publication No. 1350, 1966.

Simpson, Joanne, and Dennis, Arnett S., *Cumulus Clouds and Their Modification,* National Oceanic and Atmospheric Administration Publication, May, 1972.

Tricker, Robert A., *Science of Clouds,* American Elsevier Publishing Co., 1970.

Science Parts the Iron Curtain

By William J. Cromie

**With official backing, a new spirit of cooperation
is growing between Soviet and American scientists**

Dr. James F. Holland, noted for his work on the treatment of leukemia with drugs, was invited to lecture and do research at Moscow's Institute of Clinical and Experimental Oncology. To do this, he would have to leave his job as chief of medicine and director of the Cancer Clinical Research Center at Roswell Park Memorial Institute, Buffalo, N.Y., for eight months. Unwilling to be separated from his family for that long, Holland took his wife and six children, 7 to 21 years old, with him for the trip to Russia.

Apartments are scarce in Moscow and the Hollands moved into the Hotel Ukraine in November, 1972. But, in about a month, Russian officials announced they had found a five-room apartment for the family. And, when moving day dawned, a volunteer "moving crew"

Russian and American scientists celebrate the completion of a joint experiment at Soviet Institute for High Energy Physics near Moscow.

of senior research scientists and others from the institute showed up to help the Hollands. Then the Russians produced vodka and threw a party for the Americans.

The small, neighborly incident demonstrates the rapid closing of the gap between scientists of the United States and Russia. Ten years earlier, Dr. Holland probably would not have been invited for so long a stay. Even five years earlier, Russia's working scientists would have been reluctant to socialize so warmly with Westerners.

This new attitude followed the visit to the Soviet Union by President Richard M. Nixon in May, 1972, when he and the Russian leaders signed what presidential adviser Henry A. Kissinger called "an orgy of new agreements" for greater exchanges and cooperation. Four of the agreements, covering a five-year period, call for an exchange of data and scientists and work on joint projects. They signal a dramatic change in national attitudes.

Each of the four—on science and technology, environment, medicine, and space—has a variety of individual provisions. To carry out the science and technology agreement, for example, the nations set up a Joint Commission on Scientific and Technical Cooperation and established working groups that approved projects in six areas of mutual interest and benefit: energy, agriculture, computer applications, water resources, microbiology, and chemical catalysis. Energy projects approved included work on electric power systems, transmission lines, magnetohydrodynamics, solar energy, and geothermal energy.

Russian-American cooperation in medicine and health is not all new. It is built on prior contacts and exchanges dating back to a State Department pact signed in 1958 and renewed every two years since then. Its science and technology section provided for exchanges by the National Academy of Sciences and the Soviet Academy of Science. In 1970, the Department of Health, Education, and Welfare (HEW) began to discuss a health agreement. In February, 1972, HEW and the Soviet Union's ministry of health acknowledged the desirability of joint programs on heart disease, cancer, and environmental health. The five-year agreement, signed with the others in May, calls for "joint efforts toward combating the most widespread and serious diseases." Joint studies have begun on environmental health, cardiovascular diseases, and those involving tumors. Teams of scientists from each country exchanged visits to better understand scientific capacities and research in particular fields. Combined working groups then designed projects to use each nation's abilities and serve each nation's needs to the fullest extent.

In the fight against cancer, joint working groups are meeting to develop specific projects in chemotherapy, virology, cell genetics, and immunotherapy. The work in chemotherapy and virology seems to promise the most immediate gains. Dr. Holland's visit was one example. Scientists of each country also exchanged new drugs, viruses, cell lines, and animal tumors for study and experiment.

The author:
William J. Cromie is president and editor of Universal Science News, Incorporated, and a frequent contributor to *Science Year.* He toured Russia in 1972.

American cancer researcher Dr. James F. Holland explains a point on research methods to his Russian colleagues at the Institute of Clinical and Experimental Oncology in Moscow. Later the Holland family dines in its new Moscow apartment.

"We've accepted their animal and clinical studies on these drugs," says Dr. Steven T. Carter of the National Cancer Institute (NCI). Dr. Carter adds, "We firmly believe that the best hope of making progress in the cure of solid tumors—breast, gastrointestinal, and lung cancer— is in using drugs after surgery. The Soviets have a good deal of experience in this technique."

The exchange of viruses suspected of causing cancer in humans and other primates also has brought encouragement. "Soviet scientists had made claims over the years about isolating viruses that they believe produce leukemia in primates," said Dr. Anthony M. Bruno, NCI assistant director. "In this country, we have found abundant evidence of a cause-and-effect relationship between viruses and nonprimate tumors, but we have not linked viruses and human cancer. Therefore, we were very anxious to test the Soviet viruses." While the earliest findings have not yet substantiated the Soviet claims, "...they are encouraging," Bruno said. "In human cell cultures inoculated with the Soviet viruses, we see growth of particles that resemble viruses found in cancerous monkey tissue."

Bruno finds the relationship promising. He pointed out that U.S. technology developed rapidly after World War II, and U.S. science could luxuriate in almost unlimited technological capacity. But Russia was recovering from the war's devastation. Soviet scientists had to find research approaches that were less dependent on technological sup-

port. "Although our equipment and facilities are often superior," Bruno said, "many of us have been impressed by the talent of Soviet scientists we have met."

Bruno emphasizes that both nations can benefit from scientific co-operation, but he warns that it would be misleading to look at the exchange as merely a useful trade. "The most serious mistake we could make," he says, "would be to think of this as a trade from which we want 'our share.' To think we can wisely allow Russia a pound of inspiration only if the Russians provide 16 ounces of it in return—that kind of thinking would be disastrous."

Many U.S. scientists feel that their Soviet counterparts can contribute fresh perspectives and approaches to common problems. Coming from a completely different cultural and intellectual climate, the Russian scientists may be able to propose techniques and avenues of research that never occurred to Americans. Conversely, Russian scientists may find that U.S. technology can contribute to the solution of a problem that has stymied them.

Medical men expect important gains from such interchange. A report on the joint project on heart disease said, "Some of the most elaborate and interesting medical approaches in the U.S.S.R. are unfamiliar or unknown to U.S. heart specialists and their coronary patients." In another instance, the United States received a set of Soviet-developed stapling machines that can be used to repair organs

U.S. physicist Edgar Delli and his Polish co-worker monitor their experiment at The Institute for High Energy Physics, *left.* An American group, *above,* tours an institute at Akademgorodok.

during surgery. Their use replaces the difficult sewing procedure more familiar in the West. The stapler can also be used on blood vessels. Western technicians had tried for some time to develop such a device, but they had not succeeded.

On the other hand, Soviet scientists are extremely interested in artificial heart devices developed in the United States. They have been given samples of American-made pumps and devices to aid failing hearts and lungs. The United States has also given Russian researchers the first ultrahigh-speed centrifuges to be used in a Communist country. Their use in separating cell components is especially important in cancer studies. In addition, two automatic analyzers are to be sent to Moscow and Leningrad. Russian scientists will use them in a joint study of blood fats.

One incident illustrated the warm cooperation and understanding that is building between the two countries. Mstislav V. Keldysh, president of the Soviet Academy of Science and a scientist so talented that he is widely considered a national treasure, had a serious circulatory illness. Major arteries in his abdomen and upper legs were becoming blocked. Keldysh had been examined twice by Dr. Michael E. De-Bakey, the famed Texas surgeon who developed techniques to replace damaged blood vessels with artificial ones. When Keldysh's condition worsened in January, 1973, DeBakey flew to Moscow with a surgical assistant and a nursing team. The next morning, DeBakey performed by-pass grafts around the blocked arteries. Keldysh recovered quickly and was soon back at work.

Many observers found the lack of any effort at secrecy to be the most impressive aspect of the entire drama. A leading scientist of one nation had been treated by a leading surgeon from its archrival. Yet each nation was careful—one not to lunge for a resounding propaganda triumph, the other not to hide its appreciation.

Cross-fertilization of scientific thought is already helping mathematicians—particularly those building mathematical models of the atmosphere or using mathematics in weather forecasting. Forecasters start with a value representing each atmospheric characteristic, then calculate ahead in 10-minute steps to get a 24-hour forecast. They had slower, less-modern computers, so Soviet scientists worked out ways to get forecasts in larger steps, with fewer calculations. Gurii I. Marchuk, developer of the system, introduced it in the United States in 1967. It

Scientists monitor equipment in tunnel at the Yakutsk Permafrost Institute in Siberia.

was not directly adaptable to U.S. needs, but Western meteorologists worked out a modification that is.

A National Weather Service meteorologist, John A. Brown, Jr., says, "We are now working with this modified system. And the Canadians are testing a model based on it that looks very promising. It may allow us to do the same job as Marchuk's method does, but in an easier way." If that works out, Russia might end up adopting a Canadian model based on an American modification of a mathematical improvement initially proposed by a Soviet scientist. The circle would finally have been made complete.

The Apollo-Soyuz Test Project (ASTP) is the most visible cooperative activity coming from the growing international accord. It involves the 1975 rendezvous and linking together of a U.S. Apollo spacecraft with a Russian Soyuz spacecraft. Crew members will conduct joint scientific experiments for two days.

Russian science can now share its achievements in medicine, physics, and space. The director of the Moscow Institute of Experimental Surgery, *below,* holds a weekly conference with his colleagues. A Russian scientist tests a new chemical laser, *bottom left.* Cosmos pavilion, *bottom right,* at an "economic achievement" show salutes space sciences in Russia.

Glynn S. Lunney, American technical director, called ASTP "a test of our ability to cooperate." Lunney and Professor Konstantin D. Bushuyev, his Russian counterpart, led engineers working out project details. Their shirt-sleeve conferences in both countries during 1973 produced little apparent discord. Professor Bushuyev found only one area where agreement was lacking: "Dr. Lunney drinks coffee black, and I drink coffee with cream."

The two countries had established a pattern of trading information gathered in space exploration. It included samples of lunar rocks and soils retrieved by astronauts and by robot Russian craft and information from the unmanned exploration of Mars and Venus. Cooperation in the fields of space and medicine would have occurred without the 1972 summit meetings. The agreements in the field of environmental protection, however, break new ground.

Russell E. Train, chairman of the President's Council on Environmental Quality, identified four general subjects about which he believes U.S. scientists and technicians can learn much from the Soviets —Arctic permafrost, earthquake prediction, the effect of atmospheric changes on climate, and wildlife management. The United States is particularly interested, Train said, in "their experience in dealing with permafrost problems in terms of building, highway, airport, and pipeline construction. They probably have the most extensive Arctic experience of any nation." He added, "They have had success in predicting earthquakes in the Garm-Dushanbe region of central Asia, and we plan to install some of their [seismometers and other seismic] instruments along the San Andreas Fault in California. We also have agreed to integrate our tsunami [large earthquake-produced waves] warning system with theirs."

A working group of eight American wildlife experts visited nature preserves in Russia in January, 1973, and agreed on 22 specific joint projects, most of them involving the exchange of scientists. For example, Alexsy V. Yablokov and A. A. Kirpichnikov will join Scott McVay of the Environmental Defense Fund in a study of the rare bowhead whale at Icy Cape, Alaska. Soviet scientists will also team up to study the biology and behavior of seals in the Bering Sea with a research physiologist of the University of Alaska. And, they will chart the migration of northern swans with a Maryland pathobiologist.

The wildlife group found several Soviet-developed ideas and techniques that can be applied to conservation problems in the United States. "In the protection of endangered plant species, the Russians have demonstrated greater accomplishment than we have," notes Joseph P. Linduska of the Bureau of Sport Fisheries and Wildlife. "They have brought forest insects under control without pesticides—with a simple and strictly biological system. They provide little nest boxes—artificial housing sites—all through the forest. The nests increase the population of birds and small mammals that eat insects and this keeps the insects under control."

Russian physicist, *right,* explains the operation of a device his group had prepared for joint work at the National Accelerator Laboratory, located near Chicago.

As scientific ideas and approaches are exchanged, researchers in both countries plan similar experiments so they can compare results. And senior scientists in both countries now feel it will be fruitful to spend extended periods in the others' laboratories. Such personal exchanges could produce results that might be much more difficult–or even impossible–to obtain by working separately. An experiment during 1972 and 1973 at the National Accelerator Laboratory (NAL), near Chicago, demonstrated the value of such exchange visits.

At a 1970 conference in Kiev, physicist Ernest Malamud of NAL learned of an interesting experiment in which Soviet scientists measured the way the radius of a proton increases when exposed to higher energies. The Russians used energies of 10 billion electron volts (GeV) in an accelerator at Dubna and 75 GeV in a larger machine at Serpukhov. "The natural extension of this experiment was to study what happens to protons at the kind of energy we expected to achieve at

NAL–200 to 400 GeV," said Malamud. "I proposed a joint experiment, and, in 1971, after a good deal of discussion and negotiation, we put together a group of 7 Soviet and 10 American physicists.

"While we waited for them to arrive, we tried the experiment ourselves, using thin polyethylene foil as a proton source. However, this did not work so well. The Soviets had, on the other hand, spent many years developing a technique that uses a beam of hydrogen gas moving at supersonic speed as the source. They built such an apparatus especially for the NAL accelerator and brought it with them when they came in March, 1972. This was a very important contribution." Malamud added that having the Russian device immediately available, "saved us many years."

Vladimir A. Nikitin, a senior Russian scientist, found the results gratifying. He said: "The biggest difficulty was the language problem, but we found that—as far as physics was concerned—our ways of thinking are similar. We got to know each other very well and would regard the prospect of working together again very positively. We exchanged not only equipment and techniques, but also ideas and attitudes that helped to make the experiment a success."

Although scientists speak in glowing terms about the new era of relaxation and cooperation, problems still exist. One is money. No new funds were allocated for joint Soviet-American programs in the 1974 federal budget; U.S. agencies must pay their share out of funds appropriated for other programs.

Distance, language, and communications are also problems. One veteran foreign service officer concedes that neither government is free of bureaucratic rigidity. While saying, "Their society remains under tight internal control, and access to some areas is refused or difficult," he adds that "there is inflexibility on both sides. Technically, no citizen of the Soviet Union is admissible to the United States under present Justice Department rules, and each must have a waiver secured by the State Department."

By working together, scientists should be able to overcome these problems. The staff at the National Accelerator Laboratory, for example, points to the language problem that handicapped early discussions in the international group. But the scientists soon found that technology provided them a solution. Seated side by side at a computer console, each could express himself clearly in terms the other understood. The new moves toward international cooperation seek to make that experience a symbolic goal for much of science.

For further reading:

"A Busy Week in Moscow," *Science News,* June 3, 1972.

Leighton, Lauren G., "Another View of Akademgorodok," *Bulletin of the Atomic Scientists,* April, 1972.

Walsh, John, "Soviet-American Science Accord: Could Dissent Deter Détente?," *Science,* April 6, 1973.

Science File

Science Year contributors report on the year's major developments in their respective fields. The articles in this section are arranged alphabetically by subject matter.

Agriculture

Anthropology

Archaeology
Old World
New World

Astronomy
Planetary
Stellar
High Energy
Cosmology

Biochemistry

Books of Science

Botany

Chemical Technology

Chemistry
Dynamics
Structural
Synthesis

Communications

Drugs

Ecology

Electronics

Energy

Environment

Genetics

Geoscience
Geochemistry
Geology
Geophysics
Paleontology

Medicine
Dentistry
Internal
Surgery

Meteorology

Microbiology

Neurology

Nutrition

Oceanography

Physics
Atomic and Molecular
Elementary Particles
Nuclear
Plasma
Solid State

Psychology

Science Support

Space Exploration

Transportation

Zoology

Agriculture

The use of DDT as an insecticide to protect farm crops was banned, effective on Dec. 31, 1972. Diethylstilbestrol, a growth stimulant used to increase weight gains in livestock, was banned because of a possible link with cancer (see DRUGS). The Environmental Protection Agency began a close study of aldrin and toxaphene, insecticides related to DDT. The agency also began to investigate the effects of antibiotics in poultry feeds.

Radioing weeds out. A new approach to weed control was discovered in the laboratories of Texas A&M University at College Station. Agronomist Morris G. Merkle and physicists J. Robert Wayland and Frank S. Davis found that ultrahigh-frequency (UHF) radio waves (about 2,450 megahertz) destroy weeds. The Oceanographic International Corporation built a research model and started testing it at College Station. It will treat 50 acres of cropland a day. The process is somewhat like flame weeding in that a protective shield must be placed around the crop plants while the radio waves are directed at the weeds. Experimentally, weed seedlings and weed seeds, as well as tiny parasites, such as nematodes and noxious fungi, can be destroyed by the new treatment. The radio waves are also lethal to floating aquatic weed plants that often clog rivers and streams.

Trickle irrigation, a new system of economic water management, first tried in Israel, was used on farms throughout the United States. In this system, many long, perforated plastic tubes are placed on top of the soil. Water flows through the tubes and seeps into the soil at a constantly controlled rate. The process was used from the citrus groves and vegetable farms of Florida and Texas to the orchards, vineyards, and row crops of New Jersey, Michigan, New Mexico, Arizona, and California, as well as to those in the Wenatchee and Yakima valleys of Washington. The U.S. Department of Agriculture (USDA) estimated that over 25,000 acres of high-value crops in California are now being watered with trickle irrigation systems.

New breeds. The Finnish Landrace breed of sheep, Finnsheep, introduced in 1971, are now being crossbred with

Biologists at Brookhaven National Laboratory check a tobacco plant, the first mature hybrid organism ever to be produced by the fusion of non-sex cells from two species.

Italy's Chianina beef cow, which may weigh 2 tons and stand 6½ feet at the shoulder, is being crossbred with traditional American beef-cattle breeds.

Agriculture

Continued

standard breeds. The attempt is to produce offspring with the Finnsheep capability of early, frequent, and multiple births and the standard breeds' high-quality wool and meat.

The new animals do not produce high-quality wool or meat. However, they can produce lambs once every eight months instead of once a year. In addition, 3-year-old ewes of this breed average 3 or 4 lambs at each lambing, compared with a U.S. average of slightly less than 1 for other breeds. They also lamb at 1 year of age. The most extensive studies of Finnsheep and their crossbreeding are being conducted under the direction of breeder Hudson Glimp of the USDA Animal Research Center at Clay Center, Nebr.

The latest innovation in beef cattle are Chianina crossbreeds, which were developed in 1971. Chianina cattle come from Italy. The crossbreeds developed in the United States produce better beef than the purebred cattle and are from 25 to 50 per cent heavier than Angus, Hereford, and other conventional breeds.

Plant varieties. A new technique for developing hybrid plants that by-passes pollination was revealed by botanist Peter S. Carlson, Harold H. Smith, and Rosemarie D. Dearing of the Brookhaven National Laboratory in Upton, N.Y. They fused cells from two different species of the tobacco plant, *Nicotiana glauca* and *Nicotiana langsdorffii*, and put them in a growth medium that would support only the hybrids. The resulting hybrid was identical to the same cross made by conventional means. The scientists call the new process parasexual hybridization. With it, agricultural scientists may be able to cross widely divergent species. See GENETICS.

Agronomist Charles A. Francis of the Center for Tropical Agriculture at Cali, Colombia, developed a high-lysine variety of corn, in 1972, which is superior to previous varieties in milling, storage, and rate of grain maturity. The fact that it has a high-lysine content generally increases its digestibility.

Interest in high-lysine corn for poultry, swine, and livestock feed was thus revitalized. The corn more commonly

Nature's Insecticide

Instead of becoming a normal stable fly, *below,* a pupa sprayed with a juvenile hormone, *bottom,* never completes its full development.

Modern insecticides have been partly successful in man's battle against those insects that spread disease, destroy crops, and pester man and beast, but they have many serious drawbacks. For example, the insecticides kill not only the destructive insects but also other insects, some of which are natural predators that control the pests. The insecticides may also persist in the environment, ultimately finding their way into foods and into the bodies of animals, including man.

In some cases, insect pests develop resistance to the chemical insecticides—their biochemical mechanisms evolve to make the insecticide ineffective. Man's response has been to develop more toxic and dangerous insecticides that provide only temporary insect control, but all too often also create new environmental hazards.

Our team at the U.S. Department of Agriculture laboratory at College Station, Tex., used a different approach to find new substances that attack only one species of pest, do not persist in the environment, and are not hazardous to animals. For example, we have developed two substances to control the stable fly, *Stomoxys calcitrans*, which are similar to natural substances within the insect. We call these chemical substances insect growth regulators.

The stable fly causes millions of dollars damage in the agriculture industry each year. It is a blood-sucking insect that annoys cattle and other livestock and causes them to eat less than normal. The flies lay their eggs in decaying organic matter in cattle feed lots, along sandy beaches, in open fields, and in urban areas.

The chemical substances we developed are analogues of the insect juvenile hormone. This hormone is one of three known secretions that regulate the insect's growth and metamorphosis from larva to pupa to adult.

At certain stages in this development, the hormone is normally absent. If it is sprayed on the insect during these periods, the fly's life cycle will be disrupted. If this disruption occurs, the insect develops in an abnormal way and it will eventually die.

We have successfully demonstrated this new approach in laboratory tests and under field conditions where the insect is normally found. In one test, for example, we treated a 10-foot-square grassland area in Florida and reduced the stable fly population by 99 per cent in a few weeks. In a test at a Nebraska feed lot, we reduced the stable fly population by 89 per cent.

The analogue apparently does not affect the natural predator of the stable fly, the parasitic wasp, *Muscidifurax raptor*. These wasps deposit their eggs in stable fly pupae. As the wasp larvae hatch out and grow, they destroy the tissues of the stable fly pupae.

It makes no difference whether the pupae have been exposed to the hormone analogue or not, because a wasp larva thrives in both the treated or untreated pupae. And, most important, those wasps that have consumed the treated pupae, in the same manner as those that have not, continue to lay their eggs in other stable fly pupae and destroy them.

We proved this by comparing the development of untreated stable fly pupae with pupae that had been exposed either to the juvenile hormone analogues, the wasp, or to both. Adult stable flies emerged from the untreated pupae within seven days. None of the treated pupae developed into adults, however, and when we dissected them we found dead, incompletely developed stable flies in every case.

The young parasitic wasps we found within the treated pupae, however, were perfectly normal. We found that the wasp's development was apparently unaffected by concentrations of the juvenile hormone analogues as much as 1,000 times greater than those that were needed to stop the normal development of the stable fly.

Tests made by Harry E. Smalley, veterinary toxicologist in our laboratory, show that the juvenile hormone analogues used to control the stable fly also are harmless to laboratory animals and livestock. Indeed, the major significance of this research is that it points to an exciting new approach in the control of insect pests. Species-specific agents could easily be incorporated into established pest-control programs to reduce the quantity and number of conventional, highly poisonous insecticides used, and ultimately, perhaps, replace them completely.　　[James E. Wright]

Agriculture

used, because of its superior milling and storage qualities, required high-cost protein supplements.

Important to the protein contributions in human food was the development of IR 480 rice by the International Rice Research Institute at Los Baños in the Philippines. The protein content of this new variety was increased by 20 per cent in 1972. There was no loss in yields compared to the standard IR-8, the rice strain that began the green revolution in underdeveloped nations.

Five lines of grain sorghum that are resistant to the green bug pest were developed in 1972 by plant breeders Harold L. Hackerott and Thomas L. Harvey at the USDA's Fort Hays (Kans.) Station and released by the Kansas Agricultural Experiment Station. The green bug pest is an aphid that is particularly destructive to wheat.

A variety of green pea that is resistant to Race 5 *Fusarium* wilt was also developed and tested in 1972 in the Puget Sound area of Washington. *Fusarium* wilt is a fungus that attacks the roots and stems of many fruit and vegetable plants.

William A. Hoglund, plant pathologist at the Northwestern Washington Research and Extension Unit of Washington State University at Mt. Vernon, said that a minimum of one freezing and one canning type of these peas will be marketed in 1973.

Remote sensing by satellites opened new dimensions for monitoring and viewing the earth's agricultural resources. The first Earth Resources Technology Satellite (ERTS) was launched on July 23, 1972, and its performance has exceeded expectations. It views the ground and records energy waves that cover a wide spectrum from infrared through and beyond ultraviolet. See SPACE EXPLORATION.

Its orbit keeps it constantly over the earth areas that are in daylight, and it passes over any single location every 18 days. It provides data on crop yields, changes in land use, and crop damage caused by insect pests and disease. This information is recorded as color variations. This operation should help prepare the first worldwide inventory of food resources.   [Sylvan H. Wittwer]

Anthropology

The discovery of 38 new hominid (man-like) fossils made the 1972 season in the Lake Rudolf region of Kenya an overwhelming success. The new discoveries brought to 87 the number of hominids recovered from this unusually rich area on the eastern shore of Lake Rudolf.

Richard E. F. Leakey of the National Museums of Kenya, who directed the search in this area, provisionally classifies 18 of the new specimens as genus *Australopithecus* (near-man) and 16 as *Homo* (true man). The affinity of the other 4 is unclear.

The australopithecine material includes a well-preserved juvenile mandible (jawbone) found by Musa Mbithi, a Kenyan assistant. It has five teeth and several partially developed tooth crowns still in place. The left half of another mandible, found below a volcanic tuff (porous rock), was reliably dated by the potassium-argon method at 2.61 million years old. This bone is evidence that the large form of australopithecine lived in the Lake Rudolf area before that date.

Some of the most exciting finds were bones from the postcranial skeleton (all the bones except the skull) of *Australopithecus* because they should provide valuable information on posture and locomotion and because they have been difficult to locate in the past. These include the fragmentary arm and leg bones known collectively as KNM-ER 1500, found by John Kemingech, also a Kenyan. This was the first time in east Africa that scientists can be certain that the limb bones they are dealing with come from a single australopithecine. Preliminary studies of this material suggest that *Australopithecus* may have had longer arms and shorter legs than does modern man.

Oldest skull of *Homo*. The Lake Rudolf material that Leakey attributes to the genus *Homo* includes several teeth, parts of skulls and mandibles, several arm and leg bones, and a nearly complete but badly fragmented skull. The skull, KNM-ER 1470, was found east of the Koobi Fora and Ileret regions. It had been broken into more than 150 pieces. Kenyan field aide Bernard Ngeneo sighted the fragments in sandy sediments in a deep gully about 115 feet

Anthropology

Continued

Richard E. F. Leakey compares the cast of a 2.9-million-year-old skull unearthed in Kenya in 1972 with that of a 1-million-year-old *Australopithecus*. The find shatters the theory that man descended from *Australopithecus*.

The "dentist," Belle, grooms the teeth of her friend Bandit with a dental tool she made by stripping leaves from a twig. Born in the wild, the chimpanzees, which live at the Delta Research Center in Covington, La., are the first animals that were observed to prepare and use a tool for social grooming.

below the 2.61-million-year-old tuff. When Meave Leakey, Richard's wife, and Bernard Wood, a London anatomist, reconstructed the skull, its steep sides and small brow ridges looked remarkably like those of modern man. The skull has a capacity of at least 800 cubic centimeters (cm³), larger than that of any australopithecine and just a bit short of *Homo erectus*. The estimated age of this specimen is 2.9 million years, which makes it a contemporary of *Australopithecus*—and the earliest fossil material that has so far been attributed directly to early man (*Homo*).

Analysis of hominid specimens found before 1972 produced interesting results. Ralph L. Holloway of Columbia University studied a fossil cast of the interior of an australopithecine skull recovered from the Swartkrans site in South Africa by Charles K. Brain of the Transvaal Museum, Pretoria, in 1966. The cast, presumably from the large form of australopithecine, has a capacity of only 530 cm³, the size of a large ape's brain, but Holloway found that the structure of the brain is clearly that of a hominid.

Statistical studies of hominid measurements suggest that australopithecines do not lie exactly between apes and men on the evolutionary line of development but had evolved into their own unique position. A study by Henry M. McHenry of the University of California, Davis, of a hominid upper arm bone found in the Lake Rudolf area in 1970, and one by Charles E. Oxnard of the University of Chicago of hominid foot bones from east and South Africa seem to bear this theory out.

Kow Swamp. Since 1968, archaeological work in the Kow Swamp area of southeastern Australia has produced an extensive series of at least 40 human fossil skeletons, dating from 9,000 to 10,000 years ago. A. G. Thorne of the Australian National University, Canberra, and P. G. Macumber of the Geological Survey of Victoria, Melbourne, gave a preliminary description of this material in August, 1972. They indicate that the Kow Swamp people had thick skulls, massive mandibles, markedly flattened bones of the forehead, and well-developed brow ridges.

Anthropology

Continued

Considering the ancient nature of the structure of the Kow Swamp people, the estimate that they lived 9,000 to 10,000 years ago seems surprisingly recent.

Meat-eating chimps. Anthropologists continue to look on the chimpanzee as the animal most closely related to man and thus the one whose behavioral patterns are most likely to resemble those of early hominids. More than a decade of intensive study at Gombe National Park in Tanzania, primarily by zoologist Jane van Lawick-Goodall, shows that the chimpanzee, like man, is a meat-eater. Most other apes and monkeys are almost exclusively plant-eating animals.

Geza Teleki of the University of Georgia reported in January, 1973, how chimpanzees capture, kill, eat, and share various kinds of animals. They eat young bushbucks and bush pigs, but more than half the victims seen in the last 10 years have been monkeys, especially juvenile baboons. Several males usually cooperate in a hunt; females hunt occasionally, but not in the presence of males. The chimps share flesh in an almost ritualistic fashion. Practically all of a victim, except for some of its bones, is ultimately consumed. The brain, however, appears to be the prized portion and is rarely shared.

Chimp cannibals. Studies of chimpanzees have uncovered cannibalism, another trait attributed to early man. In August, 1972, J. D. Bygott of Cambridge University, England, reported two instances in which adult male chimpanzees captured and ate a juvenile chimpanzee. One occurred in the Budongo Forest of Uganda, the other at Gombe. The Gombe incident lacked some aspects of the usual meat-sharing behavior; they played with the carcass, eating only parts of it.

Mosquito gourmets. What appeared to be an unequal representation of the various ABO blood groups among malaria victims suggested to Corinne Shear Wood at the University of California, Riverside, and her co-workers that the carrier of the disease, the mosquito *Anopheles gambiae*, may actually prefer to bite people with a specific blood type. Her tests suggest that type-O victims are selected through substances in their skin cells or sweat.   [Charles F. Merbs]

Archaeology

Old World. The theft of antiquities from archaeological sites for sale in the world's art markets increased sharply in 1972 and 1973. The prices paid for such treasures rose enormously, and so did the temptation to pillage sites for marketable objects. Meanwhile, archaeologists searched for ways to stop the illicit trade. The ancient sites in Turkey and Italy in particular suffered, since these countries are exceptionally rich in antiquities of many periods.

The disturbance of archaeological sites by robbers destroys a great deal of important evidence. Much of that which is most significant about any ancient find is determined by the context in which it is discovered, and by its relationship to other objects found at the same site. The context also sometimes determines the object's age.

The problem was highlighted in December, 1972, when Thomas Hoving, director of the Metropolitan Museum of Art in New York City, announced that the museum had acquired what some archaeologists call the finest Greek vase in existence. It is a calyx krater, or wine-mixing vessel, from about 500 B.C., the work of a potter named Euxithrous and a vase painter named Euphronios. The museum paid nearly $1 million for the krater, which supposedly came from a private collection in Europe. In May, 1973, however, Italian authorities said they had found evidence indicating that the vessel actually had been illegally taken from an Etruscan tomb in western Italy in 1971.

Africa. In November, 1972, Richard E. Leakey, director of the National Museum of Kenya, Nairobi, reported his discovery of fragments from the skull of a man who lived about 2.6 million years ago. Leakey found the fragments on the east shore of Lake Rudolf in northern Kenya. The skull bones are from the same period as man-made tools found earlier in the year in the same area by archaeologist Glynn Isaac of the University of California, Berkeley.

These artifacts, the most ancient tools yet found, are mainly pieces of volcanic lava apparently carried from some distance away and collected to be sharpened into weapons. Isaac suggests that

The well-preserved remains of a 2,100-year-old body believed to have been a Chinese noblewoman were found near the city of Changsha in southern China.

they could have been used to kill animals, such as zebras and pigs, and to cut up the meat for consumption. The tools were found in groups, apparently discarded, at five campsites that were located during the summer. Evidence indicates that from 8 to 10 individuals lived at the campsites. This evidence suggests that these prehistoric people practiced cooperative hunting and centered activities around a home base.

Shipwreck figurines. A team of divers and marine archaeologists from the University of Haifa, Israel, in January, 1973, completed the excavation of an ancient Phoenician ship that sank within sight of the coast north of Haifa about 500 B.C. The recovery was conducted by Elisha Linder of the university's Center for Maritime Studies.

The ship contained a cargo of earthen vessels and more than 200 figurines of Tanit, goddess of fertility and the only known female deity of the Phoenician world. Previously, archaeologists found only drawings and symbols of the goddess and impressions on bronze coins. The clay figurines, ranging in height from 6 to 15 inches, seemed to have been mass-produced from molds.

Russia. A Stone Age settlement, described as the largest yet found in eastern Europe, was discovered during the summer of 1972 by Russian archaeologist N. M. Shmagli 115 miles south of Kiev. The settlement is said to contain about 1,500 houses and to cover about 700 acres. The houses were built of adobe, and some of them may have had two stories. Buildings were laid out concentrically, with streets branching out from the center like wheel spokes.

This exceptionally large town belongs to the Tripolye civilization, which existed about 3000 B.C. in the Ukraine. The population of the town subsisted on primitive agriculture and even some domesticated animals. Its discovery lends further weight to the theory that the domestication of animals and development of agriculture led to village settlement and the comparatively rapid appearance of large cities. Earlier evidence of this trend has been found at sites in Turkey.

China. In August, 1972, sources in Peking, China, released information on a notable discovery dating from the Han dynasty, which ruled ancient

Greek vase from 500 B.C., bought by New York City's Metropolitan Museum, was later found to have been stolen from a site in western Italy.

Archaeology

Continued

A statue of the Egyptian sun god, Horus, as a child, was found in Lebanon.

China from 202 B.C. to A.D. 220. The tomb of a noblewoman, thought to be the Marchioness Li Tsang, was excavated in Ma Wang Tui, a suburb of Changsha, the capital of Hunan Province. The woman was about 50 when she died. Her well-preserved, mummified body had been painted red and wrapped in silk. It lay in a coffin painted in red, white, black, and yellow with designs of birds and animals. The coffin, surrounded by charcoal that may have been intended to act as insulation, rested in a sarcophagus – a bigger stone coffin – that was decorated with paintings on silk representing folk scenes of heaven, earth, and hell.

Over a thousand objects were found in the tomb, making it an unusually rich and significant discovery. These objects included pottery, figurines, lacquered bowls containing still-recognizable eggs, rice, and such fruits as pears, peaches, and melons. Three musical instruments, one with 21 strings, were of particular interest because they added to knowledge of the history of music in China.   [Judith Rodden]

New World. Robert A. McKennan and his Dartmouth College associates located 42 new archaeological sites in the eastern Brooks Range of Alaska near Arctic Village in 1972. Most of these were caribou-hunting campsites that were occupied briefly about 3000 B.C. The scientists found many tools, including spear points and stone knives.

During the summer of 1972, a University of Connecticut field party led by Jean S. Aigner excavated a large fishing village at Sandy Beach Bay in the Aleutians. They determined that people lived at this site from 3500 to 2100 B.C.

X-ray fluorescence analysis of obsidian rock, comparing sources of the rock with specimens found in archaeological excavations in California, showed that prehistoric tribes in the area traded and used obsidian widely. The study was done by Robert N. Jack of the University of California, Berkeley. The obsidian came from at least 18 different places. Most of it found at any archaeological site came from nearby sources. However, some pieces must have been transported great distances, and were ap-

A workman digs a trench in the ruins of a building at the pre-Columbian village site. Once a traditional center, the village was discovered by a joint expedition from Harvard University and the University of Arizona in September, 1972. It was hidden in a forest on the island of Cozumel off Mexico's Pacific coast.

Archaeology

Continued

parently part of the trade between different tribes and social groupings.

Archaeologist Richard George of the Carnegie Museum in Pittsburgh excavated a Monongahela Woodland village site in Westmoreland County, Pennsylvania, during the summer of 1972. The village was occupied for a few years at the beginning of the A.D. 1100s. It had a circular stockade 190 feet in diameter. From 15 to 20 circular houses were grouped around a central courtyard in the stockade. The inhabitants apparently ate fish, deer and smaller mammals, migratory birds, and mussels. They also ate corn, beans, and hickory nuts.

Patrick Munson of Indiana University in Bloomington directed excavations at a prehistoric site in Dearborn County, Indiana, that had been occupied many different times. The two lowest levels have yet to be dated. Artifacts in the middle level are from the early ceramic period, about 500 B.C. The upper levels contained pottery from about A.D. 800. However, most of the occupation levels, including a number of burial areas and large storage pits, have been identified

with an early Fort Ancient culture occupation, about A.D. 1100.

Central America excavations at the site of Teotihuacan, some 40 miles north of Mexico City, continued to produce surprises. Jorge Acosta of Mexico City's National Institute of Anthropology and History discovered a man-made stairway leading to a natural tunnel under the Pyramid of the Sun. The tunnel ends in a group of four chambers. There is evidence that the tunnel and the chambers once had elaborate plastered walls. Pottery vessels and engraved slate disks found in the chambers were dated to about A.D. 200.

The purpose of the tunnel and chambers may never be firmly established. To those who dug them and built the Pyramid of the Sun, however, they were probably thought of as caves. Caves are symbols of the origin of life, or of particular groups of people, in Mexican mythology. In Mexican murals of "paradise," rivers always flow from a cave. Oracles were also associated with caves, which symbolizes the importance of such places for receiving the words of the gods.

Archaeology

Continued

Archaeologist Shirley Gorenstein of Columbia University examines a fragment of Tarascan pottery from ruins of a community that she excavated in Mexico.

Richard E. Blanton and his associates of Hunter College, New York City, are conducting a long-term study of prehistoric settlements in the Valley of Oaxaca, Mexico, with special attention to the famous Monte Albán site near the city of Oaxaca. They have identified a wall around Monte Albán, reservoirs, and an irrigation system built east of the main hill of Monte Albán. The installations were built from 200 B.C. to A.D. 300 during a period of major population growth. The area had an estimated population of from 25,000 to 40,000, according to Blanton.

Recent research on the social and political structure of the lowland Maya in Guatemala and Mexico by Joyce Marcus of Harvard University suggests that the civilization had four regional capitals. Each regional capital had from five to eight secondary centers surrounding it, and each of these was supported by surrounding smaller sites. The ruling group in each capital apparently contracted marriage alliances with the ruling group in the secondary centers to maintain political control.

South America. Recent excavations in the seasonally flooded plains of western Venezuela by Alberta Zucchi of the Venezuelan Institute of Scientific Investigations have identified the Cano del Oso culture group that lived there between 1000 B.C. and A.D. 500. This early tribal group lived mainly by hunting and fishing, though they also had the earliest corn farming in the area. Following this occupation was another that built truncated pyramidlike mounds. They were related to the Arauquinoid people of the Orinoco River area.

Four years of archaeological investigations in the Moche Valley on the northern coast of Peru by Michael E. Moseley of Harvard University have produced evidence that urban societies began developing in that area shortly before the birth of Christ. For many years, archaeologists had thought that invaders brought urbanism into the Moche Valley about A.D. 600 from the Huari area of the highlands. Moseley now has evidence that a series of large settlements had clearly predated this invasion.   [James B. Griffin]

Astronomy

Planetary Astronomy. The discovery of deuterium in Jupiter's atmosphere in 1972 was the first time this hydrogen isotope has been found in an extraterrestrial object.

Reinhard Beer and Frederic W. Taylor of the Jet Propulsion Laboratory (JPL), Pasadena, Calif., announced in January, 1973, that they had observed deuterium, locked in the CH_3D molecule, in the spectrum of Jupiter. In March, John T. Trauger and Fred L. Roesler of the University of Wisconsin and Nathaniel P. Carleton and Wesley A. Traub of the Smithsonian Astrophysical Observatory, Cambridge, Mass., reported they had observed the spectrum of the HD (hydrogen and deuterium) molecule also on Jupiter. Deuterium was also found in interstellar space in 1972 and 1973 in the DCN molecule (deuterium, carbon, and nitrogen) and in the atomic form, D.

The discovery of deuterium is important to the cosmological research concerned with the formation, or synthesis, of the basic chemical elements. The measurement of the ratio of deuterium to

hydrogen is a most important isotopic and elemental abundance ratio if it can be shown that it is primordial, or the earliest existing. There seems little doubt that this is the case on Jupiter. The ratio (about 2×10^{-5}) is 10 times less than on earth. The earth's deuterium is probably greater relative to hydrogen because of the preferential evaporation of hydrogen, in the form of water (H_2O), from the oceans and its subsequent escape from the top of the atmosphere.

The ratio on Jupiter reinforces the theory that the universe started with a "big bang" and will continue to expand.

Echoes from Saturn's rings. JPL observers Richard M. Goldstein and George A. Morris, Jr., reported the first detection of radar echoes from Saturn's ring system in March, 1973. They did not detect echoes from Saturn itself, presumably because ammonia (NH_3) deep in the planet's atmosphere absorbed the radar signals. The strength of the echoes from the rings is particularly surprising. Goldstein and Morris believe the echoes imply that the rings must include a large component of

Astronomy

Continued

meter-sized "boulders," probably extending far beyond the visible rings.

Confusion over Titan grew after Laurence Trafton of the University of Texas announced in July, 1972, the possible detection of molecular hydrogen (H_2) on that satellite of Saturn. Trafton also showed that Titan could have as much as 30 times more methane (CH_4) in its atmosphere than earlier estimates allowed. This work, which opens up the possibility of a massive, relatively warm atmosphere on the satellite, has projected Titan from relative obscurity to a prime object in solar system studies.

Trafton's work, plus observation of Titan's thermal-emission spectrum by Robert E. Murphy, David Morrison, and Dale Cruikshank of the University of Hawaii, has led to controversy over the true nature of the satellite's atmosphere. A model conceived by Carl Sagan and George Mullen of Cornell University has a massive H_2 atmosphere. At the other end of the scale, John Caldwell, David R. Larach, and Robert E. Danielson of the Princeton University Observatory note the un-

certainty in Trafton's observations of H_2 and consider the atmosphere to consist primarily of CH_4 and its photochemical products. The latter are produced by the action of sunlight on the atmosphere. In-between are a range of models, including one by Donald M. Hunten of the Kitt Peak National Observatory in Arizona, which would require large amounts of nitrogen (N_2). This molecule, unfortunately, cannot be detected by spectroscopic means.

All of these models have very different properties in terms of atmospheric structure and evolution, but observations should be possible in the near future which will resolve these controversies.

Water, water, everywhere... Until two years ago, H_2O was known to exist only on the earth and, in very small quantities, on Mars. In 1971, water ice was discovered in Saturn's rings, and in late 1972 and early 1973, two groups independently found that large areas of the Jupiter satellites Europa, Ganymede, Calisto, and possibly Io are also covered with water ice. This identification was made by Carl B. Pilcher and

"Henderson, if there's anything you've always wanted to do, I suggest you do it sometime in the next 23 hours and 57 seconds."

Astronomy

Tom B. McCord of the Massachusetts Institute of Technology (M.I.T.), Cambridge, Mass., and Stephen T. Ridgway of Kitt Peak; and also by Uwe Fink, Nicolass Dekkars, and Harold P. Larson of the Lunar and Planetary Laboratory in Tucson, Ariz. It is in accord with the current idea that these low-density satellites are about 75 per cent water ice.

Venera 8 and Venus. The Russian spacecraft, Venera 8, operated for 50 minutes after landing on Venus on July 22, 1972. The mission differed from earlier ones in that the spacecraft landed in Venusian daylight. As a result, it was able to show that visible light diffuses through the planet's extensive atmosphere and cloud system to the surface.

Measurements by Venera 8's gamma ray spectrometer indicate that the soil contains 4 per cent potassium, 0.002 per cent uranium, and 0.00065 per cent thorium, a mix of radioactive elements very similar to that in granites on earth. Such rocks are formed at high temperatures, and their presence on Venus has been interpreted as implying that Venus, like the earth, has undergone extensive internal change since it was formed.

The clouds Venera 8 had to pass through to reach the surface continue to mystify planetary scientists. Nevertheless, in early 1973, Andrew T. Young of JPL suggested that the upper visible clouds are a highly concentrated form of sulfuric acid, H_2SO_4. The presence of considerable amounts of H_2SO_4 would answer another important question: Why is there such an apparent lack of observable sulfur compounds in the planet's atmosphere? At temperatures as high as those on Venus, the normally abundant and relatively volatile sulfur should be changed into a gas and put into the atmosphere as carbonyl sulfide (COS). But sensitive spectroscopic measurements always fail to detect COS.

Ronald Prinn of M.I.T. apparently provided the answer in March by noting that COS should be broken down by sunlight very rapidly in the upper regions of the Venusian atmosphere; its photochemical derivatives, together with H_2O, will lead to the rapid formation of H_2SO_4. If Young's conjecture and Prinn's theory are correct, the upper clouds of Venus must be seen as an exceedingly acidic photochemical smog covering the planet. [Michael J. S. Belton]

Stellar Astronomy.

In 1972 and 1973, astronomers recorded one of the brightest supernovae ever seen, detected nuclear reactions in solar outbursts, observed the most distant known object in the universe, and studied strong radio emissions from an X-ray star.

A star explodes. On May 13, 1972, Charles T. Kowal of the California Institute of Technology discovered the fourth brightest extragalactic supernova ever observed. He found it in the galaxy NGC-5253, about 13 million light-years from the earth. Hampered by the low altitude of the supernova as viewed from North American observatories, astronomers made conflicting measurements of its fading light as it cooled.

Thomas A. Lee and other researchers at the University of Arizona's Catalina Observatory found that near-infrared emission from the supernova held virtually steady as the visible light decreased, indicating that the infrared light was produced as synchrotron radiation caused by relativistic electrons (those traveling at close to the speed of light) passing through the star's magnetic field.

But a team headed by Robert P. Kirshner of Hale Observatories in California found evidence of a gradual decline of the infrared light. Instead of indicating synchrotron radiation, their measurements were consistent with ordinary thermal emission from a hot gas. The measurements indicated that the star's exploding shell became transparent as it expanded, permitting light to escape from deeper layers. Disagreements in explaining the infrared emission probably cannot be resolved until another bright supernova is observed at a higher altitude.

Solar flares. Although the sun is in the period of minimum activity in its 11-year cycle, notable solar outbursts occurred in 1972. In one region, a large sunspot group stretched 160,000 miles across the surface in August. Major flares were observed in the region on August 2 and August 4. Then, on August 7, the greatest flare in several years was seen in ordinary visible light.

The solar activity was accompanied by the most intense decrease in cosmic rays ever recorded on the earth. Such events take place when solar plasma sweeps out into interplanetary space, partly deflecting the cosmic rays that

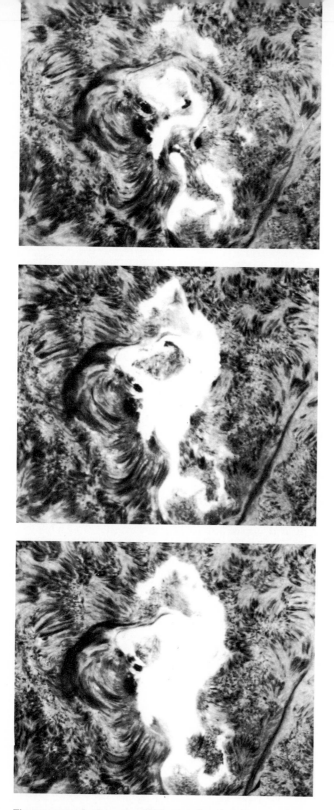

The greatest solar flare since 1960 erupted on Aug. 7, 1972, and continued for more than four hours. It set off intense X-ray, ultraviolet, and radio emissions that disrupted communications over parts of the earth.

enter the solar system from all directions in the galaxy. Thus, fewer cosmic rays reach the earth. The largest previous such incident, in July, 1959, took seven days to reach the lowest point of the cosmic-ray flow. But Martin A. Pomerantz and Shakti P. Duggal of the Bartol Research Foundation in Philadelphia reported that the cosmic-ray intensity on August 4 dropped to an even lower level in only one day.

The August solar observations included two detections of gamma-ray emission lines by the National Aeronautics and Space Administration's (NASA) Orbiting Solar Observatory-7 (OSO-7) satellite. The gamma rays were detected by an instrument built for the satellite by Edward L. Chupp and his associates at the University of New Hampshire.

The New Hampshire physicists found evidence of gamma-ray emission at energies of 0.511-, 2.22-, 4.4-, and 6.1-million electron volts (MeV). This indicates that nuclear reactions and the mutual annihilation of matter and antimatter particles occur in the flares. The 0.511-MeV line, in particular, is ascribed to the annihilation of electrons with their antimatter counterparts, positrons. Chupp believes the positrons are produced in the flares as decay products of short-lived radioactive isotopes. The isotopes are created when a beam of high-energy protons, accelerated by the flare, interacts with nuclei in the ordinary solar gas. The 2.22-MeV line was generated by deuterium nuclei that were produced by collisions between the flare protons and nearby neutrons.

Cygnus X-3, a galactic X-ray source, has also been known as a source of weak radio emission. But on Sept. 2, 1972, Philip C. Gregory of the University of Toronto discovered that its radio emission at the 2.8-centimeter wave length was more than 200 times stronger than usual. No such event had previously been found in any galactic X-ray source. See X Rays from the Sky.

Radio astronomers throughout the world monitored what turned out to be the first of at least five outbursts. During the first Cygnus X-3 outburst, the radio intensity as measured at successively longer wave lengths peaked at successively later times. Observers agreed that the outburst was probably caused by the ejection of a cloud of rela-

Astronomy

tivistic electrons from the central X-ray source, probably a binary star. Floyd W. Peterson of the National Radio Astronomy Observatory in Charlottesville, Va., found that the ejection of the electrons must have persisted for about 29 hours. See ASTRONOMY, HIGH ENERGY.

Quasar near the edge? In April, 1973, Robert F. Carswell and Peter A. Strittmatter of the University of Arizona announced that the spectrum of quasar OH-471 showed a red shift of 3.4. Interpreted in terms of the expanding universe theory, this red shift implies that the quasar is roughly 10 billion light-years from the earth, one of the most distant objects yet found. See CLOSE-UP.

OH-471 is an 18th-magnitude object, not very faint as quasars go. Thus, it should be possible to observe quasars of similar brightness even at greater distances. Since none have been found, Allan R. Sandage of Hale Observatories speculates that there may be no quasars beyond OH-471 and that it may lie near the edge of a finite universe.

New telescopes. NASA's Orbiting Astronomical Observatory-3 (Copernicus) satellite, launched on Aug. 21, 1972, carried a 32-inch reflector, the largest astronomical telescope ever put into orbit. The telescope was designed for high dispersion ultraviolet spectroscopy to study the interstellar gas in our Galaxy, the Milky Way.

The first test observations were made in March, 1973, with the 158-inch Mayall telescope at Kitt Peak National Observatory in Arizona. When it was dedicated in June, 1973, the telescope was the second largest one in the world.

The new 100-meter, fully steerable radio telescope of the Max Planck Institute for Radio Astronomy began limited operations at Effelsberg, West Germany, in late 1972. Its first observations included studies of pulsars and the detection of radio emission from the red supergiant star Betelgeuse.

Meanwhile, American astronomers warned that increasing light pollution, the stray light in the sky that comes from street lamps and other outdoor lighting, was endangering major observatories in California and southern Arizona. In June, 1972, astronomers acclaimed a new Tucson city ordinance to control outdoor lighting to protect astronomical research.   [Stephen P. Maran]

High Energy Astronomy. Robert Carswell and Peter Strittmatter of the University of Arizona discovered a quasar with a record red shift of 3.4 in April, 1973. They identified the radio source of OH-471 with an 18th-magnitude star. While typical quasars are very blue, this star is a neutral color. Nevertheless, it has the characteristic emission line spectrum of a quasar.

The discovery is important because it suggests that identifying more radio sources with faint stars of neutral colors may lead to the discovery of even more distant quasars. Because there are so many galactic stars, however, astronomers need very accurate positions (to within one arc second) for the radio sources. Such positions are only now becoming available in sufficient numbers due to new radio telescope techniques. See CLOSE-UP.

Cygnus X-3. Radio astronomers throughout the world observed a giant radio outburst from the X-ray source Cyngus X-3 in 1972. On the evening of September 2, Philip Gregory of the University of Toronto discovered strong radio emission from Cygnus X-3 with the 150-ft. radio telescope at the Algonquin Radio Observatory in Ontario. The radio emission at 2.8 centimeters was more than 200 times stronger than its normal value. The emission reached its maximum two hours after Gregory first observed it, and then declined to its usual low level during the following week. In the following months, astronomers detected four more similar bursts from Cygnus X-3.

The radio burst in Cygnus X-3 shows a surprising similarity to the bursts observed in quasars: Activity begins at the smallest wave lengths and gradually spreads to longer wave lengths. The most likely interpretation is that Cygnus X-3 somehow generated a cloud of energetic electrons that produced synchrotron radiation when they passed through the source's magnetic field. At the shortest radio wave lengths, the radiation freely escapes, but at longer wave lengths, it is absorbed within the source. As the source expands, however, it absorbs less of the radiation and longer wave lengths can then be observed.

Detailed models suggest that the energy in relativistic electrons was about 10^{40} ergs—as much energy as the sun

Quibbles Over Quasars

Quasar OQ-172, with a red shift of 3.5, was identified in June, 1973, as the most distant object yet known in space. It is about 10 billion light-years from earth.

Quasars—superbright, quasi-stellar objects that apparently lie at the farthest edges of the universe—have piqued astronomers' curiosity ever since 1963 when they were first discovered and recognized as a major constituent of the universe. More than 300 are now known. Although they have very large red shifts, indicating that they are located at great distances from the earth, they appear brighter than galaxies with comparable red shifts. Yet unlike galaxies, which show a measurable image on photographic plates, the intense light from quasars appears to be confined to an image no larger than that of ordinary stars more than a thousand billion times fainter.

The continued effort to locate more quasars resulted, in early 1973, in the discovery of several quasars with very large red shifts. Using the new 90-inch telescope at the University of Arizona's Steward Observatory on Kitt Peak, astronomers in April measured a red shift of 3.4 for the quasar OH-471. Hard on the heels of this discovery came the announcement in June, that astronomers using the 120-inch telescope at Lick Observatory in California, had observed a red shift of 3.5 for the source OQ-172.

These observations have intensified the debate over the nature of quasars and their relationship to the universe. The usual interpretation of the red shift relates it to velocity in an expanding universe. According to this explanation, when a source of light is receding from the viewer, the waves of energy it emits are lengthened so that a specific emission line is shifted toward the red side of the spectrum. The amount of displacement is directly proportional to the object's speed. Thus, the red shifts for OH-471 and OQ-172 indicate that they are receding from the earth at a velocity more than 90 per cent of the speed of light. Furthermore, the Hubble law, which relates the red shift to distance from the earth, indicates the two new quasars are about 10 billion light-years away, making them the farthest objects known.

However, not all astronomers agree with this cosmological interpretation of the quasar red shift. The apparent extraordinary luminosity and often rapid variability in radiation of quasars have led some astronomers to argue that quasars are relatively close objects, and that their red shift is due to some as yet unexplained phenomena and not to the expansion of the universe. If this is the case, it may also call into question the usual interpretation of galaxy red shifts, which has been the cornerstone of extragalactic research.

Several astronomers have found that quasars and galaxies are not randomly distributed in the sky, but appear to occur in clusters. They presume that all of the quasars and galaxies in a particular group are about the same distance from the earth. Thus there is considerable interest in several groups containing objects with very different red shifts. Halton Arp of Hale Observatories in California and Geoffrey Burbidge of the University of California, San Diego, who challenge the cosmological interpretation of quasar red shifts, suggest that the presence of a quasar near a bright galaxy indicates that the quasar may have been ejected from the galaxy.

Proponents of the cosmological interpretation argue, however, that these apparent clusters are not real, but are the chance coincidence of objects located at vastly different distances on a line in the same direction.

One proponent of the cosmological red shift, James Gunn of Hale Observatories, has observed quasars in several clusters with the 200-inch telescope on Mount Palomar. He finds that the quasar red shift is close to that of the galaxies in the same group, indicating these clusters of quasars and galaxies are at the same distance. And Jerome Kristian, also of Hale, reported in January that he detected faint galaxies underlying all quasars that are close enough and small enough that the less luminous galaxies behind them can be seen. See ASTRONOMY, COSMOLOGY.

Further support for the cosmological interpretation comes from Richard Hills of the University of California, Berkeley, and John Bahcall of Princeton University, who reported in March that the fainter quasars have the largest red shifts as expected if they are, indeed, more distant. But, Burbidge and Stephen L. O'Dell of the University of California, San Diego, argue that Hills and Bahcall drew their conclusions from a study that was dominated by a few bright objects. Burbidge and O'Dell found that there is no general relationship between the red shift and apparent brightness.

More recently, Lodewyk Woltjer of Columbia University and Giancarlo Setti of the University of Bologna in Italy pointed out that the correlation between red shift and magnitude is more significant if the analysis is confined to quasars with radio properties similar to those of radio galaxies. Radio-quiet quasars show no such relationship between brightness and red shift.

Faced with the conflicting evidence on the origin of quasar red shifts, several astronomers have suggested that there may be two different kinds of quasars—one with red shifts due to cosmological expansion, the other with red shifts caused by some other phenomenon.

Radio astronomers are also reporting exciting new discoveries. Using large radio telescopes located around the world, they are examining quasars in considerably greater detail than is possible with optical telescopes. In this Very Long Baseline (VLB) astronomy, signals received by radio telescopes located thousands of miles apart are recorded on tape recorders synchronized by atomic clocks that are accurate to a millionth of a second. The proper combination of such signals provides pictures of the radio emission with a resolution comparable to that obtained from a single radio telescope many thousands of miles in diameter. Such pictures show that the radio emission from the compact quasars is complex, and comes from two or more well-separated regions.

When astronomers first began to use VLB to explore quasars in detail, it appeared that the different components in some objects were flying apart at velocities greater than the speed of light. This has been used as an argument that the quasars must really be much closer than indicated by their red shifts. In a nearby object, the same data would indicate relatively slow motions.

However, newer observations suggest that there may not be any real motions at all. Rather, the illusion of very rapid motion may be caused by several fixed components that flash on and off between observations. Radio astronomers in many countries are now cooperating to examine the structure of quasars in unprecedented detail and it is hoped that this research may eventually settle the question of the apparent faster-than-light motions. [Kenneth I. Kellermann]

radiates in a month. The source of the electrons is unknown.

X-ray sources. In 1972 and 1973, astronomers identified the optical counterparts of several X-ray sources discovered by the Uhuru satellite (see X RAYS FROM THE SKY). In six cases, the X-ray source is a member of a double star system. In one of these, Hercules X-1, the orbit of the X-ray source has already been determined from the variations in its 1.2-second period of pulsation. The X-ray source has now been identified optically with the variable star HZ Hercules, which has been known for at least several decades.

The orbit of the optical companion was determined from an analysis of the radial velocities of the two stars. Then, by analyzing the orbits of both stars, observers determined their masses. The mass of the X-ray source, which is only a few miles in diameter, appears to be about 1.2 times that of the sun. This is less than the mass limit to white dwarfs and neutron stars.

Most current models for the source interpret the 1.2-second pulsing period as the rotation period of a magnetic neutron star. Matter flows from the optical companion along the magnetic field lines to the magnetic poles of the neutron star. As a result, the poles become hot and emit X rays. The rotation periodically takes these polar regions into and out of view, and this causes the rapid X-ray pulsations.

During the 1.7-day orbital period, the light of the optical companion varies by a factor of 4. It is strongest when the X-ray source is in front of the optical companion, because a significant part of the X rays are absorbed on the companion's surface. As a result, the side of the companion facing the X-ray source (and also facing the earth) is hotter and emits more light.

Discoveries of new pulsars continued in 1972 and 1973, and by now 90 are known. One pulsar was found in January, 1973, just outside the IC-443 supernova remnant. This is the third time a pulsar has been associated with a supernova remnant. The pulsar's age is 90,000 years, compared to a very uncertain estimate of 50,000 years for the supernova remnant.

The pulsar's eccentric location indicates that it has a velocity of about 150

Astronomy

Scientists at Louisiana State University will join the search for gravity waves with this detector, which, when put together, will weigh 6 tons. It will be suspended on a superconducting magnet, and it will operate in conjunction with a similar detector at Stanford University.

kilometers per second. This confirms tentative results based on the interstellar pulsar scintillation (a kind of twinkling) which indicates that pulsars in general appear to be traveling at high velocities. Presumably, an asymmetry in the supernova explosion, which propels the pulsar outward, causes the high velocities.

A group of scientists under Kenneth Greisen of Cornell University observed the Crab Nebula pulsar with a balloon-borne gamma-ray detector in October, 1972. They measured photon energies of more than 800 million electron volts, corresponding to a wave length of 1.5 x 10^{-5} angstrom. The behavior of the pulsar is surprisingly uniform over a wide range of wave lengths, and astronomers cannot explain this constancy of pulsar shape and spectral slope. For example, if synchrotron radiation by electrons in the pulsar's magnetic field were responsible, these electrons would lose their energy in 3 billionths of a second, and there would be no way to explain their acceleration.

Gravity waves. For several years, Joseph Weber of the University of Mary-land has been finding evidence that a source of gravitational radiation exists near the center of our Galaxy. This result is astonishing, because it indicates the outpouring of a vast amount of energy from our Galaxy; so much that even if all matter in the galactic nucleus were converted into gravity-wave energy, the phenomenon could last only a million years, far less than the age of the Galaxy.

Others have now joined the search for gravity waves. Tony Tyson of Bell Telephone Laboratories in Murray Hill, N.J., has operated a massive and highly sensitive detector in Holmdel, N.J., since November, 1972. Tyson's detector operates at 710 cycles, compared to 1,661 cycles for Weber's experiment. In December, after a month of observation, Tyson reported at a New York City conference that he had failed to detect any events that could be identified with gravity waves with the intensity claimed by Weber. This may indicate that some other effects cause the events recorded in Weber's gravitational detectors, or that gravity waves only occur at very specific frequencies.   [Lodewyk Woltjer]

Astronomy

Cosmology. A report in early 1973 on an attempt to determine the source of the microwave background radiation coming from all directions in space has reinforced the theory that this radiation is the same that filled the universe in its hot early stages, just after the big bang. Its very low temperature, within 3 degrees of absolute zero, can then be explained by cooling in the expanding universe.

However, alternative theories suppose that the radiation comes from many separate faint sources—quasars or galaxies, for example. If this is true, then to a radio telescope with angular resolution finer than the average distance between sources, the radiation will no longer look smooth and continuous, but will appear "granular" because individual sources can be seen. A familiar example of this in optical astronomy is the resolution of our own Galaxy, the Milky Way. It appears as a smooth, luminous cloud to the naked eye, but it breaks up into many faint, closely spaced stars when seen through a large telescope.

In April, 1973, Paul E. Boynton of the University of Washington and R. Bruce Partridge of Haverford College in Haverford, Pa., reported on their search for small-sized granularity in the background radiation. They used the National Radio Astronomy Observatory's 36-foot diameter dish at the Kitt Peak National Observatory in Arizona. The radio telescope operated at a wavelength of 3.5 millimeters, a frequency of 85,000 megacycles per second.

The observers compared the brightness of the radiation at more than 40 pairs of points in the sky. Each pair was separated by 3 minutes of arc. By rapidly moving the telescope beam back and forth between one pair for half an hour, they searched for very small differences in the brightness of the background radiation. But they found no significant differences in any of the pairs. This means that granularity, if it exists, must be on a smaller angular scale than they used.

If the microwave background radiation is caused by individual sources, Boynton and Partridge concluded that there must be many of them—at least as many as there are galaxies in the universe. This makes it unlikely that the radiation is coming from rare objects such as quasars or peculiar galaxies. On the other hand, measurements of micro-wave radiation coming from several normal galaxies nearby indicate that these galaxies are much too weak to account for the background radiation.

The new experiment does not prove that the background radiation is the remnant of the primeval fireball, but it strongly reduces the possible alternatives.

Quasars. Several observations in 1973 support a growing feeling among astronomers that quasars are events in the nuclei of giant galaxies rather than a unique and distinct class of astronomical objects. This idea is suggested by studies of Seyfert and N galaxies, which have small, bright, active nuclei that look like mini-quasars. Like quasars, their spectra show the emission lines and blue colors characteristic of energetic sources with high temperatures. Also, like quasars, the nuclei change brightness in times as short as a few weeks or even days. However, the nuclei of N and Seyfert galaxies are generally fainter than quasars, so that the underlying galaxies can easily be seen by astronomers.

In January, 1973, Jerome Kristian of Hale Observatories in California reported on his attempts to find similar galaxy images underlying known quasars. He examined photographs of 26 quasars taken with the 200-inch telescope during the past 10 years and found galaxy images on several.

More important, Kristian predicted whether a galaxy would be seen with each quasar, depending on the brightness and distance of the quasar. Although quasars are much smaller than galaxies, their light is smeared as it passes through the earth's atmosphere. Thus, if a quasar is bright enough, its image on a photograph will be larger than a galaxy image, and the fainter galaxy underlying it will simply not be seen.

Kristian's study produced the predicted results—where galaxies should have been seen, they were, and where they should not have been seen, they were not. There is still no evidence to contradict the idea that all quasars occur in galaxies, but most of these galaxies are not seen because their light is overwhelmed by that of the quasars.

A study of the object Ton 256, reported in May, supports the assumption that the extended images around some quasars are galaxies, and not something else of the same size and shape. Ton 256 was originally classed as a quasar, although better photographs later showed an extended halo around the bright core, like those seen in N galaxies and in Kristian's study.

Using a new sensitive spectrum scanner at Lick Observatory in California, Joseph Silk, Harding Smith, Hyron Spinrad, and George Field of the University of California, Berkeley, separately measured the light from the halolike envelope and the central core. Their observations indicate that Ton 256 has two parts. One is the giant elliptical galaxy, the other is the quasar at its center. The center has the emission lines and blue color typical of quasars, while the envelope is as big and bright as the brightest galaxies, and has the same red color expected for an elliptical galaxy at the quasar's red shift. See CLOSE-UP.

Anniversaries. Two important anniversaries occurred in 1973. Polish astronomer Nicolaus Copernicus was born 500 years ago, in 1473. Contradicting the common sense, common belief, and established authority of his time, Copernicus argued that the earth moves around the sun and that we are not at the fixed center of the universe.

Man's view of his place in the cosmos became even more decentralized in the early 1900s. Contributing to this new viewpoint were two American astronomers—Harlow Shapley, who died in October, 1972, and Edwin Hubble, who died in 1953.

Around 1917, Shapley concluded, from studying the directions in the sky of globular star clusters, that our solar system is not located at the center of our Galaxy, as people had thought, but is well off toward the edge.

In 1923, Hubble discovered the first Cepheid variable in M31, the bright spiral nebula in the constellation Andromeda. By comparing the known brightness of other Cepheids with the apparent brightness of the Cepheid in M31, Hubble showed that the spiral nebulae are faraway galaxies, and not part of our own Galaxy. Within a decade, he also showed that the universe is uniformly populated with many galaxies, as far as telescopes can see, and that the whole system is expanding. [Jerome Kristian]

Biochemistry

Several biologically important compounds were synthesized in 1972 and 1973. Other interesting developments included reports of, and controversy over, the synthesis of a "memory molecule," experiments indicating that white blood cells called macrophages can acquire new biochemical powers by engulfing certain molecules, and indications that a component of hemoglobin may be useful to fight cancer.

Vitamin B_{12} synthesized. After 11 years of combined efforts by 99 scientists from 19 countries, it was reported in January, 1973, that vitamin B_{12} had been synthesized. Major credit for the achievement went to chemists Robert B. Woodward of Harvard University and Albert Eschenmoser of the Eidgenosische Technische Hochschule in Zurich, Switzerland, the project's directors.

Vitamin B_{12}, also known as cyanocobalamin, was first isolated in 1948 and is the largest and most complex vitamin yet discovered. The body must have it for normal blood-cell formation, for growth, and to maintain nerve tissue. B_{12} deficiency causes pernicious anemia, a severe malady in which the number of young red and white blood cells and platelets is considerably reduced. However, most cases of pernicious anemia are caused not by a lack of the vitamin in the diet, but by the absence of a protein necessary for absorption of the vitamin from the intestine. Thus, physicians usually control the anemia by injecting the protein alone or in combination with the vitamin.

B_{12} is the only known vitamin that contains a metal ion, cobalt. The cobalt is surrounded by four nitrogen-containing, roughly pentagonal rings that are linked together.

The synthesis of B_{12} involves a series of complex steps, 37 of which produce a four-ringed compound. Woodward named it beta-corrnorsterone because it is the cornerstone of the final steps of the synthesis. A further 14-step sequence of reactions completes the synthesis. X-ray crystallography analysis shows that the structure of the synthetic B_{12} is identical with that of the natural vitamin.

Human parathyroid hormone. Three teams of biochemists reported synthesis

Biochemistry

Continued

Chemist Robert B. Woodward of Harvard University holds a model of vitamin B_{12}. He was a codirector of the project that synthesized the vitamin in 1973.

of the active portion of human parathyroid hormone. The hormone, also called parathormone, is normally produced by the four parathyroid glands in the neck, behind the thyroid gland. It helps maintain a constant concentration of calcium and phosphorus in the blood.

The natural hormone is initially produced as a prohormone, which is a linear chain of 106 amino acids. The prohormone is rapidly converted into its storage, or glandular, form, which is a chain of only 84 amino acids. The 84-amino acid hormone is apparently further fragmented in the body since a chain of its initial 34 amino acids is fully capable of altering calcium and phosphorus levels in tests. All 84 amino acid links and their order have been determined for the parathyroid hormone of cattle and pigs. Chains of the first 34 amino acids in the cattle hormone and the first 30 amino acids in the pig hormone have been synthesized and also can alter calcium and phosphorus levels in tests.

For the work with the human hormone, Claude D. Arnaud and Glen W. Sizemore of the Mayo Medical School

and Clinic in Rochester, Minn., collected human parathyroid-gland tissue for two years from 150 cooperating laboratories and physicians in 12 countries. From this tissue, the scientists purified fractions containing the hormone and sent them to the National Institutes of Health (NIH), Bethesda, Md. A NIH team that included H. Bryan Brewer, Jr., Thomas Fairwell, and Rosemary Ronan isolated 3.8 milligrams (about 1/7000 ounce) of the hormone from the fractions, and determined the 34-amino acid sequence of the active portion. The sequence differs from that of the comparable portion of the cattle hormone at six points and from that of the pig hormone at five points.

The NIH scientists sent their structural information to Werner Rittel and his co-workers at Ciba-Geigy, Ltd., in Basel, Switzerland. The Swiss team synthesized the active portion of the human hormone, stringing the proper amino acids together like beads on a string. Future syntheses will produce the substance for use in medical treatment and for further research.

Parallel Parathormone Portions

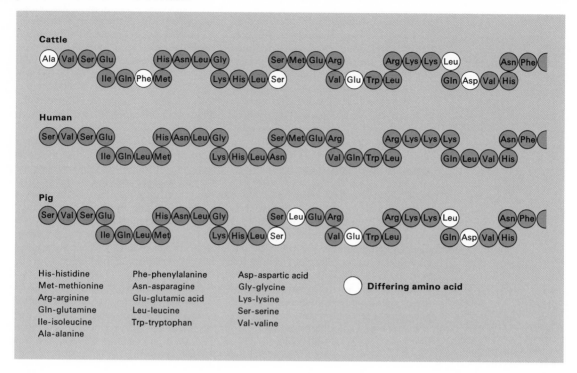

His-histidine
Met-methionine
Arg-arginine
Gln-glutamine
Ile-isoleucine
Ala-alanine

Phe-phenylalanine
Asn-asparagine
Glu-glutamic acid
Leu-leucine
Trp-tryptophan

Asp-aspartic acid
Gly-glycine
Lys-lysine
Ser-serine
Val-valine

Differing amino acid

Biochemistry

Continued

The newly determined structure for the active portion of human parathormone, which helps maintain constant blood levels of calcium and phosphorus, differs at six points from the equivalent portion of cattle parathormone and at five points from pig parathormone.

Scotophobin. In July, 1972, biochemist Georges Ungar and his colleagues at the Baylor College of Medicine in Houston reported that they had isolated, analyzed, and synthesized a substance by which a remembered experience could be transferred from one animal to another. This report was greeted with great controversy as had earlier reports of similar experiments.

In 1968, Ungar and his co-workers reported experiments in which they trained rats, which normally prefer darkness to light, to avoid darkness. Then they injected brain tissue extracts from the rats into untrained mice, which also normally prefer darkness, and the mice began to avoid darkness. Ungar claimed that the extract contained a substance made in the brain as a result of the training, and that the transfer of the substance to the mice had evoked the behavior acquired through the training. He called the substance scotophobin, from the Greek words for dark and fear.

These experiments, apparently demonstrating the transfer of learning by chemical means, evoked great interest and discussion. Other scientists repeated the experiments, some claiming success, others reporting failure.

In their latest report, Ungar and co-workers described the isolation and analysis of scotophobin from the brain extracts of trained rats. They found it to be a short chain composed of 15 linked amino acids. Because they were uncertain of the identity of the second, fifth, and eleventh amino acids, the scientists synthesized three different chains. They reported that one chain caused untrained mice to avoid the dark as effectively as natural scotophobin and also had chemical properties identical with those of the natural material.

Scientists who have criticized this work feel that the steps involving the isolation, purification, and analysis of the natural scotophobin were not carried out carefully enough. They agree, however, that further tests with the synthetic scotophobin should finally prove or disprove the Texas team's claims. See NEUROLOGY.

New roles for macrophages? The scavenging white blood cells called mac-

Biochemistry

Continued

rophages remove a variety of foreign materials from the blood by a process known as endocytosis. They use their cell membrane to engulf the material and draw it into the cell where it is digested. In December, 1972, physicians James Theodore, Julio Acevedo, and Eugene Robin of the Stanford University School of Medicine reported experiments in which macrophages acquired enzymatic activities they do not normally have.

The scientists suspended macrophages taken from rabbits in solutions containing the enzyme uricase for one hour. This enzyme oxidizes uric acid into allantoin. Next, they washed the macrophages thoroughly and placed them in solutions containing uric acid. The cells oxidized the uric acid, presumably using the uricase from the first solutions. Several experiments designed to check this conclusion seem to have confirmed it.

The investigators suggest that macrophages may be able to carry many other substances into the body, making the cells valuable tools in biochemical research and medical treatment. For example, genetic material, including a missing or nonfunctional gene, might someday be incorporated into the macrophages of persons with genetic deficiency diseases such as phenylketonuria. The acquired genes could then supply the missing ingredient needed for the patients' health.

Light versus cancer. Neurologist Ivan Diamond and his colleagues at the University of California, at San Francisco, came up with a novel way to destroy cancer cells. The scientists knew that hematoporphyrin, the nonprotein component of hemoglobin, increases the light sensitivity of cells, and that it is also absorbed preferentially by cancer cells. So they dosed isolated rat cancer cells and tumors within the rats with hematoporphyrin and exposed them to bright light. The isolated cancer cells were totally destroyed within 50 minutes. Cell kill in the intact tumors was 75 to 95 per cent after 3 to 5 days of 3- to 5-hour light exposures. The scientists suggest that similar treatment might be tried against several human cancers, including brain tumors, that are difficult to treat in other ways.

[Earl A. Evans]

In a new technique, artificial capillaries help grow masses of animal cells, *right,* for the first time to densities found in normal body tissue. A dense, dark cell mass and several portions of capillaries are clearly visible when they are magnified 340 times, *bottom.*

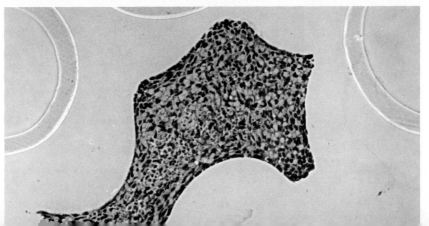

Books of Science

Here are 35 outstanding new science books suitable for general readers. The director of libraries of the Smithsonian Institution selected them from books published in 1972 and 1973.

Anthropology. *The Natural History of Man* by Carl P. Swanson. A well-written account of the known biological facts about man, his evolution, and his scientific awareness. This book combines art and text in a smooth presentation of the concept of evolution as a process, and the archaeological evidence on which it is based. The author attempts to define the uniqueness of man and regards man's development as adaptation to his environment. (Prentice-Hall, 1973. 416 pp. illus. $13.25)

Archaeology. *Viking America: The Norse Crossings and Their Legacy* by James Robert Enterline. Like a detective, the author gleans clues from archaeological finds, references in literature, and old maps to support speculations about the Norse exploration of North America. (Doubleday, 1972. 217 pp. $6.95)

Astronomy. *The New Astronomies* by Ben Bova. A summary of the new ideas in astronomy that have appeared since World War II and how they have changed the science. The author describes new equipment and techniques, telling how the radio telescope and neutrino detectors led to the discovery of quasars, pulsars, and neutron stars. (St. Martin's, 1972. 214 pp. $7.95)

Biology. *Life – the Unfinished Experiment* by S. E. Luria. The basic facts about the molecular processes involved in the development of life, from gene to cell to complex organ and finally to the individual within a species. (Scribner's, 1973. 167 pp. $7.95)

Wallace and Natural Selection by H. Lewis McKinney. This biographical study of Alfred Russel Wallace, a contemporary of Charles Darwin, tells how he developed and contributed to the theory of natural selection. The book compares the work of Darwin and Wallace and relates the influence of new theories of geology and sociology on their biological findings. (Yale, 1972. 193 pp. $12.50)

Botany. *Land Above the Trees: A Guide to American Alpine Tundra* by Ann H. Zwinger and Beatrice E. Willard. A guide to the flora of the alpine tundra, a specialized habitat that lies between the tree line and the region of perpetual snow. The book includes illustrations, pencil sketches, and lists of species of flowers that grow in America's highest mountain ranges. (Harper & Row, 1972. 489 pp. illus. $15)

Chemistry. *Chemistry for the Million* by Richard Furnald Smith is an abbreviated presentation of the science of chemistry, including a brief history and a description of the concepts of chemical composition, structure, and the properties of substances. He devotes special attention to such topics as the periodic table, inert gases, and halogens. (Scribner's, 1972. 175 pp. $7.95)

Molecules in the Service of Man by A. H. Drummond, Jr. An explanation of the chemistry of caged molecules, molecular sieves and pincers, polymers, and other unusual molecules and how they can be used in detergents, plastics, and other substances useful to man. The book speculates on the future of coal as a source of chemicals, a timely topic. (Lippincott, 1972. 173 pp. $5.95)

Computers. *The Computer from Pascal to von Neumann* by Herman H. Goldstine is a thorough but concise history of computing devices from 1600 to modern times. The author recounts the works of innovators such as Blaise Pascal, Gottfried W. Leibniz, and Charles Babbage, but concentrates on developments since World War II that led to the invention of modern computers. (Princeton, 1972. 378 pp. $12.50)

Ecology. *The Columbian Exchange: Biological and Cultural Consequences of 1492* by Alfred W. Crosby, Jr. This blend of history, ecology, and biology describes the effects of the two-way exchange of plants, animals, and diseases between the old and new worlds after Columbus discovered America. It concentrates on the wordwide changes in food supplies, the movements of people, population growth, and patterns of spreading disease. (Greenwood, 1972. 268 pp. $9.50)

The Logarithmic Century: Charting Future Shock by Ralph E. Lapp. With the aid of semilogarithmic graphs, the author demonstrates how uncontrolled exponential growth rates of industry and population will have disastrous effects on the world. He warns against the current high rate of energy and fuel consumption, stating that life in a technological society cannot exist without fuel.

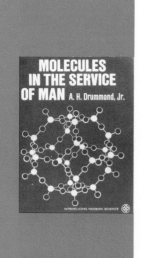

Books of Science

Continued

The book also discusses how atomic power may solve both the energy crisis and the ecological problems that have resulted from using fossil fuels. (Prentice-Hall, 1973. 263 pp. illus. $7.95)

Only One Earth: The Care and Maintenance of a Small Planet by Barbara Ward and René Dubos. The unofficial report commissioned for the United Nations Conference on the Human Environment held in Stockholm in June, 1972, is a fact-filled account of the ecological, economic, health, legal, and social problems that man faces today. The book includes a discussion of energy and matter as part of man's resources. (Norton, 1972. 394 pp. $9.95)

Tropical Forest Ecosystems in Africa and South America: A Comparative Review edited by Betty J. Meggers, Edward Ayensu, and W. Donald Duckworth. The many papers in this collection highlight some of the unique features of the two ecosystems. It identifies some of the problems of adaptation facing plants and animals, including man, in the African and South American areas that are undergoing development. (Smithsonian, 1973. 350 pp. illus. $15)

Engineering. *The Great Bridge* by David McCullough. The story of the Brooklyn Bridge and the Roebling family that built it is set within the cultural and scientific context of the late 1800s. The text is rich in explanations of the technical processes used in this spectacular engineering achievement. (Simon & Schuster, 1972. 636 pp. $10.95)

Entomology. *Insects of the World* by Walter Linsenmaier. This eminent entomologist presents a basic introduction to entomology, giving a synopsis of the orders of insects and describing the chief characteristics of many insect families, including their development and life cycles. (McGraw-Hill, 1972. 392 pp. illus. $25)

Food and Drugs. *Marihuana Users and Drug Subcultures* by Bruce D. Johnson. The author draws upon important findings of many studies about college-student drug users, attempting to define personal characteristics that may lead an individual to drug abuse. In predicting the pattern of drug involvement, he concludes that drug selling and not the use of marijuana leads to participation in the hard-drug subculture. (Wiley, 1973. 290 pp. $12.95)

200,000,000 Guinea Pigs: New Dangers in Everyday Foods, Drugs, and Cosmetics by John G. Fuller. Focusing on the inadequacies of testing and quality control, the author exposes and assesses the dangers of food additives and chemical substitutes, pesticides, some prescription drugs, over-the-counter remedies, harmful toys, hazardous household items, and other consumer products. (Putnam's, 1972. 320 pp. $7.95)

Geology. *The Restless Earth: A Report on the New Geology* by Nigel Calder. Text and illustrations show how physical evidence, including fossils, has aided in the development of theories about the processes that shape the earth, such as wandering magnetic poles, drifting continents, and the formation of mountains and ocean basins. (Viking, 1972. 168 pp. illus. $10)

Geophysics. *Earth and Its History* by Richard F. Flint. The author describes the creation of the earth from stellar gas and how natural forces molded it. (Norton, 1973. 407 pp. $9.95)

Mathematics. *Mathematical Thought from Ancient to Modern Times* by Morris Kline. A complete and scholarly history of mathematics since ancient times, with strong concentration on developments after 1800. The book contains many detailed and thorough discussions of the theories of mathematics. (Oxford, 1972. 1,238 pp. illus. $35)

Medicine. *Cancer: The Wayward Cell; Its Origins, Nature, and Treatment* by Victor Richards. Recording his observations of cancer patients and his discussions with cancer researchers, the author presents both humanistic and scientific approaches to an understanding of the disease. He presents facts about the biology and treatment of cancer in simple terms and discusses its history and the factors that contribute to the cause of cancer. (University of California Press, 1972. 308 pp. $10)

Rh: The Intimate History of a Disease and Its Conquest by David R. Zimmerman. This history of the Rh blood disease that kills unborn and newly born babies is told from the viewpoints of the various researchers who worked on its prevention. In addition to detailing each step in the research process that led to control of the disease, the author presents a realistic picture of scientists and their cooperation and competition in the

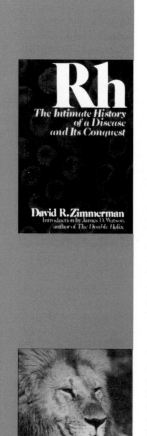

scientific world. (Macmillan, 1973. 371 pp. illus. $8.95)

Oceanography. *The Edge of an Unfamiliar World: A History of Oceanography* by Susan Schlee. The author gives an account of all the elements – people, concepts, and instruments – that contributed to the development of the science of oceanography. She describes major exploratory expeditions and analyzes the intricate relationship between the government and science. (Dutton, 1973. 398 pp. illus. $10.95)

Into the Hidden Environment: The Oceans by Keith Critchlow. A simplified account of the oceans and seas tells how they were formed, how they were explored, and what lives in them. (Viking, 1973. 125 pp. illus. $10.95)

Physics. *Albert Einstein: Creator and Rebel* by Banesh Hoffman. This popular version of Einstein's physics is woven into a personal portrait of the man. The book shows how Einstein's keen scientific and philosophical intuitions led him to establish his revolutionary theories. (Viking, 1972. 272 pp. illus. $8.95)

The Coming Age of Solar Energy by D. S. Halacy, Jr. The author describes ways for harnessing solar power, from using the oceans to trap energy to developing solar cells for powering spacecraft. (Harper & Row, 1973. 231 pp. illus. $7.95)

The Nature of Matter by Otto R. Frisch. One of the discoverers of nuclear fission explains atomic theory in simple terms. This book concentrates on subatomic particles and how they orbit around the atom's nucleus and on the new laws of the conservation of energy and matter. (Dutton, 1973. 216 pp. illus. $7.95)

The Nature of Time by G. J. Whitrow. The author touches on the various notions man has held about the concept of time – from the principles of the various inventions man has made to measure time (including atomic vibrations), to the concept of biological clocks. The book also discusses Einstein's theory of time and relativity. (Holt, Rinehart & Winston, 1973. 191 pp. $6.95)

Zoology. *Gifts of an Eagle* by Kent Durden. The author and his father, after capturing an infant golden eagle with government permission, report on the life, personality, and idiosyncracies of the bird, which they observed closely for a period of 16 years. (Simon & Schuster, 1972. 160 pp. $5.95)

Let Them Live: A Worldwide Survey of Animals Threatened with Extinction by Kai Curry-Lindahl. Continent by continent, region by region, the author catalogs the environmental changes being made by both nature and man that threaten various animal species with extinction. The book also summarizes the species that have disappeared or are in danger of disappearing on each continent. (Morrow, 1972. 394 pp. $9.95)

The Marvels of Animal Behavior edited by Thomas B. Allen. This collection of 22 articles gives insights into the nature of animal societies – such as bees, elephants, gorillas, and whales – the methods by which these species communicate, and how they interact with each other and their environment. (National Geographic, 1972. 422 pp. illus. $12.95)

Octopus and Squid: The Soft Intelligence by Jacques-Yves Cousteau and Philippe Diole. The famous marine biologists explore the biological and ethological facts about the life and environment of the octopus, the squid, and other cephalopods. They also recount myths about these shy but basically friendly creatures as well as real experiences with them. (Doubleday, 1973. 302 pp. illus. $9.95)

The Polar Worlds by Richard Perry. The author describes, simply and readably, the animals of the Arctic and the Antarctic, telling how they live, develop, and adapt in these cold climates. (Taplinger, 1973. 316 pp. $7.95)

Serengeti Lion: A Study of Predator-Prey Relationships by George B. Schaller. This 1973 National Book Award winner is a readable and thorough study of the life, habits, and hunting methods of the lions that live on Tanzania's Serengeti Plain. The author attempts to give the reader a better understanding of predators, such as the lion, and the useful role that they play in the management of wildlife. (University of Chicago Press, 1972. 480 pp. illus. $12.50)

Smarter Than Man? Intelligence in Whales, Dolphins, and Humans by Karl-Erik Fichtelius and Sverre Sjölander. Basing their arguments on studies of brain weight, animal behavior, and the role of culture and physical environment in determining intelligence, the authors question seriously the general assumption that man is the most intelligent of all the animals. (Pantheon, 1973. 205 pp. illus. $6.95)   [Russell Shank]

Botany

Herbert G. Baker and Irene Baker of the University of California, Berkeley, announced in February, 1973, that the nectar of flowers contains amino acids, the organic compounds that are the building blocks of proteins. Prior to this, scientists had always assumed that insects gathered this sweet liquid from flowers as an energy source because of the sugars it contains. However, the Bakers found significant concentrations of amino acids in the nectar of 266 species of plants growing in California.

They reported that the nectar produced by flowers that attract carrion and dung flies was richest in amino acids. Thus nectar may be an important source of protein for these insects. Flowers visited primarily by butterflies also have nectar that is rich in amino acids, though butterflies, like bees, usually obtain amino acids from pollen.

When the scientists analyzed many species of flowers, they found a wide range of different amino acids in their nectar. *Dianthus barbatus*, the common sweet William, a favorite of butterflies, produces at least 12 amino acids.

The fact that greater concentrations of amino acids are found in the more specialized pollination systems has considerable evolutionary significance. In other words, the plants that developed later in evolutionary history are the ones with the greatest number of amino acids.

Growth regulation. Working together, botanists R. N. Strange of the University College of London and Harvey Smith and John R. Majer of the University of Birmingham, England, reported in July, 1972, that they had discovered a growth stimulant within plants that promotes the growth of a destructive fungus. The fungus, *Fusarium graminearum*, causes a disease called head blight that discolors the plant and breaks down the protein in cereals.

The scientists found that, in wheat, the stimulant is primarily in the anthers, or male portions of the plant. Chemical analysis showed it to be choline. When the researchers tested pure commercial choline on the fungus, this chemical made it grow and eventually destroy the wheat plant. The scientists are continuing their research to determine the comparative concentrations of choline in extracts from the wheat heads and other parts of the plant.

Charles A. Beasley of the University of California, Riverside, showed, in March 1973, that hormones can be used to artificially regulate the growth of unfertilized cotton ovules. He removed young ovules from the ovaries of flowers and floated them on the surface of a liquid containing glucose. The ovules to which he added no hormone growth substances shriveled and turned brown within two weeks. But when he added indoleacetic acid or gibberellic acid, which are growth hormones, the ovules grew, remained white, and produced cotton fibers. When he added kinetin, another chemical that promotes growth, the ovules remained white and increased in size, but they produced no fibers. The weight of the ovules increased as more kinetin was added to the liquid.

Herbicide resistance. Moshe J. Pinthus, Yaacov Eshel, and Yalon Shchori, botanists at the Hebrew University in Israel, showed in September, 1972, that selective breeding of plants that have been treated with mutagenic materials makes crops more resistant to herbicides, or weedkillers. The development is significant because herbicides that are used in weed control sometimes damage the crops they are intended to protect.

The botanists induced mutations in a group of spring wheat plants and in a group of tomato plants by soaking the seeds at room temperature in a solution of ethyl methane sulfonate. The seedlings of 50,000 treated wheat plants were then subjected to the herbicide terbutryn. A total of 588 relatively resistant seedlings resulted. A screening of 20,000 tomato seedlings growing in soil treated with diphenamid produced 117 resistant seedlings. Lines were then selected for further breeding from the various resistant seedlings.

William G. Smith of Yale University reported in 1972 that six common species of trees in New Haven, Conn., contained large amounts of lead and mercury. Scientists assume the lead comes from the combustion of automobile fuels and the mercury from the combustion of fossil fuels in factories and homes.

Smith found the highest lead concentrations in the twigs of sugar maple, eastern hemlock, yew, and Norway spruce, and in leaves of pin oak and Norway maple. He also found above normal concentrations of mercury in

A new grass that grows in the shaded forests of Brazil and is related to the bamboo plant was discovered by botanists Thomas Soderstrom and Cleofé Calderón of the Smithsonian Institution.

Botany

Continued

these plants. Smith suggested that woody plants may play an unexpectedly important role in the short- and long-term heavy metal recycling that goes on in urban areas. The leaves of deciduous trees return the metal to the soil in a single season, but lead and mercury concentrations in twigs and other tissue of perennial trees may not be returned to the soil for many years.

The birds and bees. In a study of pollination in the mountains of southern Mexico, botanist Robert W. Cruden of the University of Iowa reported in 1972 that hummingbirds pollinate more plants that grow at higher elevations than do bees. Cruden also found that bees do not pollinate as much during the annual rainy season, but hummingbirds remain active all the time.

There is apparently little difference between the effectiveness of bees and hummingbirds as pollinators in good weather, at altitudes where both are active. However, the types of plants pollinated by hummingbirds tend to grow at higher altitudes, where weather conditions are less favorable for the bees.

Cruden also noted that the plants adapted to hummingbird pollination continued flowering during the rainy season, while the ones that were adapted to bee pollination did not.

Fossil plants. A. Chandrasekharam of the University of Alberta in Edmonton, Canada, discovered carbonized impressions of fossil leaves, in 1972, in rocks found at Genessee, Alberta. About 60 per cent of the fossils consist of deciduous angiosperm leaves. Angiosperms are plants that have their seeds enclosed in ovaries or fruits. The fossil leaves are remarkably similar in size and configuration to those of living trees of the same genera. Woody structures may be so well preserved in fossil impressions that their most minute structure can be seen, but it is remarkable for the delicate spongy tissues of flowering plant leaves to be so well preserved.

Francis T. C. Ting of the geology department at the University of North Dakota, also in 1972, discovered the first example of fossil peat in lignite rocks near Medora, N. Dak. It was well preserved.   [William C. Steere]

Chemical Technology

Chemical innovators worked on developing biodegradable and sunlight-degradable plastics, and on synthetic natural gas and other energy-related technologies in 1972 and 1973. Copper extraction and waste-reclamation processes also received much attention.

Plastics technology is trying to devise products that will decompose quickly when discarded. Two developments are especially promising. Japan's Mitsui Toatsu Chemicals has developed a disintegrating polystyrene that is safe for food packaging. The material, called Toporex-ed, is similar to conventional polystyrene in appearance, physical properties, nontoxicity, and other characteristics. But a nontoxic, degradable compound incorporated into the polystyrene molecule is destroyed by the sun's ultraviolet rays. Because it is part of the molecule, the added component cannot migrate into the food in the container.

Finland's Amerplast, located in Tampere, announced commercial production of biodegradable polyethylene bags in mid-1973. The process involves mixing 0.1 per cent of a powdered iron compound with polyethylene before polymerization. The technology was developed by Gerald Scott, chemistry professor at Aston University, Birmingham, England. It is most successful with transparent bags; work continued on developing effective degradable, multicolored plastic bags. Scott has also devised a sunlight-degradable polyvinyl chloride. Thus, photodegradable forms of all four major packaging plastics – polyethylene, polypropylene, polystyrene, and polyvinyl chloride – are now available.

The first commercial outlet in the United States for biodegradable polystyrene developed by Bio-Degradable Plastics, Inc., Boise, Ida., is cold-drink-cup lids. They are used by Der Wienerschnitzel, a fast-food chain. Possible future uses for the process appear to be in styrene-based dairy containers, egg cartons, meat trays, and throwaway plates.

Fuel from chemistry. The search for new and efficient energy-production methods continued as the U.S. fuel shortage mounted. Substitute or synthetic natural gas (SNG) quickly became the focal point of much industrial activity, as

Particles of dust having some unique magnetic properties can be put into tanker's petroleum cargo to identify the guilty ship when an oil spill occurs.

companies sought to rush development of this source of fuel. SNG consists essentially of methane, CH_4. Like natural gas, it has a heating value of about 1,000 British thermal units (B.T.U.) per cubic foot. About 30 SNG plants were in some stage of planning or construction early in 1973 in the U.S.

It is easier to produce SNG from liquid hydrocarbons, such as naphtha or crude oil, than from coal. The liquids have a higher ratio of hydrogen to carbon and are easier to process. In addition, the technology for transforming liquids has become further advanced than that for converting coal.

The most recent process to generate SNG from heavy liquid hydrocarbons has been developed by the Institute of Gas Technology (IGT), Chicago, as an offshoot of its work to produce gas from coal. The IGT process produces methane from powdered coal in a hydrogen-rich atmosphere at 1300° F. and 1,000 pounds of pressure per square inch. Particles in a bed of solids pick up coatings of unreacted carbon. The solids circulate continuously to a second vessel, where the carbon reacts with steam and oxygen to produce the hydrogen that is needed for the first step.

The SNG recovered is cooled and purified; no further treatment is required. It has a heating value of about 900 B.T.U. per cubic foot.

Cleaning up copper. Strict air-pollution-control laws are forcing rapid changes on the U.S. copper industry. Traditionally, copper has been extracted from sulfur-rich ore concentrates. The three steps – roasting, smelting, and converting – each require heat input and release unwelcome fumes of sulfur dioxide (SO_2). Removing the SO_2 from the effluent is expensive, so copper producers are trying two new ways to avoid polluting the atmosphere.

In the first, ore concentrates are heated in one or a series of enclosed vessels so that the combined gases given off are relatively rich in SO_2, which can be converted to sulfur or sulfuric acid.

The second approach uses hydrometallurgy (wet processes with chemical and electrical techniques) rather than pyrometallurgy (processes that depend on heat to accelerate reactions) to extract copper metal. Starting with similar sulfur-rich ore concentrates, Anaconda

Company dissolves their copper content in ammonia, filters the solution to give a pure liquid, then extracts the copper from the ammonia solution with an organic solvent. The copper is finally recovered from the cathode of an electrolytic cell. The sulfur that has been separated from the copper ends up as ammonium sulfate or gypsum.

The Cymet process of Cyprus Metallurgical Processes Corporation is somewhat more complex. Finely ground copper-ore concentrates are added to an iron chloride solution, which dissolves the copper. After a series of intermediate steps, copper powder is precipitated from a copper chloride solution. The powder is separated and purified by an electro-refining process.

Putting wastes to use. Efforts to convert wastes to useful products held increasing appeal as an antipollution technique. Burning garbage to generate power is to be tested in suburban Buffalo, N.Y. A demonstration unit will combine combustion with pyrolysis, or decomposition by high heat. It can destroy solid wastes at up to 3000° F. while producing steam to generate power. Most of the demonstration project is being financed by the Environmental Protection Agency. The agency believes that costs of the system can be held to about $4 to $6 per ton of waste handled.

Using municipal solid wastes to produce fuel oil and other by-products will get a large-scale tryout near San Diego, Calif., in 1974. A new plant will process 200 tons of waste per day using a flash pyrolysis process that reaches 900° to 1000° F. In preparation for the pyrolysis, the waste is ground, sorted, and reground to about the consistency of a household vacuum cleaner's contents. Each ton of waste is expected to yield about a barrel of low-sulfur fuel oil. Glass, metals, and char will also be recovered.

Sewer pipe made from a composite of a polymer and crushed, discarded bottles has been installed in the Huntington, N.Y., municipal sewer system. The lightweight pipe is made by crushing discarded bottles (without washing them or removing labels or metal), mixing the pieces with a liquid plastic, and heating the mixture in a mold to harden it. The resulting composite is strong, light, and highly resistant to corrosion. A pipe section about 40 inches long and 8 inches in

Chemical Technology

Continued

Clear polyester film kept windowpanes that were shattered by an explosion in place. The window on the left was not covered by the film and was blown out by the blast. The film, which takes only minutes to apply, has been used to prevent injury from flying glass during the civil strife in Northern Ireland.

diameter uses the equivalent of about 118 beer bottles.

Fly ash, combustion residue so fine it often goes up the stack with gases from coal- and oil-burning power plants, has almost made the transition from a waste product to a by-product.

Ash is now used as a partial replacement for cement in concrete products and as fill material for roads and construction sites. Many new applications promise to gobble up ever-increasing quantities of this material. These include uses in mine-fire control, in anti-skid winter road surfaces, as fillers for products such as fertilizer and rubber, for soil stabilization, and in the manufacture of basic cement.

One particularly interesting portion of fly ash is thin-walled, glassy balls of calcined clay in which gases have been trapped during cooling. The balls, called cenospheres, can withstand extremely high pressure. They can be used to provide buoyancy in deep-ocean operations, as a closed-pore insulation in a space shuttle design, and for making lightweight floating concrete.

Mineral processing on the ocean bottom is a target of research at the University of Wisconsin, Madison. Hydroclones – cone-shaped devices that separate heavy solids from liquid and lighter solids by centrifugal force – are being modified to operate at ocean-bottom pressures. Thus, mineral concentrates can be recovered and pumped to the surface without having to bring up large quantities of waste that would have to be redumped after separation.

Plastic degrader. Britain's Coloroll Ltd., Nelson, Lancashire, has found that all major plastics can be made biodegradable by adding a common industrial chemical. The material stimulates enzymatic decomposition within about six months when the plastic is buried in at least 8 inches of soil.

Ingenious applications. Lone Star Industries, Greenwich, Conn., has a quick, low-cost way to break up ores and other hard solids that are usually ground mechanically. The material is loaded into a confined chamber and put under moderate pressure from steam or some other compressible fluid. A fast-acting

Chemical Technology

Continued

discharge valve is opened in less than 15 milliseconds, and the sudden drop in pressure sets up shock waves that fracture pieces of rock. Additional shock waves, some of them sonic, shatter the particles still more as they rush from the chamber.

Solids from a liquid slurry pose special handling problems when they must be dried, but Calmic Engineering Company of Great Britain has a new answer. The slurry goes into a chamber containing hollow, inch-long cylinders of silicone rubber. Hot air blown into the bottom of the chamber through a grid evaporates the liquid, leaving the solids coating the cylindrical rubber beads. The air flow blows the beads against a grid at the top of the chamber, and the impact breaks the dried powder coating. The air carries the powder away for recovery, and the rubber beads fall down into the chamber to begin another cycle.

Two devices for separating extremely small solid or liquid particles from gas streams promise to help control pollution. Battelle-Northwest, Richland, Wash., is developing a new electrostatic precipitator that passes the gas through a charger to give the particles negative charges, then through a nonconductive, dry, open-weave fiber bed.

Using the nonconducting collector allows the charged particles to help build up an electrostatic space charge. This apparently helps the fiber trap the particles. Such devices usually add an electric charge to particles flowing in a tube so they travel to the tube wall for removal. Battelle reports that the particles are captured sooner in its process. The fiber bed is periodically washed to remove collected material.

Versatile material. A general-purpose composite material based on sulfur dioxide, lime, and polyolefins has been developed by Japan's Lion Fat & Oil Company. It may be used as a building material or as a cardboard substitute. And when discarded, it can be burned without creating air pollution. The company has worked out technology for making the composite that can dovetail neatly with flue-gas desulfurization processes based on limestone treatment. Electric utility companies are testing such processes.   [Frederick C. Price]

Chemistry

Chemical Dynamics. Researchers made impressive gains in 1972 and 1973 toward describing the exchange reaction between an atom, A, and a diatomic molecule, BC, to form a new molecule AB and an atom C, $A + BC \rightarrow AB + C$, in the gas phase. These advances were made possible by increasingly realistic computer models of these collisions.

In this procedure, the researcher specifies an interaction potential, or force field, that controls the motions of the three nuclei. Then he starts the reactant molecules, A and BC, toward one another, and follows their motions (trajectories). In a nonreactive collision, the molecules collide and rebound with the same identity, $A + BC$. When A and BC collide and come apart as $AB + C$ an exchange reaction has taken place.

This process might be roughly regarded as shooting a marble, A, toward a pair of marbles, BC, over a surface that is not flat. The tilt and steepness of the surface is determined by the electrical forces holding atoms A, B, and C together or pushing them apart. It is called the potential energy surface of the reaction.

Molecules in a gas have a wide range of speeds, and shake and tumble in every possible way. To approximate collisions in such disorder, the initial starting conditions, such as the velocities and orientation of the molecules, are chosen at random from a large group of molecules having the correct range of conditions for that temperature. The computer calculates thousands of trajectories for these reactant molecules, and accumulates statistics.

Several groups are now active in this field, including Norman C. Blais at the Los Alamos (N. Mex.) Scientific Laboratory, Donald L. Bunker at the University of California at Irvine, Martin Karplus and his co-workers at Harvard University, and John C. Polanyi and co-workers at the University of Toronto. Polanyi's group has investigated a wide class of surfaces for energy-producing (exothermic) reactions to determine what features govern various aspects of the reaction. Polanyi found that the reactants generally come together in an entrance valley, climb a small ridge (the activation barrier) to a saddle point region (col) at its top. Then the products pass down the other side and separate through an exit valley. If the downhill portion in which most of the energy is released is in the entrance valley where the reactants come together, the potential energy surface is termed "attractive." If most of the energy is released when products separate in the exit valley, the surface is called "repulsive."

Trajectory calculations show that attractive surfaces cause most of the energy to show up as vibration in the newly formed AB molecule. Repulsive surfaces, however, cause most of the energy to propel AB and C apart at high speed. If the attacking atom, A, is more massive than the departing atom, C, even repulsive energy surfaces can cause a large fraction of the energy to excite vibrations in AB. In this case, the repulsive energy separating B from C is released while A is still approaching BC so that B recoils from C to join A. If A is approaching BC from the end, B's recoil causes only vibrations when A and B join. When A approaches from the side, B's motion causes AB to vibrate and rotate.

The angle through which the products fly apart also reveals the reaction dynamics. The scattering is termed "forward" if both products are ejected in the direction in which the attacking atom, A, moved. Attractive surfaces favor forward scattering while repulsive surfaces favor "backward" scattering.

Most reactive collisions are caused by a single encounter. However, in attractive surfaces the exit valley is not steep and AB and C may come back together in a second collision. These snarled trajectories not only reduce the internal excitation, but also broaden the angular distribution of the product molecules. On repulsive surfaces, the exit valley is usually so steep that AB and C fly apart before a second encounter can occur.

Modifying reaction rates. In energy-absorbing (endothermic) reactions, AB + C→A + BC, are the opposite of the exothermic reactions. The entrance valley is not downhill. For these reactions to occur, the reactant molecules, AB and C, must be given the energy to climb the activation barrier. In 1973, Arnold C. Wahl of the Argonne National Labora-

A Repulsive Surface

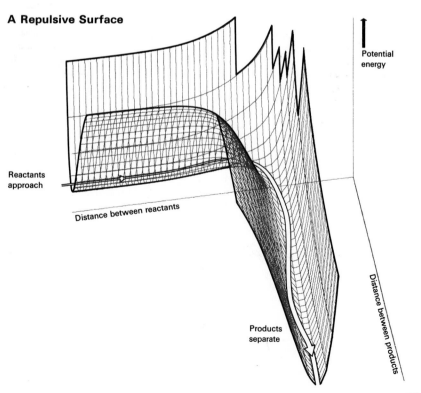

Computer-model surface displays a reacting chemical system. As the reactant molecules approach, the system moves down the entrance valley, rises over a small potential barrier, and then falls rapidly down the steep exit valley as the product molecules move apart. The weaving path shows that potential energy released by the reaction causes the product molecules to vibrate.

Potential energy

Reactants approach

Distance between reactants

Products separate

Distance between products

tory near Chicago examined the endothermic reaction between a lithium atom and a hydrogen molecule $Li+H_2 \rightarrow LiH + H$. Wahl's trajectory calculations indicate, however, that the energy must be in vibrations of the bond under attack for such reactions to take place.

Chemists know that an increase of 10° C. often doubles the rate of a chemical reaction of gases. Wahl's trajectory calculations suggest that this increased rate results when the higher temperature lifts more molecules to higher vibrational levels, where they can react.

Chemists can, in principle, increase vibrations directly in many cases by infrared irradiation, and this suggests that reaction rates can be significantly modified in this manner. In 1972, Truman J. Odiorne, Philip R. Brooks, and Jerome V. V. Kasper of Rice University in Houston dramatically illustrated this. They caused a hundredfold increase in the slightly endothermic reaction $K + HCl \rightarrow KCl + H$ by irradiating the HCl with the output of an HCl laser, exciting its vibrations. In 1973, several other experimenters used infrared lasers to increase reaction rates.

Tunneling. All these computer results use Newton's laws of motion. However, we should expect quantum mechanical complications to become more apparent in reactions with light atoms. One of the most unusual predictions of quantum mechanics is "tunneling," the ability of a particle to occasionally pass right through a barrier even though it lacks the energy to climb over the barrier.

In the past, accurate quantum calculations had been limited to one-dimensional collisions, where atoms A, B, and C are in a line. However, in 1973, Roberta P. Saxon and John C. Light at the University of Chicago, and Aron Kuppermann and co-workers at the California Institute of Technology, Pasadena, carried out calculations for planar (two-dimensional) encounters for the reaction $H + H_2 \rightarrow H_2 + H$.

They find that tunneling is an important reaction mechanism near room temperature. The reactants pass through the crest of the activation ridge from entrance to exit valley. This suggests that non-quantum trajectory calculations may be inadequate for many reactions involving the formation or rupture of hydrogen bonds.   [Richard N. Zare]

Structural Chemistry. Research continued to emphasize problems of biological importance. In 1972 and 1973, work directed by Alexander Rich at the Massachusetts Institute of Technology (M.I.T.) was particularly significant. It led to a better understanding of the process by which the genetic information contained in a molecule of deoxyribonucleic acid (DNA) is transferred to the next generation of living cells.

DNA is the molecule in a cell's nucleus responsible for ensuring that cell division produces identical daughter cells. It carries out two crucial functions: It must reproduce itself exactly, and it must dictate the exact chemical composition of the proteins being produced by the growing cell. In the latter process, various molecules of ribonucleic acid (RNA) serve as intermediaries, passing along information as to the arrangement of atoms in the DNA molecule to the proteins growing outside the cell nucleus.

Like DNA, RNA is a repeating sequence of sugar-base-phosphate units. The sugar is ribose in RNA and deoxyribose in DNA. The bases are either adenine (A), cytosine (C), guanine (G), or uracil (U). In DNA, uracil is replaced by thymine, T. The sequence of bases in a particular DNA molecule, and its RNA intermediary, dictates the sequence of amino acids that are combined to produce the protein.

DNA may contain over a million sugar-base-phosphate units, and its structure is based on a double helix – two DNA chains wound about one another and held together by hydrogen bonds A···T and G···C between the bases. On the other hand, the various RNA molecules range from less than a hundred units to many thousands, and they form a single chain, with little regularity in their structure.

In order to better understand how RNA transfers information, Rich's group carried out a two-pronged attack on the molecular structure of RNA. Both attacks are based on X-ray diffraction studies of crystals.

Crystals and heavy crystals. In one phase of the program, the group prepared crystals of transfer RNA (tRNA) isolated from yeast cultures. The tRNA selects and holds in place the correct amino acid molecule when it is added to the growing protein. The tRNA is a small

Chemistry

Continued

A small two-unit part of the double helix is formed by fragments of RNA in crystalline form. X-ray studies of the precise structural positions of the atoms found that the guanine phosphate cytosine (GpC) units were paired end-for-end and had multiple hydrogen bonds.

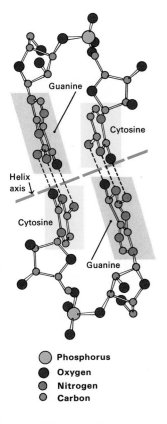

- Guanine
- Cytosine
- Helix axis
- Cytosine
- Guanine

- ● Phosphorus
- ● Oxygen
- ● Nitrogen
- ● Carbon

RNA molecule, containing only about 75 sugar-base-phosphate units. Yet chemists cannot interpret its X-ray diffraction pattern by standard methods.

Accordingly, additional tRNA crystals were prepared that contained the heavy atoms platinum, osmium, and samarium. By comparing the diffraction patterns produced by these heavy-atom crystals with that produced by tRNA itself, Rich was able to make an improved map of the electron density in the tRNA crystals. He identified phosphate groups as the regions of high-electron density, and traced out the complete sugar-base-phosphate chain.

The tRNA molecule turns out to be an L-shaped single chain, with the longer leg about 77 angstrom units (A) long (1A=10^{-8} centimeters). In four regions, the chain folds back upon itself to form short segments of a double helix, held together by hydrogen bonds between base pairs A$\cdots$U and G$\cdots$C. In most regions the molecule is about 20A thick. The four double-helix regions correspond to the stem and the three loops of a cloverleaf structure that had been previously proposed from other evidence. However, the overall geometry of the molecule is quite different from any that had been considered.

Fragmentary evidence. Further studies on tRNA crystals will undoubtedly lead to a picture of the accurate arrangement of atoms in the molecular chain. At the present time, however, the crystals are not of high enough quality to permit the researchers to locate individual atoms with a high degree of accuracy. So Rich's group also carried out structure studies on fragments of RNA that are small enough to ensure precise atomic positions, and these should correspond to their locations in the larger RNA molecule.

The two fragments they chose for study were dinucleotides, each of which contained two sugar-base units and one phosphate group. The M.I.T. group first grew crystals of the sodium salts of GpC (p=phosphate) and ApU. Then they exposed these crystals to X rays and recorded the diffraction patterns. The patterns had about 3,000 spots of varying intensities. By interpreting the positions and intensities of these spots, they arrived at an atom-by-atom picture of the two dinucleotide molecules.

The dinucleotides GpC and ApU are especially interesting because they are "self-complementary." One molecule of GpC, for example, can form two sets of hydrogen bonds G$\cdots$C and C$\cdots$G with a second molecule of GpC turned end-for-end, the same coupling that occurs in the double-helix structures of DNA and tRNA. Indeed, the crystal structures of both GpC and ApU feature a small, two-unit portion of the double helix. The helix is approximately 20A in diameter.

Since the positions of the dinucleotide atoms have been determined with an accuracy of 0.01 A, the results of these studies can now be used as guides in interpreting the more approximate results obtained for tRNA. The structure of GpC and ApU will also serve as accurate templates, or patterns, for use by researchers in studying the structures and reactions of the other RNA molecules. This should provide the precise molecular basis of how RNA transfers information in living systems.

The hydrated proton. It has long been known that a hydrogen ion does not exist in aqueous solution as a bare proton, H$^+$. Instead, it becomes attached to a water molecule to form a hydronium ion, H$_3$0$^+$. Crystal-structure studies in 1973 have now shown that the hydronium ion also is found in the solid state.

Much of the crystal structure work on hydrated proton compounds is being carried out in the laboratories of Ivar Olovsson of the University of Uppsala, Sweden. Working at temperatures as low as −180°C. (−292°F.), Olovsson's group has obtained X-ray diffraction patterns from crystals of hydrates of several strong acids, such as sulfuric (H$_2$SO$_4 \cdot$H$_2$0), hydrochloric (HCl-2H$_2$0 and HCl-3H$_2$0), and perchloric (HCl0$_4 \cdot$2½H$_2$0).

They found H$_3$0$^+$ in some of these crystals and more complicated species, such as H$_5$0$_2^+$, in others. The H$_3$0$^+$ ion is nonplanar, the three hydrogen atoms forming a triangular base of a shallow pyramid with the oxygen atom at the vertex. In the H$_5$0$_2^+$ ion, one hydrogen atom lies between H$_2$0 groups. The 0$\cdots$H$\cdots$0 distance, about 2.4 to 2.5 A, corresponds to a very strong hydrogen bond. However, it is not yet clear whether or not the central hydrogen atom lies exactly midway between the two oxygen atoms. [Richard E. Marsh]

New Synthesis Procedure
For Needed Drugs

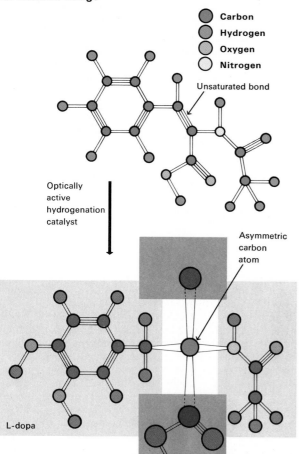

Carbon
Hydrogen
Oxygen
Nitrogen

Unsaturated bond

Optically
active
hydrogenation
catalyst

Asymmetric
carbon
atom

L-dopa

Certain optically pure substituted amino acids, such as the drug L-dopa, are directly synthesized by making a catalyst that is optically active to hydrogenate the unsaturated bond. This forces the resulting molecule into only one of the two possible configurations around the asymmetric carbon atom. Only the L configuration of dopa is biologically active.

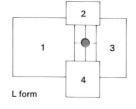

L form

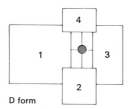

D form

Chemical Synthesis. The brilliant total synthesis of vitamin B_{12} by Robert B. Woodward of Harvard University and Albert Eschenmoser of the Eidgenosische Technische Hochschule in Zurich, Switzerland, highlighted several notable achievements in chemical synthesis in 1973. See BIOCHEMISTRY.

Synthesizing L structures. Molecules that contain asymmetric carbon atoms are called chiral, or optically active. They twist the vibrational planes of light waves to the right (*dextro*, D) or left (*levo*, L). Chemists have long sought economical ways to commercially synthesize pure D and L structures. The chiral chemicals required as starting materials to manufacture drugs and biochemicals are usually obtained by degrading naturally occurring compounds by expensive enzymatic processes, such as fermentation.

This difficulty was removed in 1973 by the direct synthesis of isomers of the α-amino acids L-phenylalanine and L-dopa that are nearly optically pure. L-α phenylalanine is used in low-calorie sweeteners, and L-dopa is useful in treating Parkinson's disease.

A team of chemists headed by William S. Knowles of the Monsanto Company, St. Louis, used a new trick to accomplish this feat. They prepared phosphines having asymmetric phosphorus atoms and chemically bonded these optically active compounds to rhodium to produce an optically active hydrogenation catalyst. Then they used these catalysts with hydrogen gas to hydrogenate D-α-acylamino acrylic acids, producing the L-α acylamino acids with about 90 per cent optical purity. That is, they had 90 per cent L, and 10 per cent D configuration. The researchers said they can achieve complete optical purity of the L-α-amino acids by simple purification procedures.

A second method. Donald J. Cram and co-workers at the University of California, Los Angeles (UCLA), reported on other methods of optical isomer separation. This group used a chiral "crown" ether that complexes with only one of the D or L molecules.

Crown ethers tend to complex with an ammonium ion, even if it already has large groups bonded to its nitrogen atom. The chemists used asymmetric amines having an $-NH_3^+$ group in

Chemistry

Continued

place of the normal $-NH_2$ group. The $-NH_3^+$ is held to the exposed oxygen atoms in the ether by electrostatic forces and hydrogen bonding.

The formation of a complex with only one of the D or L configurations is called "chiral recognition." It arises because the crown ether has a cavity that will contain only a D or L substituted ammonium ion.

To separate the optical isomers of an ammonium ion carrying an asymmetric substituent, the UCLA chemists mixed a water solution of the D and L substituted ammonium salt with a hydrocarbon solution of the ether. Only the preferred optical isomer of the substituted ammonium ion easily enters the hydrocarbon phase, because it becomes a hydrocarbon-soluble complex of the crown ether.

The equal mixture of L and D isomers found in the normal ammonium ions is enriched to 62 per cent of the preferred isomer after a single separation step. Repeated decomposition of the complex, followed by separation, causes the preferred optical isomer to become increasingly pure. After 10 or more extraction steps, it is optically pure. This method is competitive with the Monsanto hydrogenation process and together they should provide economical supplies of pure L-α-amino acids for biochemical and dietary preparations.

Closing a chapter. Columbia University researchers completed a search in 1973 that began a century earlier. In the mid-1800s, the English chemist James Dewar and the German Albert Ladenburg both proposed three-dimensional structures for benzene (C_6H_6). Their proposals were later disproved and the correct structure determined by such contemporaries as the German chemist August Kekulé. The Ladenburg benzene structure is called "prismane" because of its prismatic structure. All efforts to synthesize prismane had failed. However, Thomas Katz and his coworkers at Columbia University reported its synthesis in 1973. This completes an important chapter in organic chemistry; the Kekulé, Dewar, and Ladenburg C_6H_6 structures have all been synthesized. [M. Fred Hawthorne]

Communications

Demand for more communication capacity continued unabated the world over. The number of telephones in the world increased in 1972 to more than 300 million. In the United States, long-distance traffic continued to grow at more than 10 per cent per year. The fastest growing component was communication between computers and other digital business machines. Such data communications over the voice telephone network use modems (modulator-demodulators) to convert streams of digital data to continuous (analog) signals and vice versa.

Data communications may eventually have its own network for unconverted digital data. Carriers began scrambling to enter this growing field. The Bell System filed an application with the Federal Communications Commission (FCC) for permission to operate a new data service called the Digital Data Service. In April, 1973, Data Transmission Company, Vienna, Va., broke ground for the first, seven-city segment of its network. A new Canadian digital system called Data Route first went into service in April, 1973.

Canadian communications satellite Anik I is made ready for its launch from Cape Kennedy pad.

Data communication today relies on linking sender to receiver for the duration of the "call." Future data communication may be through "message switching," instead. Most data that needs to be transmitted occurs in bursts, or packets. Each burst of data would be switched rapidly to any available network for transmission. Great Britain pioneered packet transmission in the late 1960s with a demonstration network and tests. Packet Communications, Waltham, Mass., proposed a new service based on this idea to the FCC in January, 1973.

A different proposal envisions national, regional, and local loops, all interconnected, carrying high-speed conveyor-like data belts. Each message would find its way from source to destination by transferring from loop to loop as required.

Still, most of the growth in communication facilities was in telephony, broadcasting, and other voice services. In April, 1973, the Bell System announced plans for field trials of a new transmission medium – the millimeter wave guide, capable of carrying some 230,000 telephone conversations at once.

279

Gas Discharge Television Panel

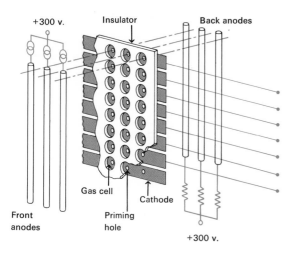

A thin, experimental gas-discharge panel, *right,* may replace the bulky cathode-ray tubes in television sets. Cathode current to a back anode primes a gas cell through the tiny priming hole, *above.* A front anode can then ionize the gas so it glows. Increased current strength brightens the glow so that variations in the current can give the bright and dark areas a TV picture requires.

Communications

Continued

Domestic satellite. Ten years after launching Telstar, the first active communications satellite, the United States appeared ready to start a domestic satellite system. The FCC has paved the way for a number of common carriers to own and maintain communications satellites.

The FCC action followed Canada's successful launching of its second communications satellite, Anik II, in April, 1973. Commercial service began in June.

Anik I, the first of Canada's satellites, was launched in November, 1972, and began service in January, 1973. Some U.S. common carriers are leasing circuits from these satellites until the U.S. domestic satellites begin operating.

Optical communications. Many scientists expect optical communications to fill much of the growing communications needs. The transmission medium would be optical fibers, perhaps of glass or quartz. Judging by the research progress and accomplishments in 1972 and 1973, they may be right.

Signal losses in optical fibers can now be held almost routinely to less than 10 decibels (db) per kilometer (km) in cer-

tain parts of the optical spectrum. Up to mid-1973, the best value announced was from the Corning Glass Works, Corning, N.Y. It reported loss of only about 2 db/km in one segment of the spectrum. If water were this transparent, we could see all the way to the bottom of the deepest ocean.

Optical fibers of this sort have two components: a cylindrical core and a cladding, or layer that surrounds it. The cladding's index of refraction must be less than that of the core so that a ray of light that strikes the core-cladding interface is reflected back into the core. The fiber thus guides the ray. Two problems limit the fiber's ability to carry information over long distances. Absorption or scattering cause loss of signal, and dispersion causes the train of light pulses to smear as they travel along the fiber.

Pure core. To reduce absorption, the core material must be as pure as possible. Also, to reduce light scattering, it must be as uniform as possible and free of bubbles and imperfections. Quartz has low loss, but also a low index of refraction. Few materials are compatible with

Communications

Continued

it, yet have a still lower index, so they can be used as cladding.

An important step forward was the discovery that certain borosilicate glass has an index lower than fused quartz. Fibers with borosilicate glass cladding were prepared and studied at Bell Telephone Laboratories in 1972.

A completely different and ingenious way of constructing optical fibers made of a single material has also been demonstrated at Bell Laboratories. It has three components: a tube, a solid inner rod, and supporting plates inside the tube that keep the rod centered. This configuration is preserved as the assembly is heated and drawn to a fiber about as thick as a human hair. The solid-glass rod becomes the guiding core of the fiber.

Dispersion, the second handicap, is least in a fiber whose index of refraction decreases gradually from its central axis toward the edges. Japanese scientists experimenting with such fibers have reported significant improvement in their transmission characteristics.

Two notable anniversaries. A quarter of a century ago, physicists John Bardeen, Walter H. Brattain, and William Shockley discovered the transistor effect. They were awarded the Nobel prize for Physics in 1956. Also 25 years ago, mathematician Claude E. Shannon published his first paper on information theory, now a classic.

The transistor is now used in products ranging from tiny hearing aids, radios, and heart pacemakers to large computers, electronic switching systems, and space capsules.

Shannon's theory of information also had an impressive impact on telecommunications. It still provides a theoretical framework to tell the communications scientist and engineer what is physically possible, and to guide his attempts to design better communications systems. It tells, for example, how far practical communication systems are from the ideal. Attempts to approach closer to the ideal still continue. New codes permit a closer approach to Shannon's upper limit on the information rate, and improved methods have been proposed for receiving digital signals over noisy channels.　　　[Solomon J. Buchsbaum]

Aluminum-coated disk, similar to an LP record, is a competitor for home videotape equipment. It carries 90 minutes of color TV programming. The disk's "pick-up" is a helium-neon laser beam.

Drugs

The Food and Drug Administration (FDA) banned the synthetic hormone diethylstilbestrol (DES), which was being used to fatten farm animals intended as human food. The animals got DES in their food or from a long-acting pellet implanted under their skin. On Aug. 4, 1972, the FDA prohibited DES as an additive in cattle feed and on April 27, 1973, rescinded approval of its use as an implant in cattle and sheep. This leaves no approved veterinary uses for DES in animals. The FDA stopped its use in poultry in 1959 and has never allowed it in swine.

The latest bans were based on the strict provisions of the Delaney Amendment of the Federal Food, Drug, and Cosmetic Act. It forbids administering any cancer-producing drug to animals if residues of such a drug can be detected in any edible part of the animal.

Although several studies had shown that DES causes mammary tumors in certain strains of mice, residues had not been found in livestock fed or implanted with it. In July, 1972, however, the U.S. Department of Agriculture announced that a new, more sensitive method of analysis could detect DES in the livers of steers fed this drug. In April, 1973, it reported the same findings for animals with implants. The amounts found were about one-half part per billion. Since companies licensed to manufacture and use DES as a growth promoter could not show that residues were not present, DES was banned. In his rulings, however, the FDA commissioner said there was no evidence that DES was a public health hazard because no human harm had been demonstrated in more than 17 years of its use.

Methadone restrictions. On March 15, 1973, the FDA and other government agencies tightened the regulations governing the use of methadone in treating heroin addicts. This synthetic narcotic analgesic, an addicting pain reliever, has been used increasingly during the last 10 years to treat heroin addicts, either to gradually take them off the drug so they suffer no withdrawal effects, or as a substitute for heroin.

The new rules limit the distribution of methadone to hospitals, approved treatment-rehabilitation centers, and specially authorized community pharmacies in remote areas. Persons entering methadone maintenance programs will be supervised more closely, and dispensing the drug for home use will be sharply curtailed. This system was adopted to curb a growing black market.

Prostaglandins. In October, 1972, the Committee on the Safety of Medicines in the United Kingdom (the British equivalent of the FDA) approved two prostaglandins – Prostin E_2 and Prostin F_2-alpha – for use in selected hospitals and clinics to induce labor and terminate pregnancy. The clearance of the first two prostaglandins for routine, nonexperimental use may herald the introduction to clinical medicine of other prostaglandins. The prostaglandins, a family of fatty acids containing 20 carbon atoms, were described in 1933, although their structure was not known until 1962.

More than a dozen closely related prostaglandins have now been found in many tissues and organs in man and other mammals. Among these are extracts of the prostate gland, which gave rise to their name. Several methods for synthesizing these substances have been developed and a new, rich source is the sea whip, a species of soft coral found in the Caribbean Sea.

Although similar chemically, individual prostaglandins often have quite specific – even opposite – effects on different biological systems. They can stimulate or inhibit contractions of the uterus and the gastrointestinal tract, elevate or reduce blood pressure by changing the muscular tone of blood vessel walls, enhance or depress neural activity in the central and peripheral nervous system, and alter many metabolic and endocrinal functions.

Results of intensive investigations now being conducted with these substances suggest that they may play a part in regulating many basic physiological processes and in determining an organism's response to external changes such as infection and injury. In 1970, John R. Vane and his colleagues at the Royal College of Surgeons, London, showed that aspirin may reduce pain, lower fever, and lessen inflammation by inhibiting the synthesis of prostaglandins, which might cause these symptoms.

The naturally occurring prostaglandins and scores of synthetic analogues, or close chemical relatives, are being

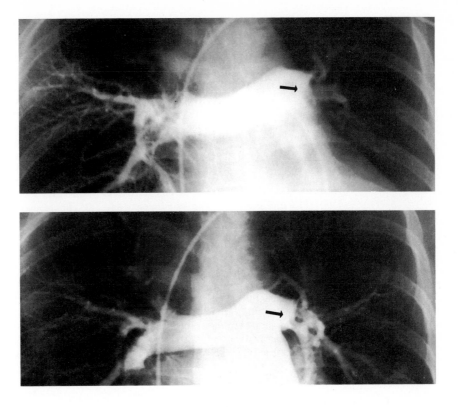

When a blood clot plugs a lung's blood vessel, *top*, an injection of the experimental drug urokinase dissolves the clot so blood can flow to the lungs again.

Drugs

Continued

tested for their potential as therapeutic agents. For example, two prostaglandins reduce acid secretion by the stomach and thus might benefit patients with peptic ulcer. Other prostaglandins relax the muscles in the bronchioles, the air passages in the lungs, and could aid asthma sufferers.

FDA's new role. On July 1, 1972, the Division of Biologics Standards (DBS), which regulates the production of biological products, was transferred from the National Institutes of Health to the FDA and renamed the Bureau of Biologics. This action followed charges that the DBS was seriously mismanaged.

The FDA initiated a comprehensive review of the efficacy, safety, and labeling of all biological products that have been licensed for human use during the past 70 years. Advisory panels of experts were being formed in 1973 to set standards and develop methods for evaluating various categories of these products, including bacterial vaccines, viral and rickettsial vaccines, antitoxins, toxoids and toxins, antivenoms, and blood and blood derivatives.

New drugs introduced in 1973 included:

- Hyperstat (diazoxide), a potent antihypertensive for emergency reduction of blood pressure in acute cases of high blood pressure. It may cause the body to retain sodium and water and raise blood sugar levels, but it often lowers dangerously high blood pressure when injected rapidly into a vein.

- Cleocin (clindamycin phosphate), a new antibiotic that is effective against several pathogens that cause serious infections. Cleocin appears especially useful in treating infections caused by anaerobic bacteria, those that thrive without oxygen, such as *Bacteroides* most commonly found in patients after surgery or during drug or radiation therapy that impairs their normal defense mechanisms against infection.

- Ancobon (flucytosine), a new oral drug for treating extremely serious internal fungus infections. It has been used against the fungi *Candida* and *Cryptococci* when the only other systemic antifungal agent, amphotericin B, was ineffective. [Arthur Hull Hayes, Jr.]

Ecology

Spiders are among the most important insect-eating predators in nature. Yet, few attempts have been made to measure the ecological significance of spider predation. Ecologists Bennett C. Moulder and David E. Reichle of the Oak Ridge (Tenn.) National Laboratory reported in the fall of 1972 on their study of energy transfer in forest-floor food chains dominated by spiders.

The ecologists ranked spiders from five species into three weight categories: small (less than 1 milligram), medium (1 to 10 milligrams), and large (more than 10 milligrams). They conducted their study in a forest area where trees had previously been tagged with a long-lived radioisotope, cesium-137.

Since the radioactive substance spread throughout the trees, the leaves that fell to the ground were radioactive. As they decayed into the soil or were eaten by insects, the radioactive tracer spread throughout the forest food chain.

Once a week for a year, the scientists measured radioactivity in forest spiders and their prey. This enabled them to tell what the groups of spiders ate, because the radioisotope concentrations in prey species were similar to those found in the spiders that ate them.

At the same time, the ecologists conducted laboratory experiments in which crickets tagged with radiocesium were fed to forest spiders. By checking radioactive concentrations in the spiders, the scientists were able to determine the amount of food energy, or calories, the spiders ingested and assimilated. Also, the rate at which the spiders eliminated the radioactive cesium reflected their rate of metabolism, the process of converting food to energy.

Moulder and Reichle found that medium-sized spiders, the group with the greatest population, exerted the greatest impact of the three groups. They accounted for 57 per cent of the total prey killed, 59 per cent of prey eaten, and 55 per cent of the energy output of spiders. Large spiders, much less numerous and with a slightly lower metabolic rate, were the least important of the three groups. They accounted for 17 per cent of prey killed, 14 per cent of food eaten, and 13 per cent of the energy expended. Small spiders, which were more numerous than the large-spider group, were responsible for 26 per cent of the prey

eaten, and 32 per cent of the total energy output. As a group, spiders annually consumed 43.8 per cent of the mean annual number of invertebrates, mostly insects, on the forest floor.

The scientists also found that how much the spiders ate varied according to the season. Small spiders ate the greatest amount of food in summer and the least in winter. Medium-sized spiders ate the greatest amount in the spring and the least in the winter. Large spiders consumed the most food during summer and the least during fall.

There was also a seasonal pattern to the amounts of energy expended by the spiders. In fall and winter, small spiders took in more calories than they used. In the spring, they consumed the same number of calories as their energy output, but during the summer they used more energy than they took in. This meant that the small spiders could store food, or grow, or both only during the fall and the winter.

Medium-sized spiders took in more calories than they used during all four seasons, growing continuously throughout their lifetimes. Large spiders consumed calories in excess of their energy losses in winter, spring, and summer. They had an energy deficit in the fall, probably because of the reproductive cycle. Some of the spider species tested die after mating and reproduction in the fall, and those spiders that are weakened and dying eat less.

In addition to providing new and detailed information on food chain processes in ecosystems, this study demonstrated that spiders grouped according to size functioned similarly, even though they represented five different species.

Desert reptiles. Scientists have long attributed the presence of large reptiles in deserts to their ability to conserve body water. In late spring, 1972, ecologist Brian Green of Adelaide University in South Australia reported on his studies of water loss by the sand goanna (*Varanus gouldii*), a large carnivorous lizard, weighing up to 4½ pounds. Green studied these reptiles at two sites in South Australia, Kangaroo Island and Calperum Station, near Renmark.

He captured and injected lizards with a radioactive saline solution. Before releasing them, he marked the animals. He periodically recaptured them, took

Ecology

blood samples, then gave them another injection of radioactive saline and took a second blood sample. By measuring the difference in radioactive concentrations between the first and second blood samples, he was able to determine the amount of water the reptile lost during the release period. Green also measured temperature and humidity inside and outside the animals' burrows in order to relate water loss to local conditions as well as to the animals' activity.

At the Kangaroo Island station, the daily water losses ranged from 0.186 ounce per 2.2 pounds of body weight during winter to 0.74 ounce per 2.2 pounds of body weight during summer. The rates of water loss at Calperum ranged from 0.267 ounce per 2.2 pounds of body weight during winter to 1.36 ounces per 2.2 pounds of body weight during summer.

Green's studies showed that the most active lizards at Calperum lost as much or more water than goannas at Kangaroo Island even though they were only half as active as those at Kangaroo Island. Green concluded that this was probably

due to the hotter, drier environment at Calperum. Prior to his field study, he had placed animals from the two field localities under the same laboratory conditions and found no difference in the rates of water loss. Green's study also showed that the least active animals at Calperum Station could effectively regulate their water loss by remaining in their humid burrows.

Site fixity. In late spring, 1972, turtle ecologists Archie and Marjorie H. Carr of the University of Florida provided new information on the homing mechanisms of the green sea turtle, after 11 years of tagging and studying turtles on the Caribbean coast of Costa Rica.

They found that the turtles tend to clump their nesting sites, building the largest number of egg nests in one 3-mile stretch of beach near Tortuguero. Each turtle does not come back to a specific spot on the beach, as had been supposed, nor do the turtles scatter in random fashion along the entire 22-mile stretch of turtle beach rookery.

The Carrs questioned how turtles can consistently return to a 3-mile-long nest-

Two forestry students from West Virginia University set up a series of solar cells to measure light intensity at various places on the floor of a forest. Their data can help foresters decide what seedlings will grow best, and when, in given areas.

Ecology

ing site that is constantly being resurfaced by winds and waves and changes in beach vegetation. They believe that the nesting site is imprinted, or fixed, in the turtles when they are hatched and persists throughout the turtle's life. The Carrs call this "site fixity" and point out that by going back to a natal site, a turtle can both retain a favorable nesting place and find a mate.

Red oak study. In the fall of 1972, ecologist Harrison L. Flint of Purdue University reported on his study of cold hardiness of young red oak (*Quercus rubra*) twigs to determine differences in cold hardiness and its role in determining the range of the trees. Flint studied trees grown from seeds collected in 38 different geographic regions. The trees had been planted on a single site in Indiana in 1950 and 1952.

Flint began his study in 1968. He collected twig samples for two years in fall, winter, spring, and summer to see if hardiness had changed during the year. He removed twigs representing the latest growth and divided the twig samples from each tree into five groups. All sample groups were refrigerated overnight at about 41°F. The following morning, control groups from each sample were removed and brought back to room temperature. The other sample groups were placed in freezers preset at 41°F. The temperature was gradually lowered, and twigs from the remaining samples were removed at different freezing temperatures, ranging from 23°F. to −54.5°F. The freezing temperatures Flint selected included those that would kill the trees during the particular season in which the samples were taken.

He established an index of injury, guidelines enabling him to determine the extent of injury caused by freezing, by making a solution containing sap from the twigs and then passing an electric current through it. The greater the amounts of current conducted by the solution, the less the freeze damage.

Flint found that trees from northern origins showed more cold hardiness in the fall and winter than trees of southern origins. But these differences tended to disappear in the samples that were taken in spring and summer.

Based on a detailed statistical analysis of all the samples, Flint concluded that cold hardiness is not now an important factor in controlling the distributional range of red oaks. The wide geographic variation in twig hardiness suggests that natural selection in the red oak has kept pace with climatic changes in the areas where it grows. Twig hardiness appeared to be greater than necessary for the red oak to survive in the climates from which the sample trees were taken, meaning that the hardiness of established trees is probably no longer an important factor in the survival of the red oak. Other phenomena, such as pollination, production of seeds, and soil conditions, appear more important than cold hardiness.

Ants and trees. An unusual instance of a beneficial insect-plant relationship was reported in late summer, 1972, by ecologist David H. Janzen of the University of Chicago. In the Nigerian rain forest, a small tree, *Barteria fistulosa*, grows along trails and in heavily logged areas. Many of the trees have a colony of ants, *Pachysima aethiops*, living in the naturally hollow branches and feeding on scale insects or fungus. Certain insects, such as caterpillars, adult beetles, grasshoppers, and locusts, are currently a threat to these trees. Janzen discovered the importance of the ants in protecting the trees against these insects by comparing insect damage to trees that have no colonies of ants with trees occupied by the *Pachysima* ants.

Tree saplings with an ant colony had an average of 61 leaves, 6 living branches, and 9 leaves per branch. Saplings without ants had an average of 13 leaves, 2 living branches, and 4 leaves per branch.

The beneficial relationship between ant and tree apparently goes back to past mammal feeding pressure. Even elephants, when they lived in the area, avoided browsing among these *Barteria* saplings, although the shoot tips were palatable, because *Pachysima* ants fall on elephants or any other large object that disturbs the foliage. According to Janzen, the ants' piercing sting causes a deep, throbbing pain and sore and stiff muscles, lasting one or two days. Painful memory probably prompts mammals to avoid the tree.

The ants also are diligent in keeping the leaves of *Barteria* cleaned of plant matter that falls from taller trees. The ants literally pick up debris, carry it to the edge of the leaf, and drop it over the side.   [Stanley I. Auerbach]

Electronics

Electronics manufacturers were asked in 1973 to help solve some of the problems that automobiles present to society. For years, they had sought increases in the use of electronics in cars, but the auto industry resisted, mainly because of the cost.

New federal safety and pollution-control standards are now radically changing the picture. Most emphasis in trying to control engine emissions is now on the catalytic converter, a chemical approach. Yet, electronics is the ultimate answer, many specialists say. They believe the specified emission levels can be maintained for 50,000 miles of driving only by improvements in such electronic devices as voltage regulators, ignition systems, and fuel controls.

In somewhat of a pioneering effort, the Chrysler Corporation featured all solid state ignition systems in its 1973 cars and trucks. Other manufacturers were following suit in their 1974 models.

Safety electronics. A federal regulation requires seat-belt interlocks in all 1974 cars built or sold in the United States. The interlocks prevent a driver from starting the engine unless he and

any right-front-seat occupant weighing more than 47 pounds have fastened their lap and shoulder belts. Integrated circuits linked to seat sensors handle the signals that result as the seats are occupied and the belts are fastened.

Laser applications. The laser, once described as "a solution looking for a problem," has a number of new and intriguing applications. Bendix Recognitions Systems, Southfield, Mich., has developed a laser-based automatic baggage-sorting and handling system for airline terminals. It identifies, sorts, and routes baggage at speeds of up to 300 feet per minute. The first model was being installed in the Eastern Airlines terminal in Miami.

A label applied to each item of baggage can be coded for more than 1,000 separate flight numbers or destinations. As a bag passes by on a conveyor belt, the gas laser system reads the code. Electronic circuits relay coded data to an automatic sorting device that routes each bag to the proper spur line.

Serving the arts. A professor at the University of California's Scripps Insti-

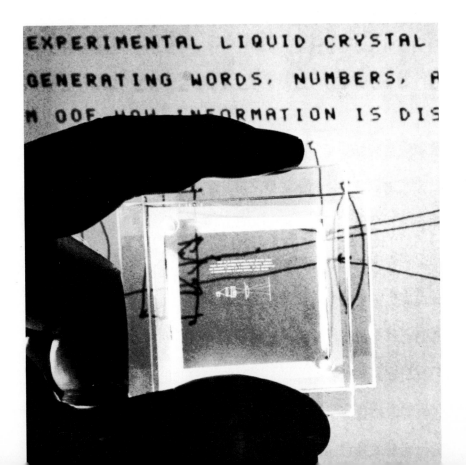

Laser lettering on a liquid crystal slide, foreground, can be projected on a wall screen, background. A small voltage erases image from the slide.

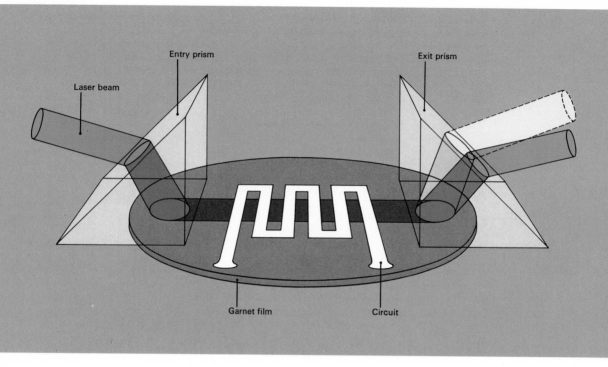

Laser beam

Entry prism

Exit prism

Garnet film

Circuit

Electronics

Continued

New electronic switch may put phone calls and other information on a beam of laser light. A magnetic field from the tiny circuit can switch beam from one position to another in the exit prism.

tution of Oceanography in La Jolla has developed a way to use a laser to clean and protect sculptured works of art (see LASERS TODAY: STILL NOT IN FULL FOCUS). The laser can also be used to record their images in three dimensions as a template, or pattern, for future guidance in restoration.

Geophysicist Walter H. Munk became interested in the problem in Venice, Italy, where pollution is slowly destroying priceless art. He found that laser bursts cleaned marble pieces that were almost blackened by carbon, soot, chalk (calcium carbonate), and other oxides. On some Venetian sculpture, the layer was about one-third of an inch thick. Munk and his co-workers found that a ruby laser readily removed the deposit and had little effect on the light marble.

Using a laser to make a hologram of a sculpture could permanently record the detailed image of the work in three dimensions. Thus statues such as Michelangelo's *Pietà*, attacked and damaged by a deranged man in 1972, could be accurately restored by comparison with the hologram-template.

Medical applications of lasers are also increasing. Russian doctors have developed a nonsurgical method of treating glaucoma, an eye ailment that often leads to blindness.

The disease starts when passages for the eye's lubricating fluids become blocked. The pressure that builds up in such cases can permanently damage the optic nerve. Surgery can be used to cut new passages, but it is risky and may scar eye tissue. Russian doctors have used a ruby laser to punch through the blockage. According to the Soviet inventor, Michael M. Krasnov, the method is painless, because the laser bursts last for only 20 billionths of a second. The treatment must be repeated at two- to eight-month intervals.

New laser. Researchers at the University of Illinois, Urbana, reported in March, 1972, on developments with a semiconductor laser that produces visible light at room temperature. The device is made of an alloy of indium, gallium, and phosphorus. Another team at RCA Laboratories in Princeton, N.J., developed a junction laser diode struc-

Electronics

Continued

ture of aluminum and gallium arsenide that emits visible red light continuously at room temperature. Such devices may lead to the development of simple and inexpensive lasers for use in communications and displays.

Laser photograph plans were announced in April, 1973, by the Associated Press (AP). The new system, which the AP is calling Laserphoto, will deliver dry, glossy prints cut and stacked at a newspaper editor's desk after they have been transmitted by wire from distant locations.

In the electronic darkroom of the new system, a laser beam will expose dry silver paper. With no liquid chemicals used, pictures will be processed by exposing them to heat. Installation of Laserphoto equipment is to begin in 1974. The process was developed by Massachusetts Institute of Technology professor William F. Schreiber.

Bubble memories. Magnetic "bubble" technology advanced in 1973. So-called magnetic bubbles are locally magnetized sites in thin plates of materials called orthoferrites. Moved around on the surface by magnetic fields, the bubbles can be used to store and process digital data. They require little power, and can store large amounts of data in a very small volume. Potentially, they could replace the bulky tape, disk, and drum memories that computers now use.

But the spontaneous appearance of mysterious "hard" bubbles had plagued operation of the bubble memories. They had different properties than the desired "soft" bubbles representing the data bits, and so made the memory inoperable.

Bell Labs scientists discovered in 1973, however, two ways they could suppress the hard bubbles by tailoring the orthoferrite material. In one technique, they formed a sandwich composed of a high-magnetization material atop a low-magnetization material grown on a nonmagnetic substrate, or foundation. In the other technique, they implanted hydrogen ions in the surface of the ferrite.

In April, 1973, Bell Labs announced it had developed a prototype, 1.15-million-bit bubble memory. The entire modular memory is formed from individual 20-kilobit bubble "chips."

Motorized wheel chair that is controlled by the occupant's eye movements is being tested for persons who are unable to use their arms and legs.

Electronics

Continued

Scientists at IBM Corporation's Thomas J. Watson Research Center, Yorktown Heights, N. Y., found they can make bubble memories of thin films of amorphous materials on glass and plastic substrates, instead of the more complex single-crystal garnet materials Bell Labs used. The IBM team developed a way to evaporate amorphous films of gadolinium-cobalt and gadolinium-iron with the necessary magnetic properties onto various substrates. The technique could lead to simple manufacture of large bubble memories. See PHYSICS, SOLID STATE.

Computer technology progressed at both the large and small ends of the scale. Low-cost, limited-capacity "micro-computers," and a superfast, large capacity system, reportedly the largest and fastest general-purpose machine ever built, were introduced in 1973.

Microcomputers have capabilities between disk calculators and minicomputers. Many new companies offered microcomputers. Many of them were priced between $50 and $100. The little devices open up whole new areas for computer applications – areas where even the smallest minicomputer available today can provide much more capability than is needed.

The microcomputers are typically being incorporated in "smart" terminals, units that communicate with larger computers but have some processing power of their own. They are useful in retail stores, bank-teller stations, gas stations, and many industrial operations where sophistication and high speed are not essential.

At the other end of the scale, a new, large-scale computer was announced in March, 1973, by Amdahl Corporation of Sunnyvale, Calif. Gene M. Amdahl said the Amdahl A System would be faster than the IBM 370/165.

The machine is designed around a family of very fast large-scale integrated circuits of a type called emitter-coupled logic. It has a system cycle time of less than 25 billionths of a second. The standard main memory consists of two modules, each storing 1 million 8-bit bytes. Six more modules can be added to the unit, bringing the total to 8 million bytes.   [Samuel Weber]

Energy

The energy crisis continued unabated into 1973. Its relief appears clearly to require new technology, public education, and changes in laws and economic and regulatory policies.

Increasing marginal domestic oil and gas production posed technical and economic problems. The rich Alaskan and offshore resources were under legal constraints and delays because of growing environmental concern.

Oil imports, too, that had been available in the past were in jeopardy because of complex factors. Japan and Western Europe are using more Middle East oil, and the 1973 devaluation of the dollar added to the U.S. problem. Growing Arab sophistication in oil negotiations and recognition that oil in the ground may be a better investment than U.S. dollars in the bank were also factors.

Technology thus sought new energy sources and ways to avoid restrictions on old ones. Some electric utilities installed gas turbines to meet peak summer power demands. Without a practical way to remove sulfur from combustion gases, certain industries and coal-fired power stations switched to oil fuel to meet emission standards. These new uses for gas and oil add to an already serious supply problem. It manifested itself in early 1973, when some schools were closed and some airline schedules were curtailed by fuel oil shortages.

As refineries scrambled to meet fuel oil demands, gasoline inventories were depleted. In late spring, independent gas stations began closing as the supply dwindled and prices rose.

Synthetic natural gas. Of the fossil fuels now in use, natural gas has most appeal. It is the most versatile and economical, and it pollutes the least. But gas supplies are also rapidly diminishing. For that reason, technologists are seeking ways to produce synthetic, or supplemental, natural gas (SNG). A panel of the National Academy of Sciences' National Research Council called, in January, 1973, for more emphasis on developing SNG processes.

SNG can be produced from any fossil fuel. Carbon from the fuel is combined with water at high temperature. Hydrogen from the water combines with the

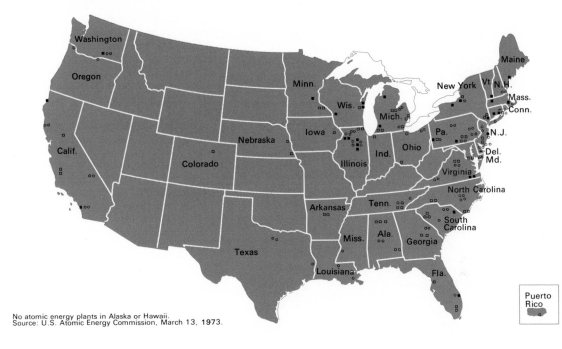

- ■ Operable (30)
  (Kilowatt capacity 15,568,300)
- □ Being built (62)
  (Kilowatt capacity 55,024,100)
- ○ Planned (67)
  (Kilowatt capacity 77,254,700)

No atomic energy plants in Alaska or Hawaii.
Source: U.S. Atomic Energy Commission, March 13, 1973.

Energy

Continued

carbon to form CH₄, or methane, the basic component of SNG.

There are three gasification processes for naphtha, that part of petroleum that boils between 175° C. and 240° C. (about 350° F. and 460° F.). Each uses a catalyst to desulfurize naphtha vapor. Several naphtha-based SNG plants are under construction in the United States. However, this process is considered by many observers to be a stopgap response to the energy shortage because nearly all of the naphtha will have to be imported.

Gasification of coal is more difficult. By some estimates, U.S. coal reserves can last a few hundred years at present rates of consumption. But most U.S. coal can be used only in a gasification process. Either its low heat value makes it impractical to transport and unsuitable for use in existing furnaces or its sulfur content is too high. Sulfur can be removed following gasification.

Five major coal-gasification processes have been developed, four of them in the United States. But only one, the Lurgi process developed in West Germany, has been tested in a small pilot plant.

In the Lurgi gasifier, coal in a moving bed works its way down toward a combustion area while a jet of steam and air or oxygen moves up to the same region. As the coal moves down, it is dried, and volatile components are driven off. By the time it reaches combustion temperatures it has turned into coke. In the combustion area, the carbon in the coke is gasified, turning into methane as it reacts with the steam and air. The El Paso Natural Gas Company has proposed to build a Lurgi gasification plant near Farmington, N. Mex.

Other attempts to find new sources of energy included the development of practical ways to harness the sun's rays and the geothermal heat beneath the earth's surface. See THE POWER ABOVE, THE POWER BELOW.

Home uses. While specialists sought new energy sources, science also looked for thriftier, more efficient patterns of energy use. Household uses account for 20 per cent of the U.S. energy budget. Home space heating and cooling received particular attention. Studies by John Moyers, an engineer at the Atomic

Traditional energy sources are left intact by users of the Pedicar. Senator Alan Cranston (D., Calif.) pedaled one from the Senate Office Building to the Capitol and back in a trial run in April. The Pedicar was invented by a former helicopter engineer, Robert Bundschuh, and is produced by Environmental Tran-Sport Corporation in Windsor, Conn. It weighs 125 pounds and costs $550.

Energy Commission's Oak Ridge (Tenn.) National Laboratory, ranked the efficiency of energy use in window air conditioners. It ranged from 4.7 to 12.2 British thermal units (B.T.U.) of cooling capacity per watt-hour of electricity. Air conditioning is a fast-growing market for electricity, and U.S. power demands are highest in the summer. Buying efficient air conditioners would help to reduce the need for new electric generating capacity.

A Rand Corporation study concluded that electric home heating should be discouraged in favor of gas heating because of electricity production's thermal losses. But critics of the Rand conclusion pointed out that future gas supplies may be derived from coal by conversion processes that also waste fuel. Electric heat utilizes plentiful coal and uranium resources. And using the relatively efficient, electric-powered heat pump in the home offsets electricity production's thermal inefficiency.

Technology assessment, which tries to anticipate all effects of a new technology before it is established, gained importance in 1973. Recognition of the pitfalls of basing energy policy decisions on only direct, first-order effects of a given technology is partly responsible.

Sulfur emissions from coal-burning power stations illustrated the need to uncover second- and third-order effects. The first-order effect of removing sulfur from coal before or after combustion is beneficial—local air quality improves. Electric utilities are finding, however, that new, second-order problems arise: Reduced sulfur dioxide in stack gas makes electrostatic precipitators less effective in removing particulate matter from smoke. Large land areas may be needed to store the sulfur chemicals removed from coal. Low-sulfur coals make more ash and produce less heat.

British researchers Rosemary Prince and F. F. Ross describe a potential third-order effect. They suggest that yields and nutritional value of agricultural products may already be limited in part by sulfur availability. If so, airborne sulfur released from tall power-plant stacks may be a nearly ideal way to distribute it to rural areas. According to British experience, it can be done while ground level, urban-area concentrations are reduced.   [Calvin C. Burwell]

Environment

Efforts to clean the air met a major obstacle in 1973 when U.S. automakers announced that they could not meet deadlines for strict auto exhaust emission controls set under federal law. The Environmental Protection Agency (EPA) and William D. Ruckelshaus, its administrator, granted a one-year extension of the 1975 deadline on April 11, 1973.

The 1970 Clean Air Act requires automakers to reduce polluting emissions of hydrocarbons and carbon monoxide by 90 per cent from their 1970 levels by 1975. Those two pollutants result from incomplete combustion. Another pollutant, nitrogen oxide, which leads to smog formation, was to be cut 90 per cent by 1976. The law allows a one-year extension of these deadlines. This was the extension that Ruckelshaus granted in April.

All four major U.S. auto manufacturers are investigating the same devices for pollution control – catalytic converters. They were to be attached between the automobile's engine and the muffler. Chemicals called catalysts enable the converter to complete the combustion of the exhaust gases and reduce harmful nitrogen oxides to less noxious nitrogen compounds. Carmakers are attracted to the converters because the devices require no change in basic manufacturing processes. Catalytic mufflers will be expensive, however, and will increase fuel consumption. Some observers also predict that greater fuel consumption may cause many mileage-conscious car owners to remove the devices. (See TRANSPORTATION, Close-Up.)

Other approaches. The National Academy of Sciences and many independent scientists have sharply criticized the auto industry for relying on this expensive, relatively inefficient means of meeting pollution-control limits. Many scientists believe that redesigning the engines, although more expensive for the industry in the short run, would eventually result in cheaper, as well as cleaner, automobiles.

Several changes in automobile engines are possible. The least extensive is a variation on a two-stage combustion of fuel, sometimes called the stratified-charge engine. Fuel is first partly burned with very little air, reducing the formation of

Delegates to the United Nations environmental meeting scorned autos and other vehicles that pollute the air by using bicycles to get around the streets of Stockholm.

The "Cost" of Straightening Streams

A natural section of Shoal Creek in Montgomery County, Illinois, contrasts sharply with a nearby channelized section.

Lovely, sparkling Crow Creek, curling across the Tennessee-Alabama state line, became a focus of the environmentalist opposition to government stream channelization in 1973.

It had been a fine place to snag bass, sunfish, or occasionally rainbow trout. But now and then, Crow Creek would spill over its banks and inundate farmland and railroad tracks on its flood plain. The remedy: Straighten, widen, and deepen 44 miles of its channel. Bulldozers and dredging buckets started the job in the summer of 1971.

The channelization seems to have worked, but the price was high: Dredging cost more than $1 million. And biologists from the Academy of Natural Sciences in Philadelphia, walking along Crow Creek's new, bare, eroding banks in March, 1972, found the stream nearly lifeless. Whole populations of fish, insects, aquatic plants, mussels, and crustaceans were simply gone. Crow Creek, they said, was an "ecological disaster" that may take a generation to heal. In the spring of 1973, other biologists found only slight signs of recovery.

Crow Creek is no isolated case. Streams and rivers across the country have recently been channelized or are scheduled for it. This is causing a great deal of controversy.

At the heart of the controversy is a collision of values. In one view, channelization improves on nature. It protects crops planted on ground that may flood, and it helps drain swamps and marshes, creating profitable new farmland. Environmentalists, however, complain that such action destroys scenic stream and wetland ecosystems, habitats essential to many plants and animals.

Between 1950 and the year 2000, observers expect federal agencies will have helped to alter fully 35,000 miles of streambed. The Soil Conservation Service (SCS) has approved plans for about 21,000 miles of work and already completed one-third of that. Thus, the SCS, once considered the conservationist's friend, has become a major foe in the stream channelization fight.

True, 35,000 miles represent only 1 per cent of the nation's 3.5 million miles of streams and rivers. But most of the

planned channelization–and most of the controversy–is concentrated in the southeastern states.

Testifying before a House subcommittee in March, 1973, scientists from several federal agencies and from the Academy of Natural Sciences outlined what amounted to the biological case against stream channelization. Apart from the direct and immediate impact, it appears that stream straightening and the loss of protective forest canopy have a host of subtle, long-term effects.

Tree removal means lost insects and debris, such as autumn leaves, that form essential links in a stream's food chains. The loss of shade allows water temperatures to rise beyond tolerable levels for many desirable fish. Straightening the stream speeds the flow of water, increasing its power to cut away at unstable banks that are no longer held firm by deeply rooted vegetation.

Erosion accelerates and the water grows muddier downstream. The sediment screens out sunlight needed for photosynthesis and smothers the colonies of bottom-dwelling animals in healthy streams. It covers fish eggs and injures fish by chafing their gills.

Limnologist Robin Vannote from Philadelphia told the subcommittee that it may take half a century or more for a channelized stream to regain all its original vitality. Other scientists point out that the channelization disruption also reaches out to surrounding wetlands.

Environmental groups suggest alternatives, but the momentum of these federal programs is high. Many–although not all–have strong local support.

The stream channelization conflict repeats the basic theme of the bitter fight over the indiscriminate use of pesticides. Each had agriculture and agriculturally oriented business united on one side, environmentalists on the other. Like other forms of agricultural technology, altering natural drainage patterns has its biological costs as well as its economic benefits. And the full magnitude of these costs is just becoming known.

But the benefits are far easier to measure than the costs. Until the scenic and biological value of a forest swamp can be measured as exactly as bushels of soybeans or board feet of cypress, conflicts over Crow Creek and hundreds of other streams will go on.   [Robert Gillette]

nitrogen oxides. The remaining fuel is then burned at a lower temperature with a great amount of air, completing the combustion and reducing the amount of unburned hydrocarbons and carbon monoxide released. Japanese automakers are expected to try different versions of this general concept. Some 1973 autos made by Honda use it, and they meet the 1975 U.S. air-pollution standards. A similar technique may be applied to Wankel engines. Wankel Japanese Mazda autos, which use a kind of afterburner, have also met 1975 emission standards.

Diesel engines, which are relatively common in European passenger cars, may also be able to meet the 1975 and 1976 standards. General Motors Corporation was already marketing a diesel auto in Europe that meets the emission standards in the United States.

More radical approaches to auto-pollution control are proposed as cheaper and more efficient in the long run. Included is one that will abandon the internal-combustion engine completely. Among the substitutes that are being considered are gas- or steam-turbines and electric engines.

Some observers have suggested that a varied mixture of vehicle types could serve society better than the auto industry's present, single-type program. They suggest, for example, electric vehicles for city use, short-range freight deliveries, and public buses; steam or turbine cars for suburban or intercity use, and so on. Such proposals generally include much stronger financial support for buses and other forms of mass transit.

Several of these alternatives are under study in Japan. The Japanese government indicated it has no plans to relax its own 1975 and 1976 pollution standards. They had been closely patterned after those of the United States, but Japanese automakers apparently will be able to meet them.

International efforts to protect the environment may have begun a new era in June, 1972, when the first United Nations (UN) Conference on the Human Environment completed its work in Stockholm, Sweden. Participants and observers generally agreed the international conference had accomplished most of what was planned. Its principal achievement was the establishment of a new Environment Program, which

Environment

Continued

would draw environmental activities of existing UN agencies together. It is to conduct a worldwide information and monitoring program called Earthwatch, which is designed to gather, compile, and distribute more information about the global environment and to warn of impending environmental disasters. In December, 1972, the UN General Assembly approved all the recommendations that the conference made.

Headquarters for the Environment Program is being established in Nairobi, Kenya. This choice reflected the very strong interest that the countries of Africa, Asia, and Latin America are taking in the new UN program. In General Assembly votes leading to the establishment of the Environment Program, these Third World countries, aligned with neither the United States nor the Communist bloc, voted together in a rare show of unanimity. In part, this reflected their suspicion that the more industrialized countries will use environmental worries to slow the development of potential economic competitors in the poorer nations of the Southern Hemi-

sphere. The UN established a 58-member governing council to set policy for the Environment Program. Of the 58 seats, 39 were allotted to Third World nations. The action gave this group a decisive voice in the program's activities.

Atomic safety. Public hearings on the risk of serious accidents in nuclear plants forced the Atomic Energy Commission (AEC) to retreat from previous positions and to require greater plant safety. On April 16, 1973, the AEC regulatory staff concluded testimony recommending design changes that would enhance the safety of nuclear plants, but would also reduce their power output. The drop in power would be as much as 20 per cent for some existing plants, according to the AEC, and would cost $193 million.

In June, 1973, the U.S. Court of Appeals for the District of Columbia delivered another severe setback to the nation's nuclear-power program. It ordered the AEC to prepare an environmental impact statement for its proposed new generation of nuclear fast-breeder reactors. These are power plants that generate more fuel than they burn. They

"You've got emphysema."

are needed to soften the impact of uranium shortages expected in the 1980s. The court decision, in a suit brought by the Scientists' Institute for Public Information, opens this program to scrutiny by the public and by federal agencies. The decision has strong impact on nuclear power as a whole; it will cause delays and may force cancellation of the program to build fast-breeders. The suit by the scientists' institute cited many unresolved safety questions and the difficulty of disposing of nuclear wastes. These problems were mentioned by the court in support of its decision.

Opposition grows. Despite such moves to quiet criticism, opposition to nuclear power continued to grow. Bills designed to halt the construction of nuclear power plants were introduced in the U.S. Senate and several state legislatures.

In Sweden, the parliament defeated an effort to halt the construction of all nuclear power plants until unresolved safety questions are settled. However, the parliament withdrew approval for the government's long-range nuclear power plans. In proportion to Sweden's size, these plans were more ambitious than those of any other nation.

Lead pollution. Members of the New York City Public Health Department were among scientists who found lead in canned evaporated milk widely used for infant formulas, and in canned juices for infants and children. But the U.S. Food and Drug Administration (FDA) took little action. Inspection showed about one-half part of lead per million in the milk and juices. This concentration is 10 times the World Health Organization standards for lead in drinking water.

In late 1972, the FDA announced plans to propose that the ratio of one-half part per million would be the most lead allowed in milk. By July, 1973, however, not even this modest step had been formally taken. No action was taken on the lead in canned juices. In both cases, the lead dissolves from solder that is used in cans of a type long-since abandoned by other segments of the food industry.

Ingested lead can result in plumbism, or lead poisoning. Severe cases cause convulsions and brain damage. Extreme cases cause death.   [Sheldon Novick]

Genetics

In the two years since they were discovered, new chromosome-staining methods have allowed geneticists to identify all of the 20 chromosome pairs of the laboratory mouse *Mus musculus*. This has also made it possible to assign genes to their proper chromosomes.

Genes that are on the same chromosome tend to remain together in inheritance. The closer together they are, the less often they are separated by crossing over (an exchange of parts between chromosomes of a pair), and therefore the more they are linked in inheritance. From breeding experiments, geneticists have found 19 linkage-groups of from 2 to 24 mouse genes and have arranged the genes within the groups in their proper order. Scientists at the Jackson Laboratory in Bar Harbor, Me., have been particularly active in this work.

The problem has been to determine on which chromosome a linkage group belongs. All 20 pairs of mouse chromosomes, except one small one, had looked alike when viewed through a microscope. But the new techniques, called quinacrine-fluorescence staining and modified Giemsa staining, produce patterns of light and dark bands unique to each chromosome.

By X-raying the mice, the scientists broke their chromosomes which then could reattach in new ways called translocations. This then changes the pattern of inheritance of the genes on the translocated chromosomes. By correlating the new patterns with the chromosomes that can be seen to have been translocated by the new staining techniques, scientists have now determined which chromosome more than 200 genes are located on.

Gene number. In June, 1973, Burke Judd and his colleagues at the University of Texas, Austin, reported evidence that scientists may be able to count the number of genes of the fruit fly, *Drosophila melanogaster*. No one has had any idea how many genes there are in any organism except in very simple ones, such as viruses and bacteria.

The Texas scientists had previously determined that in one small region of one fruit fly chromosome there are 16 genes capable of mutating to cause either death or an anatomical defect. The chromo-

Genetics

somes in the cells of the fruit fly salivary glands are very large, showing characteristic bands when viewed through a microscope. In these cells, the chromosome region the scientists studied can be seen to have exactly 16 bands, strongly suggesting that a chromosome has the same number of bands as genes. The number of bands on all four of the fruit fly's chromosomes total a little over 5,000. Strangely, if this is also the total number of genes, it accounts for only about 1/30 of the fly's deoxyribonucleic acid (DNA), or genetic material, in the chromosomes.

Normally, living things have two of each gene, one on each member of a pair of chromosomes. Individuals who have too many or too few representatives of some genes have severe physical abnormalities and mental retardation.

Does this mean that every gene has to be present exactly twice? Experiments performed with fruit flies by Laurence Sandler and Dan Lindsley and their associates at the University of Washington, Seattle, and the University of California, San Diego, indicate that the answer is no.

The scientists exposed the flies to X rays, producing translocations in the reproductive cells. By a combination of breeding tests and microscopic analysis of the salivary gland chromosomes in the offspring of these flies they could identify the precise points where the chromosomes had been broken to produce the translocations. By crossing two translocated strains whose chromosome breakpoints are very close together, they could produce flies that had a duplication or deficiency in the region between the two breakpoints. Thus this region of the chromosome was represented three times or only once, instead of the normal two times. By a large number of such crosses, the scientists systematically tested almost the entire length of all the chromosomes in this manner.

As long as the duplicated or deficient region was very small, the offspring flies were essentially normal, indicating that gene deficiencies and duplications are harmful only in large numbers. Evidently, there is a cumulative effect, because each error by itself is too small to produce any noticeable change.

Chromosome likenesses, *right,* show bands that appear on actual chromosomes when they are stained by new techniques. Chromosomes 9 and 19, *below,* and an abnormal joining between them can be identified in this way.

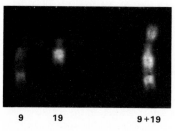

9   19   9 + 19

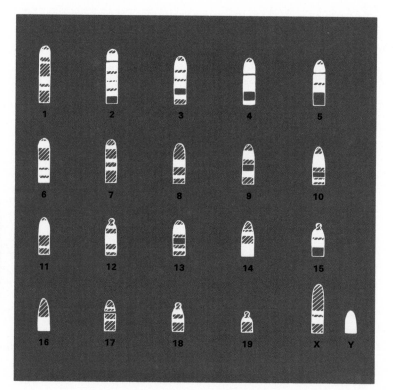

Genetics

Continued

Molecular evolution in a test tube. Molecular biologist Sol Spiegelman and his associates at Columbia University in New York City have isolated and determined the structure of a molecule that offers unique opportunities to investigate evolution at the dawn of life. Even before there were cells, there were nucleic acids, the molecules of which genes are made. These molecules, capable of reproducing themselves, evolved in several ways. They could change their subunits, called nucleotides, and they also added new subunits, becoming longer. Added length meant more genes performing more functions. Ultimately, all this led to cells and the complex life forms we know today.

The Spiegelman group isolated and determined the structure of a small ribonucleic acid (RNA) molecule that can reproduce itself in a test tube. The molecule needs only free nucleotides and an enzyme known as replicase to reproduce. The RNA "organism" then multiplies and evolves, adapting itself to such new environments as changes in chemical concentrations or poisons added to the medium. By studying the specific nucleotide changes that occur when the organism adapts to a new environment, Spiegelman hopes to learn more about how replication occurs. By studying evolution in this cell-free system as the RNA adapts, he may also get some hints as to what evolutionary pressures and steps led to the creation of cells.

Plant hybridization by cell fusion. In August, 1972, botanist Peter S. Carlson and his associates at the Brookhaven National Laboratory in Upton, N.Y., reported that they had hybridized two species of tobacco without pollination.

They first grew cultures of cells from each of the two species of plants. When plant cells are cultivated free of their rigid cell walls, it is not difficult to get them to fuse. They fused cells from the two parent cultures, forming a single hybrid cell. The single cell then grew into a mature plant. Cell fusion should also make it possible to hybridize plants that will not crossbreed naturally. Perhaps we shall someday grow combination plants, such as one with a lettuce top and a carrot bottom. [James F. Crow]

Geoscience

Geochemistry. Data gathered from the Apollo 15, 16, and 17 landing sites was evaluated in 1972 and 1973, and it provided a great deal of information concerning the age and chemistry of the lunar highlands. Many of these results were presented at the Fourth Lunar Science Conference, which was held in March, 1973, in Houston. The conference included approximately 325 reports involving the work of 650 investigators.

Age determinations are among the most interesting results of the site analyses. According to some lunar evolution models, the anorthosites (rocks rich in plagioclase feldspar) known to be abundant in the lunar highlands are primitive crustal rock that should be as old as the moon itself, about 4.6 billion years. The first evidence that this might not be true came in 1972. Potassium-argon dating set the age of the Hadley Rille anorthosite, or "genesis rock," at 4.1 billion years. A single age from a single site could obviously be misleading, but additional dating now indicates ages younger than 4.6 billion years for other highland anorthosites.

Laboratories that studied anorthosites from the Apollo 16 site in the Descartes Highlands reported potassium-argon ages of 3.8 to 3.9 billion years. Because the argon decay product of potassium is a gas, scientists at first had some doubt as to whether these ages date the time the rocks crystallized or the time of a later shock event that caused the rocks to lose all of their gases.

However, later dating by the rubidium-strontium and uranium-lead decay methods, neither of which produce a gas, indicates that the time of crystallization has been measured rather than a shock event. The uranium-lead measurements from several laboratories indicated that Apollo 16 anorthosites are younger than 4.6 billion years, although the small amount of uranium makes these samples difficult to date precisely.

Mitsunobu Tatsumoto and his colleagues at the U.S. Geological Survey determined an age of 3.6 billion years for an Apollo 16 anorthosite. Using the uranium-lead method, they established an age of 3.9 billion years for the Apollo 15 genesis rock. Scientists do not fully

299

A channel on Mars more than 50 miles long was spotted in pictures from the orbiting Mariner 9. Geochemists believe this feature was once formed by running water.

Geoscience

Continued

understand the reason for the differences between the potassium-argon and uranium-lead ages of these rocks. However, the differences may be caused by experimental uncertainties.

The anorthosites contain so little rubidium that they cannot be dated accurately by the rubidium-strontium method. However, plagioclase-rich basaltic rocks that are probably related to the anorthosites have been dated. Scientists found an age of 3.8 billion years for one such rock at the Apollo 16 site. The evidence is thus quite strong that the highland rocks found at the Apollo 16 sites crystallized about 3.9 to 4 billion years ago and are not primitive lunar crustal rocks.

Gerald Wasserburg and his colleagues Fouad Tera and Dimitri Papanastassiou at the California Institute of Technology (Caltech) noted that many rocks from the Apollo 14, 15, 16, and 17 sites appear to have crystallized around 3.9 to 4 billion years ago. To explain the surprising sameness in age at such widespread sites, they reasoned that a lunar "cataclysm" may have occurred at that time. They

believe that the Sea of Rains may have been formed at that time, and that debris from this cataclysm was widely scattered across the face of the moon. If this is true, rocks more than 4 billion years old may well be present elsewhere on the surface of the moon. In this regard, it is well to remember that we have as yet collected no samples from the far side of the moon.

Orange soil. Another significant aspect of lunar research concerned the age of the orange soil found at the Littrow Crater, where Apollo 17 landed. Scientists selected this site in the belief that it might contain some recently produced lavas, which would show that lunar volcanoes continued to erupt after 3.1 billion years ago, the age of the lavas found at the Apollo 12 landing site in the Ocean of Storms. Those are the youngest lavas found so far. When the Apollo 17 astronauts found an orange-colored soil at their landing site, it was thought that this might be oxidized iron from recent volcanic activity.

Laboratory tests showed that the orange color is due to glass rather than

Geoscience

Continued

The alternating bands of black and white in this Mariner 9 photograph of Mars' south polar cap may represent layers of dust and frost.

oxidized iron, while potassium-argon and uranium-lead age measurements indicated that the constituents of the soil are about 3.8 billion years old. These ages are similar to the ages of samples from the Apollo 16 site, and dispel the idea of recent volcanic activity at the Apollo 17 site.

Organic matter. Organic chemists continued to study traces of organic matter that are found in lunar soils. When the soils are put in hot water for 20 hours, small amounts of amino acids are detected. Experiments have shown that these acids are not caused by contamination of the samples on the earth. Another set of experiments has shown that the amino acids did not come from any type of rocket exhaust.

Sidney W. Fox and Kaoru Harada of the University of Miami and P. Edgar Hare of the Geophysical Laboratory of the Carnegie Institution of Washington noted that the ratio of amino acids to total carbon in the lunar soils is close to that observed in the Murchison meteorite, which fell on Sept. 28, 1969, near Murchison, Australia. The ratio differs

greatly from that which might result from terrestrial contamination. The scientists believe, therefore, that the amino acids must have been formed by natural processes on the moon.

It is also significant that the lunar amino acids are found only after the soils are heated in water. This shows that the organic matter is present only as insoluble amino acid precursors rather than as fully formed amino acids.

Free amino acids would decompose on the lunar surface, but the precursors are now shown to be stable enough to escape destruction. In contrast to the earth, the moon has no water, so lunar organic matter would not have evolved beyond the precursor stage. But the finding shows that amino acids, which are so important to life, could have been produced on several other objects in the early solar system. As such, the finding is important to scientists attempting to understand the origins of life.

Venus. On July 22, 1972, the Russian spacecraft Venera 8 soft-landed on the planet and transmitted data to earth for 50 minutes. Aside from measuring the

Geoscience

Continued

Photomicrographs of the orange lunar soil brought back by the Apollo 17 astronauts show that it is composed of glass, not oxidized iron as had first been thought.

high surface temperature (470°C., 878° F.) and pressure (90 atmospheres), it also detected the abundance of some radioactive elements in the surface rocks. This was the first chemical analysis of surface rocks on Venus. The experiment gave values of 4 per cent for potassium, 0.002 per cent for uranium, and 0.0065 per cent for thorium, about 10 times higher than the concentrations of the same elements that have been observed on the moon. However, these concentrations are similar to the amounts of these elements that are found in various samples of granite on earth.

These findings recall the Mariner 9 results showing that dust on Mars has a chemical composition somewhat similar to terrestrial granite. Granite is a highly differentiated rock with respect to the mass composition of the earth.

The data now available suggests that Mars and Venus, like the earth, have undergone extensive chemical differentiation. So extensive was this differentiation that rocks at the surfaces of these planets differ from those that are found in their interiors.

Tektites mystery. Tektites are small, rounded, glassy objects that are found in several places throughout the world. They are strewn in fields, and are generally thought to result from meteorite impacts. For example, the tektite field in Moravia is associated with the Ries-Kessel impact crater, while the Ivory Coast tektite field is associated with the Lake Bosumptwi meteorite crater that is found in Ghana.

There is one large tektite field covering western Australia and southeastern Asia, however, that has no obvious impact crater. Some scientists have suggested that this field might have resulted from something other than terrestrial impact. Geochemist Billy P. Glass of the University of Delaware has identified X-ray diffraction patterns for the minerals zircon, corundum, rutile, monazite, and quartz in a tektite specimen from this field.

These are all common minerals in terrestrial sediments, and this strongly suggests that the tektites from this field also originated from terrestrial impact. However, the origin of the Australian tektites will remain a mystery unless an impact crater that could serve as their source is identified.   [George R. Tilton]

Geology. Many geologists turned their attention in 1972 and 1973 to rocks on the continents that might once have been part of the deep-sea floor. The new studies may provide the information about the nature of such rocks that geologists had sought during the 1950s and 1960s. At that time, they launched a massive attempt to sample the crust beneath the sea floor. The plan, known as Project Mohole, was organized to drill all the way through the oceanic crust and into the upper mantle of the earth. The project included plans for a seaborne drilling platform the size of a football field. Mohole ran into both engineering and political difficulties, however, and the project was abandoned in 1966.

A less-ambitious drilling project, called the Joint Oceanographic Institutions for Deep Earth Sampling (JOIDES), started in 1968, has been spectacularly successful in sampling oceanic sediments. These drilled rock samples and other rocks dredged from ocean seamounts and ridges are the only direct samples that geologists have from the present-day sea floor. But the JOIDES drill has gone only a few feet into the hard rocks beneath the sediments on the ocean floor. Because these dredged and drilled samples come from the uppermost part of the oceanic crust, there is still no direct information about the lower oceanic crust or the mantle.

In 1968, geologist Ian Gass at the University of Leeds, England, convincingly argued that a certain rare association of three types of rock on the island of Cyprus indicated that they were a part of the oceanic crust that had been raised above sea level. The Cyprus rocks are thin banded red sedimentary rocks that probably were once soft muds; lava flows that had a peculiar pillowlike form; and massive rocks that were probably igneous rocks that cooled from a melt, then were greatly altered later by contact with hot water.

Intensive study indicates that the Cyprus rocks match known features of the sea floor in many ways. However, there are some features of the Cyprus rocks that do not match. For example, there are two sets of the pillow lavas on Cyprus, and the upper set does not match the chemistry of the most typical oceanic lavas.

Kirkjufell volcano on Iceland's Heimaey Island, which was dormant for more than 1,000 years, erupted in January, 1973, burying homes and other buildings with its ashes and forcing most of Heimaey's 5,000 residents to move to mainland safety. Molten lava poured from a 1½-mile-long crack.

Geoscience

Study during 1972 and 1973 indicates that there are probably many more regions of sea floor exposed on land. As geologists in other parts of the world became aware of the Gass proposal, reports of similar finds have come in from the Pacific Islands, from the West Coast of the United States, from the Maritime Provinces of Canada, and from the Middle East.

No one geologist has seen all of these sites because it is difficult to travel to some of them. But published reports indicate that these areas seem to have much in common, and the rocks there resemble many of the rocks dredged from the deep sea. Over the next few years, extensive work should determine whether or not these areas are actually uplifted segments of the sea floor.

Submarine geology. A more dramatic way of studying the oceanic rocks has been proposed by geologists from France and the United States. They suggest taking a small unmanned submarine 2 miles below the sea surface to the crest of the Mid-Atlantic Ridge, where they hope to make visual observations, obtain rock samples, and take photographs.

The manned dives would be hazardous and expensive, but observations by a trained observer in a steerable submarine raises completely new possibilities for undersea geological exploration. An area has been selected for the underwater exploration, and some of the most detailed surveys ever made from surface ships began in 1973 as preliminaries to the manned diving program.

Resources. The threat of an energy crisis has intensified the search for oil and natural gas. Geologists used a wide range of geological, geochemical, and geophysical techniques in this search during 1972 and 1973. They expect that the search for oil will become more efficient and less disturbing to the environment as their understanding of earth processes improves.

A few geologists during the year pointed out that oil and gas are only the most obvious of many resources that are becoming more difficult to find. In May, 1973, the U.S. Geological Survey released a 722-page report detailing the estimated future supply of all of the mineral resources needed by our highly industrialized society. We have an abundant supply of some, such as coal, limestone, gypsum, salt, and potassium, and the supply promises to remain abundant for at least the remainder of this century. But others, such as petroleum and copper, are slowly becoming scarce as the richer and the more easily obtained deposits are exhausted. The geological search has, therefore, turned to deeper oil wells, to deeper mines, and to places where there is simply less of the material to exchange for the effort of getting it out.

Copper ores mined in 1900 contained almost 10 times as much copper per ton of rock as do today's copper ores. To slow the trend to leaner and leaner deposits, geologists are now beginning to look for more unconventional sources of copper deposits.

Some evidence exists that copper deposits might have been formed on the sea floor, although the famous copper mines of the western United States were not. However, the copper deposits on the island of Cyprus, mined by the ancient Greeks and still producing today, occur at the top of pillow lavas that resemble deep-sea rocks. Perhaps the copper mines of Cyprus were originally rocks on the open sea floor.

Several indirect lines of evidence suggest that seawater circulates through the hot, young lavas near the mid-ocean ridge. If the seawater dissolves some of the metals from a large amount of hot rock, the heated seawater could deposit those metals where it bubbles up again to the cold ocean floor. Small amounts of such metal-rich deposits have been found near the mid-ocean ridge. Some of these deposits encrust rocks, and some are found in the sediments just above the pillow lavas.

It is also possible that the manganese nodules found at great distances from the mid-ocean ridge might owe their origin to the hot igneous rocks on the ridge. Thick metal deposits have been found on the hot, geologically young floor of the Red Sea. And one of the JOIDES drill cores taken from the Atlantic contained an accumulation of pure copper. The prospects of deep-ocean mining have therefore become more interesting to geologists. Which, if any, of these deposits will be useful is not known, but none of the new possibilities can be ignored.

Diamond pipes. Evidence is accumulating that diamonds are brought to the

Geoscience

Continued

earth's surface by rocketlike eruptions. Diamonds are a phase of carbon that typically is formed at temperatures and pressures comparable to those 90 miles below the earth's surface. They are usually found in subsurface columns of cemented fragments a few hundred feet in diameter that are known as "pipes."

According to a new theory developed by Thomas R. McGetchin of the Massachusetts Institute of Technology, a combination of gases in their fluid phase—probably chiefly water—under extremely high pressure in the earth's upper mantle cause these eruptions that form the pipelike columns. When the pressure becomes great enough to force a passage up through the crust, the material begins fighting its way upward at a relatively modest speed.

Origin of the continents. For many years, it has been customary to think of continents as fixed, permanent objects of great age. However, recent evidence of continental drift, sea floor spreading, and plate tectonics shows that the continents move. And it has also been calculated that the present-day rate of adding material to the continents, if continued over the 4.5-billion-year history of the earth, would equal the present volume of the continents. Therefore, most geologists are convinced that the continents were built up gradually by the addition of crust to their edges.

Richard Armstrong of the University of British Columbia proposed a dissenting view in 1972. He suggests that the continents achieved their present volume early in the earth's history, and continental material has been recycled through the earth's mantle since that time. In Armstrong's view, the "new" lavas added to the continent are actually recycled continental rock.

Judith Schaeffer, an undergraduate student at Princeton University, calculated in 1973 the recycling rates of the major chemical elements between the mantle and continents. Her work shows that Armstrong's theory is probably correct, and that the structure of the continental masses is more easily understood in terms of the efficiency of recycling than in the terms of more traditional geology.   [Kenneth S. Deffeyes]

"It's one of two things—either the great god of the inner earth, Timbuku, is angry with our last virgin sacrifice, or the enormous pressure of a formation of molten rock is breaking through a weak spot in the earth's crust."

Geoscience

Continued

Geophysics. Scientists learned more about what causes earthquakes and how to predict them in 1972 and 1973. Relative motions of the large lithospheric plates that make up the earth's outer shell appear to be the basic cause of earthquakes. Movements of the plates generate seismic waves.

Russian seismologists studying small earthquakes at Garm, in central Asia, in 1968 and 1969, noticed changes in the relative velocities of the two types of seismic body waves, compressional waves and shear waves, prior to a major earthquake. Ordinarily, a compressional wave travels much faster through the earth than a shear wave, arriving at a seismograph station much more quickly than the latter. From weeks to years before a major earthquake, this lead time is markedly reduced. But it returns to normal a few hours before the earthquake takes place.

Scientists from the Lamont-Doherty Geological Observatory of Columbia University observed this effect in the United States in studies of earthquakes in the Adirondack Mountains. Studies of the 1971 San Fernando Valley earthquake by scientists from the California Institute of Technology (Caltech) also revealed a sharp change in the ratio of compressional to shear velocity. The change began 3½ years before the earthquake, and the ratio returned to normal just prior to the quake.

One Russian scientist has suggested that the change in the ratio is caused by an increase in shear-wave velocity. However, scientists from Caltech and Columbia believe that they can demonstrate that a decrease in the velocity of compressional waves causes the narrowing. If this is so, then it will be of some importance to future earthquake predictions, because compressional waves can easily be generated artificially, and can be monitored to determine if changes are taking place.

Dilatancy. The explanation for the observed wave-velocity change lies in the physics of dilatancy. Before an earthquake occurs, when rocks are stressed to half their breaking strength, they may dilate, or form a new series of cracks within themselves. The new crack spaces will reduce the compressional-wave velocity through the rock. Although the stress may continue to rise, the velocity of compressional waves depends on the rate at which ground water fills the newly formed cracks. When the cracks become filled and the compressional velocity returns to normal, the stressed rock suddenly fails, and the earthquake occurs.

Dilatancy initially delays the earthquake by reducing the fluid pressure on the fault, then triggers it when the fluid pressure is restored. The time delay between the two depends on the rate the cracks form and fill with water.

Christopher H. Scholz, Lynn R. Sykes, and Yash P. Aggarwal of Columbia University point out that this model is consistent with other preliminary effects of an earthquake. They add that every one of these effects found in field observations has been reported in rock-mechanics studies in the laboratory. Among these are long-term crustal movements, particularly uplifts, and changes in electrical resistivity.

An increase in the amount of radon in deep well waters, detected by running samples of water through a radiation counter, provides a more striking effect. This radioactive element has a half-life of 3.8 days, so its sudden increase in deep waters can only be related to an increase in the surface area of the surrounding rock caused by dilation or the increased flow of ground water. The dilatancy model predicts both effects.

This model shows great promise for earthquake prediction, although it cannot yet be demonstrated that these preliminary effects occur with earthquakes that take place deeper in the earth than 5 miles. However, this depth range includes virtually all of the world's most destructive earthquakes.

Several earthquakes have been triggered by pumping fluids into deep wells. This apparently lubricates the plates and releases the accumulated pressures. The effects of this high-pressure pumping were studied in 1972 and 1973 at the Rangeley oil field in Colorado. Scientists also suspect that underground water plays a role in creating earthquakes.

Plate tectonics. Further advances were made during the year in understanding the details of plate tectonics, particularly with regard to the development of surface features of the earth. Michael Carr and Richard Stoiber of Dartmouth College studied the relation-

Moving Plates on the Pacific Floor

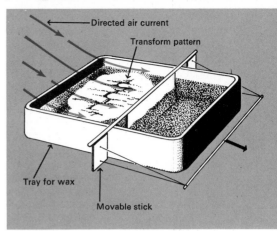

- ← Directed air current
- Transform pattern
- Tray for wax
- Movable stick

Photograph of the pattern formed in a tray of wax, *left,* closely resembles the pattern of transform faults and other features on the spreading ocean floor. The apparatus, *above,* that produced this pattern is designed to show how separation of the earth plates causes these features to develop at the plate boundaries.

Geoscience

Continued

ship between volcanic activity in Central America and the seismic waves caused by the underthrust of the Cocos plate that lies under the ocean west of Central America. The underthrust plate appears to be divided by cross fractures into a set of slabs dipping at varying angles. The volcanoes appear to lie in linear segments above these slabs, and the segments are offset at the cross fractures. Managua, Nicaragua, which was destroyed by a destructive earthquake on Dec. 23, 1972, is located along one.

According to Carr and Stoiber, the energy-release pattern of shallow earthquakes (down to about 45 miles deep) shows well-defined peaks near the cross fractures. The frequency of intermediate earthquakes (between 45 and 190 miles deep) from 1961 to 1962 correlates with volcanoes that had large eruptions during the same time period and to a lesser extent with the total volume of material ejected by the volcanoes.

Records made by the Shell Oil Company off Vancouver Island in 1972 indicate movement in the sediments, and wells drilled later showed high pore pressures in the vicinity of the thrust fault plane. Peter Vail of Esso Production Research Corporation also reported in 1972 that data across the ocean trench and the continental margin of Central America revealed a pattern of rock overlapping that resulted in a thickening of the earth's crust. Vail estimated that the overlapping would eventually result in rock formations similar to those found in Alpine mountain systems.

An up-down map. The National Oceanic and Atmospheric Administration published a map showing vertical movements of the earth's surface in the United States. The maximum rates of vertical movement are about the same as the horizontal plate movements. The vertical motions, which were determined by highly precise leveling equipment, are oscillatory. The vertical movement that can be determined from changes in the rocks is much slower than that observed by the leveling equipment. The relationships between vertical and horizontal movements are not understood, but both are caused by internal forces.   [Charles L. Drake]

Geoscience

Paleontology. The fossil remains of the largest dinosaur yet found were discovered in November, 1971, and are currently being excavated by James A. Jensen of the Brigham Young University Museum at Provo, Utah. The nearly complete fossil skeleton of the new species of sauropod dinosaur was preserved in rocks of the Morrison formation at Dry Mesa, Colo., near Grand Junction. These rocks were formed during the Jurassic Period, which indicates that the fossil is about 150 million years old.

Until now, the world's largest known land animal had been a sauropod called *Brachiosaurus* found in Tanzania, East Africa. A skeleton mounted in a museum in East Berlin measures 42 feet in standing height and 80 feet in overall length. The animal is estimated to have weighed more than 50 tons when alive.

The Dry Mesa sauropod species appears to have been much larger. Although the remains are not yet completely excavated, the shortest vertebra in its lengthy neck is 3 feet long, while the longest is nearly 5 feet. From these bones, a conservative estimate is that this animal stood more than 55 feet tall, which is almost one-third again as tall as *Brachiosaurus*.

Other unusual remains among the 18 tons of fossil bones and rock collected at Dry Mesa include a large carnivorous dinosaur unlike any previously discovered. It may represent a new family of extinct giant reptiles. Two rare turtles, a feathered reptile that was not a bird, a Pterosaur (flying reptile), and two new medium-sized dinosaurs were also found.

Warm-blooded dinosaurs. In July, 1972, Robert T. Bakker of Harvard University published evidence suggesting that some dinosaurs were warm-blooded, rather than cold-blooded animals, as has always been assumed.

He pointed out that the locomotor apparatus of the large dinosaurs, such as the duck-billed dinosaurs, the Stegosaurs, and the horned dinosaurs, was anatomically similar to that of modern large mammals, such as elephants and rhinoceroses. The legs of the large dinosaurs were vertical beneath their bodies, and Bakker says that they probably could move about as fast as elephants. If

Dinosaur bones found at Dry Mesa, Colo., are encased in plaster for the journey to a museum. The large block at the left houses a shoulder blade of *Brachiosaurus*.

Geoscience

Continued

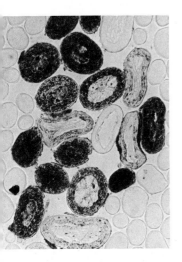

A photomicrograph cross section of hair in which the cellular structure is still clearly seen, is from an antelope that lived in 35,000 B.C. This and other well preserved animal and vegetable matter was found in Border Cave between Swaziland and Natal, South Africa.

so, then the energy the large dinosaurs needed to move at high speeds could only have been supplied by a metabolism similar to that of large mammals.

Warm-blooded animals must eat more than cold-blooded ones to fuel their high rate of metabolism. Bakker cited studies of modern animal communities that indicate that a given number of prey can sustain proportionately more predators among cold-blooded animals than among warm-blooded ones. Studies of large dinosaur fossils also suggest a low predator-to-prey ratio, which compares favorably with warm-blooded animal communities. He also pointed out that large warm-blooded animals that live in warm climates are nearly hairless. Dinosaurs were also hairless. In other words, the largest dinosaurs, which were reptiles, resembled the largest mammals more closely than they did lizards.

Early leaf architecture. Leo J. Hickey and James A. Doyle of the Smithsonian Institution carefully studied the structure of fossil angiosperms, or flowering plants, in 1972 and 1973. On the basis of their study, they classified angiosperm leaves into four levels of organization, and showed how the plants evolved from the earliest angiosperms of the middle Cretaceous Period, about 100 million years old, to such modern plants as oak and beech trees.

The researchers based the levels on the degree of regularity with which structural elements, such as the shape of the leaf edges, vein pattern, and leaf gland placement, were arranged. First-rank, or the most primitive, leaves showed the least regularity. They had simple leaf edges and irregular veins. Such leaves are found today only in a few rather rare groups of plants. Second- and third-rank leaves have greater organization. Fourth-rank leaves, such as those of the modern trees, have regular patterns of veins, and more complex lobed or serrated leaf edges.

Previous examination of fossil pollen from trees of the middle Cretaceous Period revealed the primitive nature of these plants, yet the superficial study of their leaf shapes did not. In fact, earlier comparisons of the middle Cretaceous leaves with those of today suggested that essentially modern families, if not genera, of angiosperms were already present in the early fossil record.

Applying their criteria of structural levels of leaf organization to the fossil record, Hickey and Doyle found that angiosperm pollen and leaves underwent closely parallel evolutionary trends during their early histories. The rise and expansion of the flowering plants apparently occurred in early Cretaceous time, about 130 million years ago, and diversification continued through the middle Cretaceous Period.

Origin of Metazoans. During the past few years, studies have provided new estimates on when and how multicellular organisms developed. The turning point from single-celled to multicellular organisms occurred during late Precambrian times, between 1,500 and 650 million years ago, and scientists in the United States and elsewhere can now outline the major events in this sequence. Several recent reviews of their progress were published in 1972 and 1973 by paleontologist William Schopf of the University of California, Los Angeles.

Schopf reported that the first multicelled organisms, or *Metazoa*, appear in the fossil record rather suddenly, just prior to the beginning of the Cambrian Period, some 600 million years ago. Evidence of earlier organisms comes from a series of fossil-bearing rocks of Precambrian age. The cellular evolution of early life forms can be interpreted from these fossils, and were summarized by Schopf:

1. One-celled, free-floating organisms with a formed nucleus were derived from nonnucleated ancestors more than 1.3-billion years ago. These were asexually reproducing algae.

2. Sexual reproduction, in which the number of chromosomes in the daughter cells was doubled temporarily and then reduced again by cell division, seems to have occurred about 1 billion years ago. Diversity and rate of cell evolution markedly increased.

3. Between 900 and 700 million years ago, these algal cells began to retain the temporarily doubled chromosome condition, so that this became the dominant state of the organisms. *Metazoa* began to develop then.

4. Evolution increased the complexity of the metazoic organisms, and hard body parts developed. These advances seem to have been related to increasing the organism's motion and foodgathering capabilities.   [Vincent J. Maglio]

Medicine

Dentistry. Technological advances in 1972 and 1973 produced several new coatings that bond to the enamel of the tooth and protect it from decay. These composite dental resins consist of an epoxy resin product and a form of acrylic acid, plus reinforcing fillers such as glass beads or quartz fillers.

Dentists can use the new materials to seal pits and fissures when the teeth first grow in, thus preventing decay. The new materials can also be used to restore teeth eroded near the gum where work is normally difficult. And, they can be used to coat malformed teeth to improve their appearance.

New fillings. In January, 1973, the Laboratory of the Government Chemist in London announced the development of a new translucent tooth-filling material. The new fillings, made by mixing a special aluminosilicate glass powder with a solution of polyacrylic acid, have several advantages over conventional enamel fillings made with phosphoric acid. The polyacrylic acid is much milder than phosphoric acid, which can damage the tooth if a cavity is not lined before it is filled. Also, the new materials are adhesive, eliminating the need for extensive drilling to secure the filling.

The new filling materials, now undergoing clinical trials in Great Britain, are expected to be useful in repairing the natural cracks that develop on molar biting surfaces, and in treating tiny cavities that develop at the gum line where drilling is undesirable.

Artificial teeth. In October, 1972, Peter A. Neff of the Georgetown University School of Dentistry in Washington, D.C., reported on an artificial plastic tooth made of Vacalon that, when implanted in baboons, functions much as a natural tooth. New tissue grew around the implants and bone grew through vents in the implant roots, as would occur with natural teeth.

Neff plans continued study of the implants. To be successful, the surrounding tissue must not only accept the artificial material, but the tooth must also continue to function under all types of natural stresses. His research may lead to the development of similar implants for human beings.   [Robert J. Hilton]

Plastic Vacalon tooth appears to have "taken root" six months after it was implanted in a baboon by Georgetown University researcher. Closer look confirms that bone and tissue are growing around it as if it were a normal tooth.

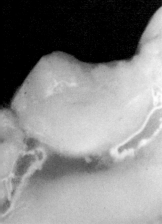

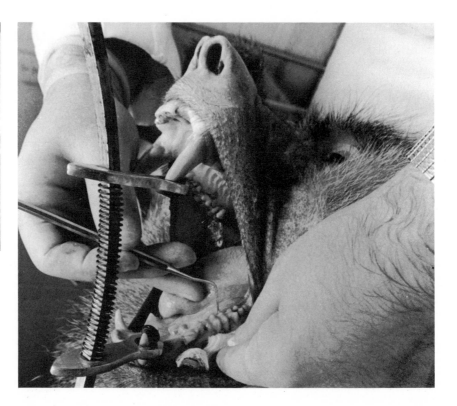

Medicine

Continued

Internal Medicine. In April, 1973, Dr. Elwin E. Fraley of the University of Minnesota School of Medicine reported the discovery of what seems to be the first definite link between a virus and a human cancer. He identified the virus as responsible for bladder cancer, a disease that strikes 20,000 Americans each year and kills about half that many.

The virus is similar to some known to cause cancer in lower animals, but it is unlike those that cause flu and other infectious diseases. It was named EFMU, for Elliott (Dr. Arthur Elliott, who worked with Fraley in isolating the virus), Fraley, and Minnesota University.

Viewed through an electron microscope, EFMU resembles "C-type" viruses that cause such cancers as leukemia in cats, breast tumors in monkeys, and Rous sarcoma in chickens. However, Fraley believes EFMU is an entirely new kind of virus, not a true C-type.

Fraley and his colleagues have studied about 25 patients with bladder cancers. Final results are in on 14, and in 12 cases, the results support the thesis that the virus causes cancer. Identification of the virus as the cause of bladder cancer raises hope that a mass-screening method can be devised to detect this cancer in its early stages or that a protective vaccine can be developed.

In November, 1972, a team of virologists headed by Bernard Roizman of the University of Chicago announced that they had found a fragment of deoxyribonucleic acid (DNA) from herpes simplex 2 virus in a human cervical cancer. Along with this viral genetic material, they found viral transfer ribonucleic acid (tRNA), a product of the virus' genes. This was the first time herpes simplex DNA and its tRNA product have been found in a human cancer. It adds to the mounting evidence that herpes viruses may be involved in cervical cancer.

Cancer diagnosis. In November, 1972, Dr. V. William Steward of the Pritzker School of Medicine at the University of Chicago reported he had successfully used a proton beam to photograph a breast cancer in a living patient. The new method requires a smaller dose for diagnostic use than conventional X-ray techniques, and it gives better visualization.

DR. JOHN SMILMAN M.D.
PEDIATRICIAN

DR. WILLIAM BARNS M.D.
INTERNIST

DR. SAMUEL O. MOSS M.D.
EAR NOSE & THROAT

DR. VICTOR B. ROBBINS M.D.
DERMATOLOGIST

DR. HERBERT WINTON M.D.
SIDE EFFECTS

PROFESSIONAL BUILDING

Drawing by Richter; ©1973 The New Yorker Magazine Inc.

The Food and Drug Administration is currently checking a new diagnostic test for certain types of cancer before approving it for public use. The test, developed in the mid-1960s by Dr. Phillip Gold of McGill University in Montreal, examines a patient's blood for the presence of carcino-embryonic antigen (CEA), which is secreted by certain malignant tumors. The CEA test's present value lies in its usefulness as an adjunct to other tests. A level of more than 2.5 billionths of a gram of CEA per cubic centimeter of blood is considered suspicious and worth further checking.

Doctors are also beginning to use CEA to monitor the treatment of cancer patients. CEA levels generally rise weeks or months before a patient shows a clinically detectable tumor. A high CEA level would lead to further tests; whereas, a normal CEA might forestall unnecessary surgery or other treatment.

Diet and cancer. Cancers of the digestive tract appear to be increasing throughout the world. In July, 1972, Dr. H. Marvin Pollard, gastroenterologist at the University of Michigan, told the Seventh National Cancer Conference in Los Angeles that cancer of the colon and rectum, and cancer of the pancreas are increasing among Americans. He added that the incidence of cancer of the esophagus in black Americans has tripled in the last 30 years. In Curaçao and certain areas of Iran and South Africa, esophageal cancer is also increasing at epidemic rates.

Excessive consumption of certain meats and alcohol, and a deficiency of the trace element molybdenum in the diet may contribute to causing these digestive tract cancers. Dr. John W. Berg, epidemiologist at the National Cancer Institute, noted that studies in France and the United States have linked the use of tobacco plus heavy alcohol intake to cancer of the esophagus. Researchers found that plants in parts of South Africa, including those used as food, contained very little molybdenum. And analyses of water supplies and cancer-incidence data in the United States also showed high rates of esophageal cancer in areas where the water is low in molybdenum.

Berg also reported that clues to the possible causes of cancers of the stomach and bowel have come from a study of Japanese immigrants. Stomach cancer is extremely common in Japan, while bowel cancer is rare. Among native Americans, on the other hand, bowel cancer has been on the rise and stomach cancer has been steadily decreasing. The bowel cancer rate among Japanese who emigrate to the United States soon approaches that of native Americans, and the high rate of stomach cancer also continues. Among second-generation Japanese-Americans, the stomach cancer rate is lower, but the bowel cancer rate remains high. Berg said an early insult or injury to the stomach, probably in the form of irritating foods, may make the individual sensitive to factors that later induce stomach cancer.

In studying the rise of bowel cancer among second-generation Japanese-Americans, Berg explored their consumption of 119 food items. To his surprise, he found six meat items that were associated with a high bowel cancer risk. He had expected to find no more than one or two. Both Argentina and Scotland, which also consume a lot of meat, also have high bowel cancer rates. This does not mean that eating meat causes bowel cancer, Berg emphasized, but something added to the meat may be a contributing factor.

Acupuncture. In February, 1973, Dr. Alfred Peng of Columbia University's anesthesiology department reported that acupuncture produced moderate to considerable improvement in 8 of 10 patients treated for nerve deafness. The patients had been deaf from 5 to 31 years. Five were described as cured or considerably improved after the treatments. Three were moderately improved, and two showed no improvement. Peng cautioned that the findings could not support any broad conclusions about the value of acupuncture.

Cardiovascular disease. A team of California doctors reported in November, 1972, that carbon monoxide from automobile exhaust fumes may aggravate pre-existing cardiac conditions. The team, headed by cardiologist Wilbert S. Aronow of the Long Beach Veterans Administration Hospital, obtained their evidence by driving 10 men with heart disease through heavy freeway traffic in Los Angeles. During the drive, the doctors monitored the volunteers' heart rate and blood pressure and recorded their electrocardiograms. When the 10 men,

Medicine

Continued

aged 40 to 56, breathed the polluted freeway air, they showed higher levels of carbon monoxide in their blood, slower heart rates, lower blood pressures, decreased lung functions, and poorer exercise performances. Electrocardiographic changes in three patients showed that their hearts needed more oxygen than their narrowed arteries could deliver.

Three weeks later, Aronow and his team repeated the study. This time, each patient wore a mask through which he breathed compressed air from a tank. None of the patients showed any worsening of their heart condition when they breathed the purified air.

Despite the dangers of automobile exhaust fumes, Aronow reports that the carbon monoxide levels found in the blood during the freeway study were lower than those found after heavy cigarette smoking. He said the freeway study also raises questions about the effects of carbon monoxide on healthy persons, and suggests the need for more research.

Atherosclerosis. Dr. Draga Vesselinovitch of the University of Chicago reported in February, 1973, that a low-fat,

low-cholesterol diet can reverse atherosclerosis (the accumulation of fatty deposits in artery walls) in monkeys. She said the findings might also apply to humans. The diet lowered serum cholesterol and decreased the size and number of fatty deposits in the artery walls. Similar studies by Dr. Mark Armstrong and his co-workers at the University of Iowa have provided additional evidence that even advanced stages of the disease can be reversed to some degree if low serum cholesterol levels are sustained for a long period.

Hormones. Dr. Robert Hiatt of Columbia University's College of Physicians and Surgeons in New York City reported in February that Columbia researchers believe they have isolated a previously unknown hormone. Preliminary tests show that the substance, called coherin, stimulates coordinated contractions of the intestine. This makes it potentially useful against a wide range of disorders that involve a breakdown in such coordination.

Traditionally, scientists have thought these contractions, known as peristalsis,

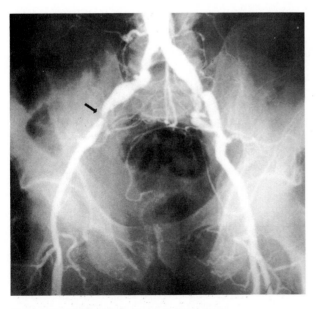

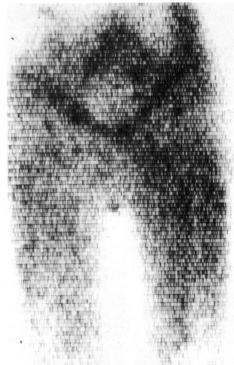

The scintigram, *right*, using radioactive isotopes, shows lowered blood flow to the calves and right thigh. It was compared with the arteriogram, *above*, which shows that both iliac arteries, particularly the right one, *arrow*, are constricted. Scintigraphy helps in evaluating patients for by-pass surgery.

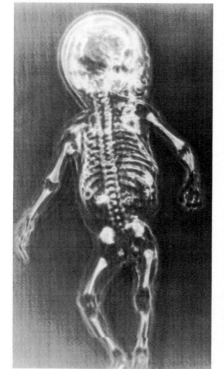

A much sharper picture of a 17-week-old dead fetus, *right,* than the usual X-ray picture, *far right,* was taken with an ultrasonic camera developed by the Stanford Research Institute. The camera uses sound waves, which are safer than X rays.

Medicine

Continued

are controlled by the autonomic and sympathetic nervous systems. This has led to skepticism regarding the role of coherin. However, a parallel study by Dr. Jacques Grenier at Strasbourg University in France supports the Columbia findings. Grenier reports that live segments of dog intestine, isolated from any nerve connections, contract in the natural manner under coherin stimulation.

So far, only small amounts of coherin have been extracted from cattle pituitary glands. This has hampered efforts to prove whether it is a hormone, and to test it more widely on man and animals.

Renal disease. In March, researchers at the University of Chicago and Argonne National Laboratory near Chicago reported they had developed a new compact and disposable dialyzer, or artificial kidney, that is 30 per cent more efficient than others now in use. The device was invented by Finley Markley of Los Angeles, former associate physicist at Argonne. Since October, 1972, members of the university's hemodialysis unit have used the dialyzer in over 70 successful clinical tests.

In addition to being faster than other artificial kidneys, the new dialyzer has a number of other advantages:
- It requires less blood to prime it, and less blood is outside the body while it is being used.
- It has a larger membrane area than other dialyzers.
- It can be operated without a blood pump because it relies on the patient's own blood pressure.
- It is disposable.
- It is less likely than others to fail through a blood leak in the membrane.
- Its estimated cost is comparable to other dialyzers.

Influenza vaccine. Claude Hanoun of the Pasteur Institute in Paris announced in February, 1973, that he had developed a vaccine he predicts will be effective against all influenza variations until 1979 or 1980. In the past, it has been necessary to develop a new vaccine each time a new flu strain appeared. Hanoun's vaccine was made available to French clinics in March. It has not yet been approved for use in the United States.   [Theodore F. Treuting]

Medicine

Continued

Surgery. Surgeons reported highly encouraging results in 1972 and 1973 in the nonsurgical treatment of patients with widespread and previously incurable melanoma, a highly malignant skin tumor normally treated by surgery.

Researchers believe that the immune response in such patients is too mild to stop tumor growth. For this reason, Dr. Donald L. Morton, professor of surgery at the University of California, Los Angeles, treated advanced melanoma patients by injecting Bacillus Calmette-Guérin (BCG), a weakened tuberculosis bacterium, directly into skin lesions. The BCG stimulates the body's immune forces so that they react against the tumor cells as well as against the BCG.

After injecting BCG, Morton found that the skin lesions regressed and 20 per cent of the patients were free of disease from two to five years later. This strongly suggests that the immune response destroyed the entire tumor.

The body's immune defenses frequently do not recognize malignant tumors as foreign tissue (see NEW INSIGHTS INTO IMMUNITY). Dr. Richard L. Simmons,

professor of surgery at the University of Wisconsin, has shown that this may occur because a protective layer of sialic acid on the tumor cell surface may block the initial recognition of the tumor-specific characteristics (antigens). He found that the enzyme neuraminidase will strip away this layer so that the body can react against the antigens.

Mice injected with their own tumor cells that had been treated with neuraminidase showed rapid and lasting regressions of their tumors. Apparently, once a mouse's immune responses were stimulated, these forces reacted not only against the treated tumor cells, but against the untreated ones as well. Simmons and investigators at the National Cancer Institute in Bethesda, Md., are beginning controlled clinical trials of the neuraminidase approach in human beings by injecting melanoma patients with treated melanoma cells.

Dr. Hillard F. Seigler of Duke University Medical Center in Durham, N.C., who has 175 melanoma patients under immunotherapy treatment, reports that the immune system is impaired in two-

The Organ-Transplant Scoreboard

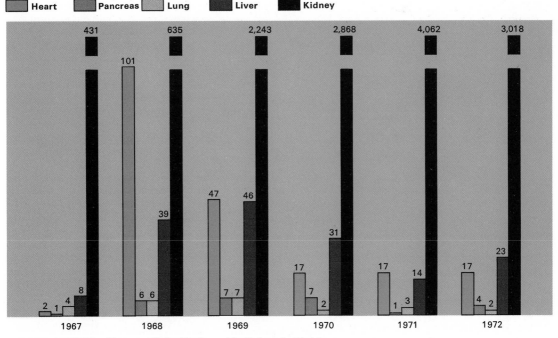

Source: American College of Surgeons—National Institutes of Health Transplant Registry

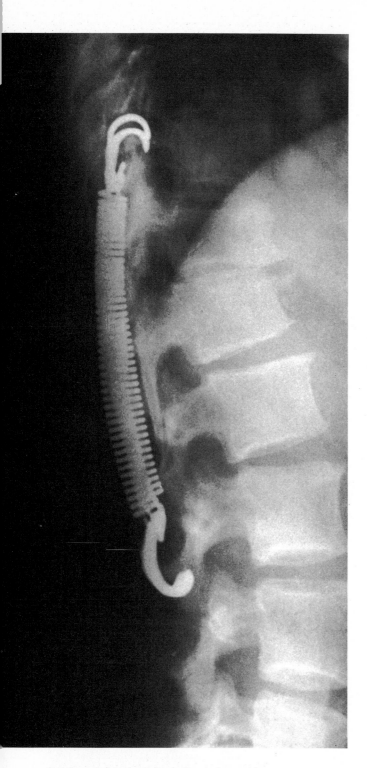

Stainless steel spring is hooked over the vertebrae arches above and below broken vertebra to stabilize fractured spine while it knits. The treatment cuts rehabilitation time from usual nine months to three.

thirds of the patients. Seigler uses BCG injections, neuraminidase, and sensitization of the patient's own lymphocytes, or white blood cells active in immunity. The lymphocytes are sensitized by taking them from the patient and exposing them to tumor cells grown in culture. The sensitized lymphocytes are then returned to the patient. Seigler's three-part program improves the immune reaction, and about 20 per cent of his patients have benefited from the program.

Hyperalimentation. In February, 1973, surgeon Stanley J. Dudrick of the University of Texas in Houston reported that he used intravenous hyperalimentation to cure 60 patients with fistulas, or abnormal connections between the intestine and the skin or other organs. This problem usually requires surgery.

Hyperalimentation allows the bowel to remain inoperative while intestinal disorders are healing. It is used for patients with severe burns, major trauma, loss of major portions of the gastrointestinal tract, peritonitis, and kidney failure. In hyperalimentation, an intravenous mixture of dextrose, amino acids, vitamins, and minerals is injected into a major blood vessel. Because the solution is highly concentrated, the smaller veins used for normal intravenous feeding cannot be used because they do not carry enough blood to dilute the mixture and would rapidly become irritated.

With hyperalimentation, 65 per cent of Dudrick's patients with fistulas healed without surgery. The overall mortality rate was 8.3 per cent compared to mortality rates of 50 per cent or more for patients previously treated surgically.

Dudrick's patients were treated in hospitals, but in 1972, two groups described home use of intravenous hyperalimentation. Dr. Belding H. Scribner and Dr. John W. Broviac of the University of Washington in Seattle trained 12 patients with severely damaged gastrointestinal tracts to infuse themselves daily for from 12 to 14 hours. The majority of the patients gained weight on the program. Dr. David M. Hume, late professor of surgery at the Medical College of Virginia in Richmond, reported he had obtained similar results.

Gallstones. After removing a patient's gall bladder, the surgeon may take X-ray pictures of the bile passages

Medicine

Continued

in the operating room, and explore the common bile duct to remove gallstones that have migrated from the gall bladder. Yet stones continue to appear in the common bile duct. In the past, when subsequent X rays showed stones in the common duct, further major surgery was generally required to remove them.

When the common bile duct is explored during gall bladder removal, a small rubber "T"-shaped tube is brought out through the abdominal wall to drain bile while the common duct is healing.

In January, 1973, Dr. Shelby J. Galloway of the department of radiology, Columbia-Presbyterian Medical Center in New York City, reported treating patients with retained gallstones by passing small catheters and stone-removing instruments through the T tube into the common bile duct. In 10 out of 13 patients, she either pulled the stones out through the T tube or pushed the stone through the lower end of the common duct into the duodenum (first part of the small intestine) from where it would be passed out of the body. There were no complications.

In September, 1972, Dr. Lawrence W. Way of the University of California School of Medicine at San Francisco reported that he treated such cases without surgery. He infused a bile salt solution (sodium cholate) through the T tube into the common bile duct of 22 patients with retained stones continuously for 14 days. The chief constituent of the gallstones is cholesterol, which dissolves in bile containing high concentrations of bile salts. The bile salt treatments eliminated the stones in 12 patients with no major complications.

Esophageal bleeding. Dr. Theodore Drapanas of Tulane Medical School reported in October, 1973, that he had used plastic tubing in successful shunting operations in 25 patients bleeding from varicose veins in the esophagus. Surgeons formerly treated such cases by sewing the high-pressure portal vein to the low-pressure inferior vena cava (IVC), creating a portacaval shunt to reduce the pressure in the portal vein system. The extensive dissection required makes the operation generally long and difficult.

Artificial lung from Wesley Memorial Hospital in Chicago is the most efficient yet developed. Its size, the smallest ever, allows it to be implanted more easily.

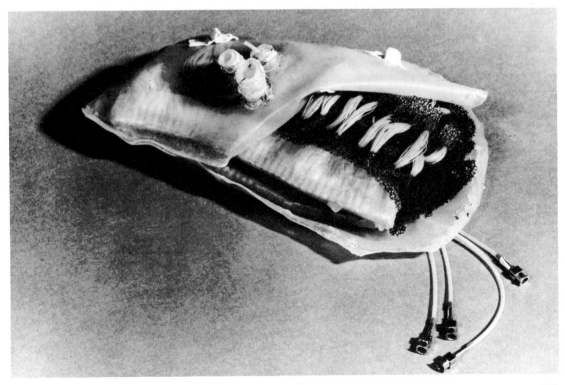

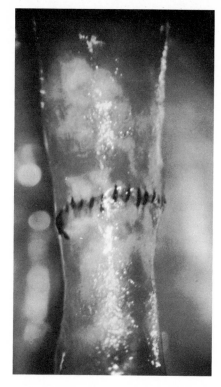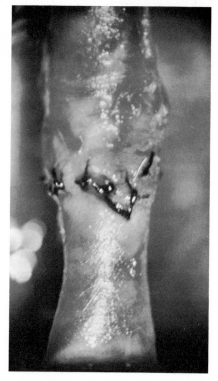

Blood vessel repaired with aid of microscope and extra-fine suture, *right*, shows a great improvement over blood vessel repaired in the usual way, *far right*.

Medicine

Continued

Drapanas simplified the surgical procedure by sewing plastic tubing between the superior mesenteric vein (a major component of the portal vein) and the IVC. This lowered the pressure in the portal vein. None of the patients have had further bleeding in the 3½ years following surgery.

Massive liver injuries are usually fatal, generally because of blood loss. Removing large portions of severely injured livers and tying off the severed blood vessels has saved many patients. But, because the operation requires an experienced team, a good blood bank, and adequate laboratory facilities, it may be impossible in small hospitals.

Dr. E. Truman Mays, professor of surgery at the University of Louisville, reported in February, 1973, that the bleeding could be controlled by simply tying off the hepatic artery, leaving the liver supplied by blood coming only from the portal vein. The liver can survive with portal blood alone until hepatic arterial collaterals develop. These collaterals are secondary vessels that branch out to compensate for the loss of the artery.

They usually develop within a few days. Twelve patients treated in this way survived. The four that died had other multiple injuries. Mays stresses that injuries that reduce portions of the liver to a pulp will still require removal of that portion of the liver.

Infant heart surgery. Several surgical teams reported encouraging results in the increasing use of profound hypothermia in pediatric heart surgery. In hypothermia, the body temperature is lowered from the normal 98.6°F. to 68°F. and all circulation is stopped. The surgeon can then work within the bloodless nonbeating small heart unimpeded by the tubes needed for cardiopulmonary by-pass. The procedure is limited by the amount of time the cooled infant brain can go without blood before suffering brain damage, generally 45 minutes.

Surgeons report good results in terms of heart repair and patient survival. The results appear to be comparable or superior to those obtained by cardiopulmonary by-pass. Hypothermia is likely to be preferred in infants with complex cardiac lesions.   [John B. Price, Jr.]

Electronics Banish Pain

Battery-operated TENS device sends electricity to flat electrodes taped to the skin near large peripheral nerves. The device is used mainly to relieve localized pain.

Pain, which is useful as a warning that something is wrong with the body, serves no further purpose when it continues and becomes chronic. But now, electronic pain-control devices can alleviate some types of pain formerly considered untreatable. In 1973, about 1,000 patients in the United States were using one such device—the Dorsal Column Stimulator (DCS)—which had been in clinical use only since 1969.

The DCS is most frequently used to relieve chronic pain resulting from multiple back operations or injuries to the spinal cord or major peripheral nerves. Another system—transcutaneous electrical nerve stimulation (TENS)—has been helpful in treating more limited, local types of pain, such as those produced by body injury or surgery, headache, arthritis, *causalgia* (a burning pain caused by nerve injury), and even cancer.

The use of electrical nerve stimulation to treat pain is not new. As early as A.D. 50, patients with pain due to gout were treated by placing the painful limb in a tub of water along with a torpedo, or electric ray, fish. Electrical nerve stimulation was used widely in the United States a century ago, but lost favor because the therapy never achieved the results expected.

In 1965, a new theory of pain, the gate theory, revived interest in the technique. The sensation of pain is carried to the brain by the smallest of our nerve fibers. The gate theory predicted that pain could be controlled if electricity were directed into larger nerve fibers to the brain. The stimulation would create signals along the large fibers that would meet the pain signals of the smaller fibers at certain "gates" in the nervous system. These gates would open only to the larger signal, thereby preventing the pain signal from reaching the brain.

The DCS, designed by neurosurgeon C. Norman Shealy of the Pain Rehabilitation Center in La Crosse, Wis., takes advantage of this phenomenon. The device consists of a stimulator combined with a small battery pack about the size of a package of cigarettes, an external transmitting antenna taped to the patient's chest, a receiver implanted in the patient's chest wall, and an electrode implanted in his body near the spinal cord. The entire unit is powered by a 9-volt transistor radio battery.

When the patient feels or anticipates pain, he turns on the battery pack. The stimulator sends a signal to the external antenna, which transmits it to the receiver. Wires tunneled under the skin carry the impulse from the receiver to the electrode, which stimulates the large nerve fibers of the spinal cord. Instead of pain, the patient then feels a pleasurable tingling or buzzing. Modifying the function of the nervous system in this way is called neuromodulation. Once the DCS is implanted, the patient need only change the battery periodically.

Because DCS and a number of other implanted devices do not work for all patients, neurosurgeons have developed screening tests that determine, before surgery, if a device will effectively relieve pain in any individual. Temple University Health Sciences Center in Philadelphia has used such a screening program for three years. As part of this program, the patient experiments with a TENS device—a small, battery-operated stimulator. This relatively simple device provides electricity to electrodes that are on the surface of the skin, rather than implanted. The electrodes stimulate large nerves close to the body surface which then act to block the pain signals at the gates.

The technique of TENS was pioneered by neurosurgeon William Sweet of Massachusetts General Hospital in Boston. With the solution of several technical problems, TENS became increasingly effective. The Temple University researchers have utilized TENS not only as a test, but as an effective pain reliever in its own right. Its greatest advantage is, of course, that the patient can use it without surgery. The patient carries a small battery-operated stimulating device with lead wires running to electrodes that are taped or painted on the skin. At a cost of $300, TENS is also considerably less expensive than the DCS, which costs $1,000.

Both the DCS and TENS have been useful in treating a problem known as central pain. This occurs when constant severe pain lasts so long that it is programmed into the brain. As a result, it continues even if the original cause is cured or removed. It appears that these new devices can gradually decondition the pain program, finally eradicating it altogether. [Charles Burton]

Meteorology

The effects and uses of weather modification became a major, worldwide concern for meteorologists in 1972 and 1973. For example, Basil J. Mason, director-general of the British Meteorological Office warned in February, 1973, that ". . . in several countries, politicians and entrepreneurs . . . are initiating and conducting major weather-modification projects without the benefit of proper scientific direction, advice, and criticism, and this may have serious repercussions on the reputation of meteorology as a science and a profession." See THE NEW RAINMAKERS.

In July, 1972, the U.S. Air Force was alleged to be using cloud seeding for military purposes in Southeast Asia, causing torrential rains and widespread flooding. This prompted a U.S. Senate foreign relations subcommittee to hold hearings on a proposed resolution calling on the United States to seek a treaty "prohibiting the use of any environmental or geophysical modification activity as a weapon of war."

Project Stormfury, a hurricane-modification experiment, became a casualty of federal budget cuts and fund impoundments in 1973. The Department of Commerce's National Oceanic and Atmospheric Administration (NOAA), which had been conducting the Stormfury experiment, announced in January, 1973, that the project would discontinue seeding operations, at least temporarily. The research aircraft used in Project Stormfury and other meteorological studies were grounded.

For the first time, cloud seeders used their capability to dissipate clouds on a lifesaving mission in January, 1973. University of Washington scientists, while on a research flight over the foothills of the Cascade Mountains, sprinkled their load of dry ice into a supercooled stratus cloud cover, causing the cloud layer to dissipate. This enabled the crew of a rescue helicopter to find and rescue three men whose small plane had crash-landed.

Meteorological satellites. The second of a new generation of operational weather satellites, NOAA-2, was launched into a nearly pole-to-pole orbit 910 miles above the earth on Oct. 15, 1972.

A student worker at the National Hail Research Experiment laboratory in Boulder, Colo., plots the movement of storms. Data plotting is part of a five-year project to study the dynamics of hailstorms and determine whether or not they can be controlled through cloud-seeding techniques.

A plastic dome protects a huge radar antenna near Fort Morgan, Colo. The new installation is one of two radar units being used by researchers to track and study potential hailstorms.

Meteorology

Continued

NOAA-2 was developed jointly by NOAA and the National Aeronautics and Space Administration (NASA).

The NOAA-2 satellite carries two new sensor systems, a very high resolution radiometer (VHRR) and a vertical temperature profile radiometer (VTPR). The VHRR enables observers to see not only clouds, but also details of the earth's surface, such as areas covered by snow. It can also measure the temperatures of sea surface areas as small as half a mile wide. The VHRR works equally well by day or night, because it measures the infrared radiation emitted by the clouds and by the earth rather than measuring reflected sunlight.

The VTPR measures the infrared radiation emitted by the carbon dioxide in the atmosphere and from this determines the vertical distribution of temperatures in the atmosphere. The satellite orbits the earth every 115 minutes. As the earth rotates beneath it, the VTPR can provide atmospheric soundings twice daily over the entire earth. These supplement the data from a network of several hundred balloon-borne weather transmitters, most of which observe the atmosphere only over the continents and mainly in the middle latitudes. The VTPR data have been used routinely for global weather analyses since Dec. 19, 1972.

On Dec. 10, 1972, NASA launched Nimbus 5, another in its series of meteorological research satellites. Flying in a nearly polar orbit, the satellite circles the earth in 107 minutes at an altitude of 690 miles. The orbit is altered slightly each day to keep the satellite in line with the sun. This allows it to observe all points on the globe twice a day, in sunlight and in darkness.

New experimental devices aboard Nimbus 5 include two microwave sensors designed to "see" through clouds. Meteorologists are supplementing satellite cameras that use visible light and infrared radiometer measurements with measurements of microwave radiations, radio waves emitted by the earth and components of the atmosphere, such as oxygen, water, and water vapor.

One of the microwave sensors aboard Nimbus 5 is an electrically scanning

321

Meteorology

microwave radiometer, a single-channel instrument that can distinguish liquid water from ice and snow through thin clouds and may reveal where thicker clouds contain high amounts of liquid water or where precipitation is falling. The other Nimbus 5 cloud-penetrating sensor, a microwave spectrometer, has five channels, three for picking up microwaves emitted by oxygen and two for detecting liquid water and water vapor. Scientists can determine the vertical temperature profile of the atmosphere—even when clouds are present—from the microwaves detected by the three oxygen channels. The other two channels provide information on water in the atmosphere and on the earth.

In addition to the microwave sensors, Nimbus 5 also carries four infrared radiometers. The surface-composition mapping radiometer measures physical characteristics of the earth's surface. The infrared temperature profile radiometer measures atmospheric temperature. The two other infrared instruments are a radiometer that measures the vertical temperature profile in the atmosphere as well as the distribution of water vapor, and a radiometer that observes cloud cover, ground temperature in clear areas, and water vapor.

Atmospheric effects of the SST. Despite the 1971 congressional decision to cancel plans for building the supersonic transport (SST), a major research program is attempting to determine the potential effects of SST operations on the atmosphere. Partly on the basis of this research, the United States will decide whether or not to resume work on the SST. The research is also expected to have international repercussions in regard to the Russian and joint British-French SST efforts.

At the request of Congress, the U.S. Department of Transportation in late 1971 funded the $7-million, two-year research effort to calculate the effects of quantities of water vapor and particulates emitted by SST aircraft. Among the questions being considered are effects on climate and on the ultraviolet radiation reaching the earth. Scientists question whether or not the earth's climate will become warmer or colder because of the water vapor added to the stratosphere by the SST. Also, ozone in the stratosphere absorbs some of the harmful ultraviolet radiation from the sun, and it has been suggested that both the water vapor and the oxides of nitrogen emitted by SST engines in the stratosphere may reduce the amount of ozone. If this were to happen, dangerous levels of ultraviolet radiation would reach the earth, and this could lead to such biologically harmful results as an increase in the frequency of skin cancers.

The effect of SST exhausts on the ozone in the stratosphere is a complicated problem involving atmospheric chemistry, dynamic meteorology, photochemistry, and radiative transfer. By mid-1973, several meteorological research centers in the United States were attempting to calculate what the effect might be. They were using mathematical models of the atmosphere, such as those developed for studies of the general circulation of the atmosphere and for numerical weather prediction.

Influences on weather patterns. NOAA announced in March, 1973, that two scientists based in Boulder, Colo., had identified an apparent link between geomagnetic storms—disturbances in the earth's magnetic field caused by electrically charged particles emitted from the sun—and low-pressure troughs in the atmosphere. This may be significant in helping scientists understand the connection that some of them suspect exists between these storms and the earth's weather patterns.

The scientists found a statistical relationship between certain types of geomagnetic activity (such as the magnetic storms associated with auroras, or northern lights) and pressure patterns, or troughs, at high altitudes above the Gulf of Alaska. These troughs deepened, or intensified, following magnetic storms. Knowing this could be a help to weather forecasters.

On Jan. 18, 1973, NOAA announced that cold-water eddies, bodies of cold water surrounded by warmer Gulf Stream water, may affect weather. Data collected by a NOAA satellite showed that the presence of eddies dissipated low-lying clouds over the Atlantic Ocean. NOAA scientists believe that these eddies may influence weather along the East Coast, but they need to collect more data before they can determine the relationship between weather and the eddies. [Jerome Spar]

Microbiology

Some of the most important work in microbiology in 1972 and 1973 dealt with the role of bacteria in reversing the effects of man's mistreatment of his environment. Also of interest is a study revealing a unique relationship between termites and bacteria.

Gobbling up oil. In September, 1972, microbial biochemists Ronald Atlas and Richard Bartha of Rutgers University in New Brunswick, N.J., reported on a way to make bacteria more effective against oil spills. From 5 to 10 million tons of oil enter the earth's oceans each year. Most of this oil comes from thoughtless actions of man.

Theoretically, the best way to get rid of it is through biological degradation, in which oil-consuming bacteria break the oil down into harmless, soluble products. Laboratory and field studies have shown, however, that the bacteria which degrade oil cannot keep up with the job that confronts them in virtually all cases of oil spills and dumping.

The Rutgers scientists found that the reason the bacteria fail to keep up is that ocean water does not contain sufficient nitrogen and phosphorus for optimum bacterial growth. When the scientists added potassium nitrate (a source of nitrogen) and potassium phosphate (a source of phosphorus) to seawater containing oil, the bacteria could degrade much more of the oil. However, the added chemicals also caused a tremendous increase in the numbers of algae in the water, creating a condition commonly called algal bloom. Algal blooms are dangerous because they deplete the oxygen that is dissolved in the water. This ultimately kills fish and plant life and causes obnoxious odors.

Atlas and Bartha next sought a way to provide nitrogen and phosphorus to the bacteria without stimulating the growth of algae. They tested the effects of several compounds that dissolve in oil (upon which the bacteria feed), but not in water (where the algae feed). They added these compounds to tanks filled with seawater contaminated with oil, and found that a combination of paraffinized urea (a nitrogen compound) and octylphosphate (a phosphorus compound) increased the rate of oil degradation by

Bacteria that were photographed living on molybdenite ore, *below*, were the first observed on any ore. Wormlike bacterium (arrow) dines on a sulfur granule, *right*. The furrow above and to the right marks the abandoned site of a previous dinner.

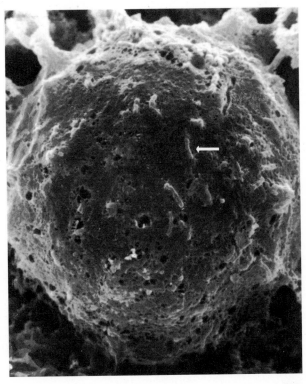

Little
But
Lethal

The viroid, a new type
of disease-causing agent,
is much smaller than
a virus' DNA (genetic
material). Magnification
is about 35,000 times.

In November, 1972, I reported that I had isolated a new type of disease-causing agent that is the first new type discovered since viruses were found in 1898. Until now, scientists have thought that all infectious diseases are caused either by microorganisms, such as bacteria, or by viruses.

Because they are so small and otherwise difficult to detect, viruses have been blamed for the many infectious diseases for which a causative agent has never been identified. In 1963, plant pathologist William B. Raymer and I decided to try to isolate the virus that caused one such malady, potato spindle tuber disease, in which potato plants produce badly elongated and spindly potatoes. Raymer and a colleague, plant pathologist Muriel O'Brien, had already shown that extracts from infected potatoes cause a disease in tomato plants in which many of their leaves turn down, curl, and die, and the growth of the entire plant is stunted. Raymer had also uncovered evidence suggesting that the agent that causes the disease might be smaller than most viruses. With this in-

formation in hand, we began working in our laboratories at the U.S. Department of Agriculture's Agricultural Research Service in Beltsville, Md.

One of our first tests was to centrifuge infectious extracts from diseased potatoes. We put the material into centrifuge tubes and spun them in a machine at very high speeds, causing the components of the material to separate into layers according primarily to size. The largest components were closest to the bottom of the tubes and the smallest closest to the top.

Next, we inoculated tomato plants with material from each layer. To our surprise, only plants inoculated with the smallest material became diseased. This meant that the infectious agent was not a virus, which would have been larger. In fact, the agent was even smaller than deoxyribonucleic acid (DNA) or ribonucleic acid (RNA) isolated from most viruses. DNA and RNA are the molecules of which the genes of all infectious agents—indeed, of all living organisms—are made. Thus, our finding suggested that the infectious agent we were seeking

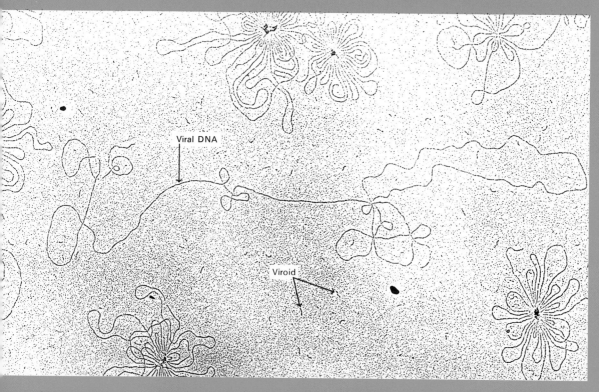

Viral DNA

Viroid

might consist solely of an extremely small molecule of DNA or RNA.

When we mixed the infectious material with an enzyme that destroys RNA, the material no longer caused disease in tomato plants. But if we mixed it with an enzyme that destroys DNA, it still caused disease. This proved that the genetic material of the agent is RNA. Further tests disclosed that no other materials are associated with the infective RNA, as they would be if the agent were a virus. Therefore, the infectious agent is a free RNA molecule.

In subsequent experiments, I found the molecular weight of the infective RNA to be 75,000 to 85,000. This is significant, because the DNA or RNA of the tiniest viruses known has a molecular weight of about 1 million. Evidently, the infectious RNA—which I call a viroid—is less than 1/12 that size.

This is enough RNA for only one very small gene. This gene may function, as do nearly all other genes, by causing a single enzyme or other protein to be made. Yet, the viroid accomplishes at least two things: It triggers the disease it causes, and it reproduces itself. The viroid gene is almost certainly too small to account for an enzyme of the size required for RNA synthesis. Evidently, then, the viroid reproduces primarily with the aid of an enzyme or enzymes of its host organism. If so, the small protein possibly made from the viroid RNA may cause the disease.

Is the potato spindle tuber agent the only viroid? Plant pathologists at the University of California, Riverside, and the Justus-Liebig University in Giessen, West Germany, found that citrus exocortis, a disease of orange trees, is also caused by a low-molecular weight RNA. However, recent experiments indicate that this RNA may be closely related to, or identical with, the potato spindle tuber viroid. But, plant pathologist Roger H. Lawson and I have found that another viroid causes a disease that stunts chrysanthemums.

I consider it likely that viroids are also responsible for certain animal diseases. One such disease might be scrapie, which kills sheep by attacking their nervous system. There also are many human diseases for which no cause has been determined. Here, too, viroids may be at work.   [Theodor O. Diener]

600 per cent and did not cause the formation of algal bloom.

The two scientists estimated that treating oil slicks with these compounds would cost less than $15 per ton of spilled oil. This is a bargain compared to the cost of cleaning up spills by current methods, which skim or soak up the oil from the water surface.

In February, 1973, microbiologist Eugene Rosenberg and biochemist David Gutnick of Tel Aviv University in Israel suggested a more novel approach for using bacteria to degrade oil. They observed that most of the oil that pollutes beaches in Israel came from the discharge of ballast from oil tankers.

After the tankers unload crude oil at a port, they take on seawater as ballast for the return trip. The water, along with the residual oil, is then pumped into the sea before the ship is loaded again. This practice probably flushes about 1 million tons of crude oil into the world's oceans each year. Rosenberg and Gutnick reasoned that this is not only a bad practice ecologically but it is also a waste of the residual oil.

They decided to attempt to solve both problems. They isolated from seawater a species of bacteria that devours crude oil and grows and multiplies quickly in the process. The scientists planned to harvest the bacteria for use as a high-protein animal feed. After three or four days of growth, there were about 90 million bacterial cells for each milligram (three one-hundred thousandths of an ounce) of crude oil used. And over half of the oil was completely degraded, with the rest emulsified, or broken into tiny bubbles that mix with water.

Next, the scientists boarded a tanker that had discharged its oil cargo and had taken on 100,000 tons of seawater for ballast. To the ballast, the scientists added about a cupful of a highly concentrated preparation of the oil-degrading bacteria. As the ship traveled to Iran to take on a new load of oil, air was bubbled through the tank to provide oxygen for the bacteria and to stir the contents. After one week at sea, the ballast was discharged. The bacteria had done their job well—there was no oil slick as is normally the case when ballast is discharged.

An unexpected and important economic fringe benefit to the oil shippers

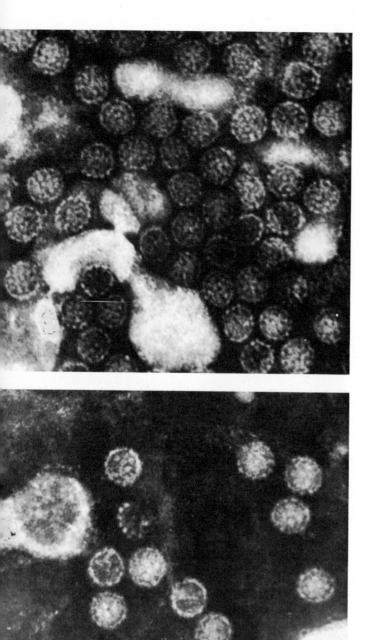

Spherical viruses found in the urine of two kidney transplant recipients are typical of those found in 45 per cent of such patients in a hospital in Pretoria, South Africa. The viruses are similar to a type long associated with cancer in animals.

also resulted from the experiment. The bacteria had done such a good job of eating the oil that the tank was much cleaner than usual. This eliminated the expensive operation of cleaning out the accumulated oily tars.

Rosenberg and Gutnick are still working on an economical way to harvest the bacteria. The scientists estimate that about 75 tons of high-protein animal feed might possibly be harvested from each "empty" 100,000-ton tank.

Mercury. Microbiologist William Spangler and his co-workers at the Midwest Research Institute in Kansas City reported experiments in January, 1973, on a possible method of detoxifying large quantities of dangerous mercury compounds that are found in many lake and river sediments.

The mercury compounds long used extensively in agriculture and industry were once considered harmless. Experiments showed that they were not retained in the tissues of human beings or in fish and birds in the human food chain. But, when mercury poisoning began to crop up in many parts of the world, scientists took another look. They found large quantities of methylmercury, one of the most poisonous forms of mercury, and one that is more dangerous because it is retained in animal tissues.

The methylmercury came from methane-forming bacteria that live in lake and river sediments. These bacteria convert the less-poisonous mercurials into the dangerous methylmercury. There seems to be no way to stop the bacteria from doing this. The only way to alleviate the problem, then, would seem to be to stop using the mercury compounds in agriculture and industry. Unfortunately, however, some lake and river sediments are already so loaded with these compounds that they could continue to release methylmercury for many hundreds of years to come.

Spangler and his colleagues noted that although methylmercury was constantly produced in the sediments, it did not accumulate there to any extent. They reasoned that perhaps other bacteria in the sediment degrade the highly toxic methylmercury.

To test their theory, they took a mixture of bacteria from the sediments and grew them in the laboratory with methylmercury as their only source of

Microbiology

Continued

food. Only those bacteria able to degrade methylmercury would survive. Thirty species of bacteria survived. These are also methane-forming, degrading the methylmercury into methane gas and inorganic mercury.

Inorganic mercury is not incorporated into animal tissues. Consequently, it is possible that the bacteria that digest methylmercury can be used to rid the poisoned bodies of water of methylmercury quickly and safely.

Termites and bacteria. In January, 1973, microbiologist John Breznak of the University of Wisconsin at Madison discovered a new cooperative association between termites and bacteria. Termites eat vast quantities of wood, but they do not have the enzymes needed to digest its cellulose, which is their major nutrient. Scientists have known for many years that bacteria living in the termite's gut digest the cellulose to sugar molecules, which the termite can use. In return, the termite provides the bacteria with a warm, moist environment.

Breznak discovered that bacteria (probably different from those that digest the cellulose) in the gut of termites convert atmospheric nitrogen gas into amino acids, which the termite uses as its source of dietary nitrogen.

To make his discovery, Breznak took advantage of an unusual capacity of the nitrogen-fixing enzyme, nitrogenase, to reduce acetylene gas to ethylene. He used a laboratory device known as a gas chromatograph to determine if ethylene was formed.

This technique was sensitive enough to detect nitrogen fixation in a single termite. To tie the activity definitely to bacteria, Breznak ground up the termite intestines along with their contents and showed that they retained their nitrogen-fixing ability.

Then, he fed other termites bacteria-killing antibiotics, and found that their nitrogen-fixing ability disappeared and they eventually starved to death. This suggests that termites can be controlled by destroying their symbiotic bacteria. Perhaps, in areas where termite damage is severe, dipping wood in an antibiotic solution would be a good way to solve the problem.   [Jerald C. Ensign]

Neurology

In July, 1972, biochemist Georges Ungar and his co-workers at the Baylor College of Medicine in Houston isolated, analyzed, and synthesized a substance that they claimed could transfer learned behavior from one animal to another. Most scientists from a wide variety of disciplines criticized this and earlier related work by the same team. But, later in 1972 and in 1973 several experimenters, working independently, reported evidence supporting Ungar's claims.

If further study proves Ungar to be right about chemical transfer of learning, it will be one of the most important developments in the history of neurology. It could help scientists answer some of the most basic questions about how the brain works in learning and memory.

Dark avoidance. Earlier work by Ungar's group had depended on the well-known preference of rats and mice for darkness rather than light. Rats were given electric shocks when they passed from a lighted compartment into a darkened one. In about a week, the rats learned to avoid the dark. Then the scientists injected chemicals extracted from the trained rats' brains into untrained mice.

Ungar and his associates reported that when they placed these mice in the apparatus that had been used to train the rats, the mice remained in the darkened compartment for only about 50 out of 180 seconds. Mice that received injections of brain extracts from untrained rats stayed in the darkened compartment for periods of from 120 to 130 seconds out of 180.

Thus, the scientists concluded that the trained rats' learned fear of the dark had been transferred to the mice by some substance in the rat brain extracts. The researchers named the substance scotophobin, from the Greek words *skotos* (dark) and *phobos* (fear).

At first, many other scientists could not duplicate, and thereby verify, these results. This was not surprising, since many steps in the experiments could vary from one laboratory to another—for example, the training of donor animals or the chemical process of extracting brain materials. All this heightened the significance of the report by Ungar

327

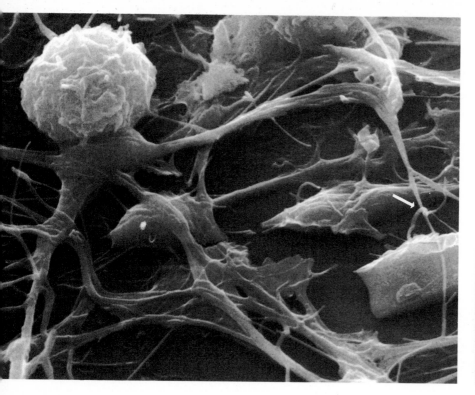

The first clear picture of what brain tissue may look like came from rat brain tissue grown in the laboratory and photographed through a scanning electron microscope. Spherical white blood cell, *left*, sits atop nerve cell that joins, or synapses, with part of another nerve cell, *below*.

Neurology

Continued

and his associates in July, 1972, that they had isolated from the brains of trained rats the specific substance that made the untrained mice avoid the dark.

Furthermore, the researchers identified the substance as a polypeptide sequence of 15 joined amino acids. Then they synthesized it. They also reported that their synthetic scotophobin induced mice to avoid the dark as effectively as natural scotophobin.

Some of Ungar's critics claim that he used very imprecise techniques to isolate, purify, and analyze the natural scotophobin. However, the methods used to synthesize scotophobin seem to conform to good chemical standards. Other critics say that scotophobin's effects may result from a physiological change, such as desensitization of the iris of the eyes in the animals that were injected.

But even a blinded mouse would only spend about 90 of 180 seconds (50 per cent, through chance) in the light. See BIOCHEMISTRY.

With fairly large quantities of scotophobin available, other scientists could check Ungar's results far more easily

than before. In November, 1972, Rodney Bryant at the Brain Research Institute of the University of Tennessee at Memphis headed a team that injected synthetic scotophobin into the brains of goldfish. They reported that the fish learned to avoid the dark more easily than did uninjected fish. They also said it was more difficult to train the injected fish to seek the dark.

Then, in December, 1972, David Malin at the University of Michigan at Ann Arbor and Helene Guttman at the University of Illinois at Chicago independently reported experiments fully confirming Ungar's prior results. They converted dark preference to dark avoidance in mice by injecting them with synthetic scotophobin.

Other experiments, almost certainly underway in 1973, should soon be reported. When all the data is in, scientists should have a much clearer view of this controversial but extremely fascinating area of study.

Cerebral hemispheres. An important study reported in 1972 by neuropsychologist Robert Nebes at the Cali-

Neurology
Continued

fornia Institute of Technology in Pasadena has more clearly defined the separate roles of the right and left cerebral hemisphere of the human brain.

Nebes studied "split-brain" patients in whom the corpus callosum, the large bundle of nerve fibers interconnecting the two cerebral hemispheres, has been cut to treat neurological disorders. Such patients show no obvious changes in personality, intelligence, or behavior under normal conditions. But, neurologists have devised special techniques of studying them that have enabled the scientists to study each half of the brain separately.

Early studies of this nature revealed that each cerebral hemisphere controls specific abilities. For example, the so-called "dominant" left cerebral hemisphere, which normally controls movement of the right side of the body, also has essentially exclusive control of communicative behavior such as speech, writing, and calculation. This has made it easier for scientists to learn about the left hemisphere than the "nondominant" right hemisphere. Yet, we have learned that the right hemisphere, which con-

trols movements of the left side of the body, seems also to be involved in the recognition and comparison of visual shapes and forms.

Nebes compared the hemispheres' abilities to infer the overall shape of a geometric figure from an examination of its scrambled constituent pieces. He projected pictures of the scrambled constituent pieces and several possible reconstructed figures through a special apparatus that allowed only one of the patients' hemispheres to receive the information. When the images were projected to the right hemisphere, the patients pointed quickly and accurately to the proper reconstructed figure. But, when the images were projected to the left hemisphere, the patients selected figures inaccurately and randomly.

As would be expected, the patients were also unable to say what they saw when the images were projected to their left hemisphere. These experiments show that the individual's ability to conceptualize the total contour of a figure is centered in his brain's right cerebral hemisphere.   [Thomas H. Meikle, Jr.]

Nutrition

Taking large quantities of vitamin C became a major food fad soon after Nobel prize winning chemist Linus Pauling claimed in 1970 that 1 to 5 grams of vitamin C a day substantially reduces the frequency and duration of colds. However, few nutritionists thought that Pauling's experimental evidence was convincing.

In September, 1972, scientists at the University of Toronto school of hygiene reported the results of their test of Pauling's claim. They used a double-blind procedure in which neither the investigators nor the subjects knew which of two groups of 500 subjects was taking 4 grams of vitamin C a day and which a placebo, a dose containing no vitamin C. The scientists measured the frequency and duration of colds and found no differences between the two groups in either frequency or duration. However, members of the vitamin C group stayed home fewer days to nurse their colds.

Although some who favor taking massive doses of vitamin C claimed that the experiment provided support for their position, most nutritionists believe it

demonstrated that the common cold cannot be effectively prevented or treated with large doses of vitamin C.

The perfect food. Many adults do not use milk or find it a satisfactory food even when it is readily available. Since 1970, several experiments have shown that many people who do not use milk have less of the enzyme lactase in the lining of the small intestine. This enzyme splits lactose, milk sugar, into the sugars galactose and glucose. If the lactose cannot be split by the enzyme, it cannot cross through the intestinal wall and enter the blood stream. Thus, lactose moves on to the large intestine, where bacteria feed on it and multiply, producing the symptoms of lactase deficiency—abdominal cramps, nausea, gas, and diarrhea.

Lactase, like many other enzymes, may be inducible—that is, the amount of the enzyme produced may increase as the amount of lactose ingested increases. If this is true, perhaps drinking milk will stimulate lactase activity. Studies on this theory have been contradictory, but evidence supporting it is accumulating.

Soybeans, the basic ingredient of meatless "meats," are converted into protein fibers, *above*, and mixed with flavors, coloring, and nutrients. Cooked and shaped, they look and taste like almost any meat or meat product, *right*, including sausages, *above right*.

Nutrition

Continued

However, Shunkichi Tamuri, Shintaro Tsuzuki, and Michio Masuda of the Tokyo Dental College reported that milk itself may contain a substance that promotes the breakdown and absorption of lactose in the small intestine.

Catch-up growth. Two studies indicated that the bodies of young children can repair damage and catch up with normal physical growth and development after a prolonged inadequate diet. D. G. Barr and his colleagues at the University Department of Pediatrics, Kinderspital, Zurich, Switzerland, studied the rehabilitation of Zurich children who had suffered intestinal disorders and poor absorption of food during infancy. They found a spectacular period of catch-up growth after the disorders had been corrected. It took only six months for the children's weight to reach normal levels. Their height and bone development were close to normal after a year, and exactly normal a year later.

In November, 1972, Zena Stein and her colleagues at the Columbia University school of public health in New York City reported their study of how the Dutch famine of 1944 and 1945 affected the mental performance of adults who lived through it as infants. They analyzed army-induction tests of mental ability and found no difference in performance between those persons who were born in cities affected by the famine and those who were not.

Catch-up behavior. Psychologist David A. Levitsky and nutritionist Richard A. Barnes of Cornell University, Ithaca, N.Y., found that early nutrition and environmental conditions affect the development of an animal's behavior. They raised several groups of rats under various environmental conditions. Among the groups were two that were kept malnourished for seven weeks. The rats in one of the two groups were handled, allowed to play with toys and other rats, and housed in pairs. The rats in the second group were kept totally isolated. After a 10-week recovery period, in which both groups were fed well, the scientists observed that the environmental conditions had had no significant effect on growth rate. But rats that had been stimulated behaved more nor-

Informative Food Labels

Nutritional labeling, which began in 1973, translates the science of nutrition into practical knowledge. The Food and Drug Administration (FDA) told manufacturers to provide extensive nutritional information on the labels of some food products and authorized such labeling on others. Relabeling must start by Dec. 31, 1973, and all labeling must meet the new standards by Jan. 1, 1975.

In the past, providing information on labels was voluntary. Much of it was deceptive, and did not disclose the nutritional components. This made it difficult to compare the nutritional value of packaged foods and plan nutritionally balanced meals.

Now, the information must follow an approved format and be presented in terms of a common household measure. It must include such data as number of calories, grams of protein, carbohydrate, and fat, and per cent of the recommended dietary allowance (RDA) provided for vitamins A, D, B_1, B_2, and niacin, and for minerals, calcium and iron.

The FDA rules were accompanied by regulations that define the minimum quality of protein required before it can be used to fulfill RDAs. For example, protein must be of sufficient quality, as determined by its ability to promote growth of experimental animals, before it may be stated on the label in terms of the per cent RDA. The regulations also define the minimum enrichment of isolated vegetable proteins, such as the protein fraction of soybeans and wheat treated to resemble animal protein, by a broad range of vitamins and minerals. And, they define the minimum nutritional content of such preparations as liquid diet meals or frozen TV dinners.

Additional regulations will probably make "percentage labeling" compulsory. This means that labels must list the proportional content of named or economically significant foodstuffs, such as the percentage of beef in "beef stew."

Tests have shown that most customers welcome nutritional labeling and use it intelligently to make better choices of food. As the nutrition education campaign planned by the FDA goes into full swing, the advantages should become even more pronounced.   [Jean Mayer]

First labels to appear under FDA's guidelines provide new nutritional information on contents.

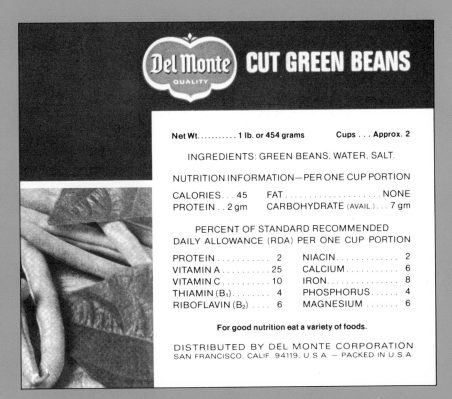

Del Monte QUALITY · **CUT GREEN BEANS**

Net Wt.......... 1 lb. or 454 grams   Cups . . . Approx. 2

INGREDIENTS: GREEN BEANS, WATER, SALT.

NUTRITION INFORMATION—PER ONE CUP PORTION

CALORIES... 45   FAT .... NONE
PROTEIN .. 2 gm   CARBOHYDRATE (AVAIL.)... 7 gm

PERCENT OF STANDARD RECOMMENDED
DAILY ALLOWANCE (RDA) PER ONE CUP PORTION

PROTEIN	2	NIACIN	2
VITAMIN A	25	CALCIUM	6
VITAMIN C	10	IRON	8
THIAMIN (B_1)	4	PHOSPHORUS	4
RIBOFLAVIN (B_2)	6	MAGNESIUM	6

For good nutrition eat a variety of foods.

DISTRIBUTED BY DEL MONTE CORPORATION
SAN FRANCISCO. CALIF. 94119, U.S.A. — PACKED IN U.S.A.

Nutrition
Continued

mally, as indicated by the way that they fought and explored, than the rats that had been isolated. This study has important implications for helping children recover from similar deprivations.

Causes of obesity. Theories of weight control have become sufficiently detailed to distinguish between short-term satiety signals, which tell a person to stop eating, and long-term weight-control mechanisms. Physiologists A. B. Steffens, G. J. Mogensen, and J. A. F. Stevenson of the University of Western Ontario in Canada and the State University of Groningen, Haren, the Netherlands, demonstrated in June, 1972, that the lateral areas of the hypothalamus monitor the products of metabolism and may initiate short-term hunger control. In a series of reports in the fall of 1972, Jaak Panksepp of Bowling Green State University in Ohio, showed that the brain cells that maintain constant long-term body weight are located in the ventromedial areas of the hypothalamus. Mechanisms that cause the weight-control apparatus to go wrong, however, are still speculative.

In 1973, experiments by physiologist Lawrence B. Oscai and his colleagues at the University of Illinois, Chicago Circle Campus, re-emphasized the role of exercise. By forcing rats to swim six hours a day six days a week, they obtained data indicating that exercise limits the number and size of fat cells in young rats; the fat cells of rats that were not exercised but maintained at a normal body weight showed no change. Jules Hirsch of Rockefeller University in New York City has proposed that limiting the number of fat cells in infancy limits their ability to expand and store fat in later life.

Many biologists believe that small cells also respond better to the pancreatic hormone insulin and that insensitivity to this hormone may be a factor in causing obesity. In November, 1972, Jonathan K. Wise and his colleagues at the Yale University school of medicine, New Haven, Conn., reported that obese persons secrete less glucagon, another pancreatic hormone that plays a role in metabolism. More research is needed, however, to learn whether this is a result or cause of obesity.   [Paul E. Araujo]

Oceanography

Budgetary cuts and fund impoundments in 1973 forced large-scale changes in federal oceanographic research. Small and varied projects were united into large, broad-spectrum programs, and the government encouraged concentration on applied, rather than basic, research. Studies relating to coastal-zone management, pollution control, and natural-resource exploration received adequate funds. But basic research into such areas as marine ecology, natural history, and physical and chemical oceanography suffered serious cutbacks.

Passage of the Coastal Zone Management Act in August, 1972, illustrated the trend toward applied research. It provides for coordinated federal and state efforts to resolve the conflicts in uses of the nation's coastline–from wilderness areas and recreational parks to port facilities and offshore mineral exploration. Further emphasizing the shift away from basic research, the National Oceanic and Atmospheric Administration (NOAA) mothballed its two most active deep-water research vessels, the *Surveyor* and the *Discoverer*, in March, 1973.

Satellites view the oceans. The first Earth Resources Technology Satellite (ERTS-1) launched on July 23, 1972, by the National Aeronautics and Space Administration (NASA) to map vast areas of the earth, revealed much information of interest to oceanographers. One ERTS-1 photograph detected acid waste being dumped off Long Island, N.Y. The satellite also photographed sediment plumes flowing out from the Hudson River and Barnegat Bay on the New Jersey coast.

In May, 1973, NASA launched the Skylab earth-orbiting laboratory. Part of its mission was to coordinate its data with information from the ERTS-1 satellite in studying thermal ocean currents and wave heights. In addition, U.S. Air Force satellite photographs, made public in March, clearly showed weather patterns, ice packs, and drifting ice.

Ocean dumping. Seventy-nine countries signed an ocean-dumping pact in London on Nov. 13, 1972. The pact bans the dumping of oil, mercury, cadmium, highly radioactive wastes, lead, zinc, and arsenic.

Researchers prepare to ride 60 feet down to the floor of the Pacific Ocean in an acrylic observation sphere, an underwater elevator built by the Naval Undersea Center off the coast of San Diego.

Oceanography

Continued

In February, 1973, NOAA reported that scientists had been investigating the feasibility and environmental safety of solid-waste disposal in the deep sea. Researchers dropped bales of garbage compressed into 35⅓ cubic-foot bundles in 650 feet of water. They retrieved the bales 3 months later and found them relatively intact. No barnacles or other encrusting organisms had as yet attached themselves, sealing off the compressed garbage, but some mobile organisms, such as brittle starfish, were found around the bales. These organisms were not harmed by the wastes, and this preliminary study concluded that such limited dumping would not have an adverse effect on the environment.

In the Atlantic Ocean, 120 miles southeast of New York City, the research submersible *Alvin* carried a team of NOAA investigators to the bottom of the Hudson Canyon, which extends from the mouth of the Hudson River to the edge of the continental shelf. They wanted to determine how effectively currents carry pollutants dumped at the head of the canyon out to the deep sea.

The scientists made eight dives in the canyon in June and seven dives in September, to depths as great as 6,000 feet. They measured current flow, gathered water samples, and studied rock formations on the canyon floor. Their findings indicate that solid wastes and pollutants deposited at the head of deep canyons may be safely carried out to the deep sea by the rapidly flowing bottom currents.

They found that rapid erosion occurs at the canyon head, just beyond the continental shelf, where strong currents scour the canyon floor. There, the bottom sediments consisted of coarse, rounded pebbles similar to those found in stream beds.

They also found a high concentration of animal life in the canyon, including a large population of crabs from 2 to 2½ feet wide of potential commercial value. But the scientists did not have enough data to determine if large-scale dumping would harm the crabs.

Deep-sea drilling. In 1972 and 1973, the research ship *Glomar Challenger* zigzagged across the Indian and Antarctic oceans sampling little-known geological

Oceanography

Continued

Turk, one of the Navy's diving sea lions, finds a torpedo on the ocean floor by homing in on its sound-generating device. The sea lion then attaches a grabber claw and haul line so that men waiting above in a boat can pull the torpedo to the surface.

features of the ocean floor. The ship's geophysical exploration of the world's oceans, known as the Deep Sea Drilling Project, is directed by Scripps Institution of Oceanography in La Jolla, Calif.

The Indian Ocean mission in September and October, 1972, sought to verify the theory that present-day Australia, Africa, South America, India, Madagascar, and New Zealand were once part of a giant continent called Gondwanaland. Scientists theorize that some 150 to 200-million years ago, the supercontinent split and the smaller blocks moved away from each other as the sea floor expanded outward from what are now the mid-oceanic ridges. Along the way, chunks of continental material, microcontinents, were thought to have broken off.

The ship's rig drilled into the mid-Indian Ocean Ridge, Ninety East Ridge, and into a supposed microcontinent called Broken Ridge in the southern Indian Ocean. Cores from Broken Ridge, however, showed that it was made of basalt, the type of rock that makes up the sea floor. Continental rock is granite. Therefore, Broken Ridge is more likely

to be an uplifted piece of the ocean floor than a chunk of continental land mass.

Final analyses of the thousands of feet of sediment core will require years of study, but some findings tend to substantiate the drifting continent theory. The mid-Indian Ocean Ridge has apparently been an active center of sea floor spreading for the last 36 million years, and sediment found near the ridge was relatively young. Sediment from the western abyss of Australia proved to be the oldest in the Indian Ocean, about 135 to 140 million years old. Thus, scientists are relatively confident that the sea floor spread in a general east-west direction during the Cretaceous Period, from 65 to 135 million years ago.

In March, 1973, the Deep Sea Drilling Project turned its attention to the Antarctic. There, the *Glomar Challenger* drilled into the sea floor, beneath the ocean current system called the West Wind Drift, or Circum-Antarctic Current, to recover fossil remains of the variety of microscopic life carried by this current. By studying the remains, scientists hope to determine long-term

Oceanography

Continued

Whirling winds and sea spray form the funnel clouds of waterspouts, the seafaring relatives of tornadoes. A recent study of the life cycles of waterspouts may help scientists understand how tornadoes, the more violent storms, develop.

changes in climate, water temperature, oxygen content, and the direction of the current. Biologically, the West Wind Drift has more plants and animals than any other ocean area in the world, and scientists credit it with having a major effect on world climates, glaciation, and even the evolution of life.

Saving fin whales. In March, 1973, Captain Jacques-Yves Cousteau reported that there is an unusually large amount of krill, or planktonic whale food, in the Antarctic because of the diminishing numbers of fin whales. Heavy fin whale fishing since 1930 has lowered the population of these mammoth plankton feeders to a point dangerously near extinction. Meanwhile, scientists at the British National Institute of Oceanography have long been studying fin whales caught by commercial fisheries and determining their ages by counting the annual rings that form on whale ear bones. They also examined the whales' reproductive systems and found that fin whales captured in 1920 matured at about 10 years of age, while those captured in 1964 matured in only

6 years. Marine biologists now believe that if the fishing of these animals can be curtailed, the abundance of food and their earlier maturation may save these ocean mammals from extinction.

Diving sea mammals. On Sept. 15, 1972, the Navy announced that a 1,200-pound pilot whale named Morgan and a 5,500-pound killer whale named Ahab had been enlisted into its Undersea Center's Deep Operational Systems program in Hawaii. Diving mammals will play an important role for the Navy in locating and recovering objects from the sea, such as torpedoes and hydrophones. They can dive deeper and search a larger area than human divers, and can home in on sound-generating devices without the need of complex sonar equipment.

Holding a hydrazine gas lift device in its mouth, the whale homes in on experimental, sound-generating torpedoes on the ocean floor at depths of more than 1,000 feet. When the whale locates the torpedo, it presses a grabber claw against it and then lets go. A trigger on the claw sets off a gas device that inflates a balloon to float the torpedo to the surface.

Oceanography

Continued

To study the recent history of coral-reef development, geologists drill sample cores from relatively young coral reefs discovered in the eastern Pacific Ocean off the coast of Panama.

Morgan was captured off Santa Catalina Island near the southern California coast in 1968 and taken to Hawaii, where he was trained. He swims alongside or behind a small boat on the way to recovery sites and then returns voluntarily to his corral from the open sea. Morgan has retrieved objects from 1,600 feet, and other whales may be able to go to twice that depth.

The Navy has trained California sea lions to retrieve objects from depths of less than 500 feet. The sea lion, riding with two men in a rubber boat, slides into the water at the approximate location of a lost object, which must have a sound-generating "pinger" attached. When the sea lion hears the pinger, it signals the men in the boat, and they give the sea lion a grabber claw and a haul line. The sea lion swims to the lost article, attaches the grabber claw, and the men haul in the object.

Man in the sea. An ambitious underwater habitat program began in Puerto Rico in September, 1972, with the first trial dives of the Puerto Rico International Undersea Laboratory. The underwater house consists of two horizontal cylinders, each 20 by 8 feet, inside a submersible barge. There are three 42-inch windows and a 9-foot-wide underwater hangar for docking a research submarine. A utility buoy moored to the submerged habitat supplies air and power. More than 39 universities have expressed interest in using the habitat for such undersea studies as fisheries development and pollution studies.

Cold-water eddies. In August, 1972, NOAA announced that oceanographers dropped two floating oil drums into a huge cold-water eddy outside the Gulf Stream and began to chart its course. The rotating body of cold water, surrounded by warmer water that had broken off from the Gulf Stream, was moving southward about 300 miles off the Georgia coast.

Oceanographers suspect that cold-water eddies rejoin the Gulf Stream. In June, they attempted to confirm this by using a research ship to track another large eddy first sighted in April, 1971. However, Hurricane Agnes thwarted their efforts. [Richard H. Chesher]

Physics

Atomic and Molecular Physics.

Significant progress was reported in 1972 and 1973 in efforts to develop shorter wave length lasers. Although the first break into the far ultraviolet came in 1970 with massive electrical discharges in hydrogen gas, these devices were not efficient enough for practical applications. Two new approaches to the development of ultraviolet lasers have now been reported, and both of them show considerable promise.

Stephen E. Harris and his co-workers at Stanford University reported in October, 1972, that they had generated ultraviolet lasing in a cadmium-argon system. They used the principle of harmonic generation. That is, if light of a frequency, f, passes through a medium in which one of its harmonics (2f, 3f, 4f, and so on) has the same velocity as f, then light of the harmonic frequency will be produced within the medium.

This technique, called phase matching, has been used before to generate coherent visible light. Infrared laser radiation is passed through a crystal such as potassium dihydrogen phosphate (KDP), or sodium barium niobate. These crystals have equal refraction indexes, and thus the same light velocity in the infrared, and at double the infrared frequency in the visible.

In the far ultraviolet, however, all crystals become opaque, and researchers must turn to other media for frequency multiplication. Harris and his co-workers demonstrated that a gaseous system can be used in this manner.

A careful mix. The index of refraction of a gas increases slowly as the frequency of the light grows from the visible into the ultraviolet. However, the index jumps upward in a phenomenon called "anomalous dispersion" in the vicinity of a resonance line, or absorption line.

Vapors of metals, such as sodium, rubidium, cesium, or cadmium, have so many of these resonances in the visible part of the spectrum that their refractive indexes in visible light are actually larger than at ultraviolet frequencies. But a mixture of a metal vapor with a rare gas whose indexes are larger in the ultraviolet than in the visible, will have equal indexes for a visible line and for its third harmonic in the ultraviolet.

The Harris group mixed cadmium vapor and argon gas in a ratio of 1 to 25 to achieve this match. In their two-step process, they passed 10,600 angstrom (A) wave length infrared radiation from a neodymium-doped yttrium aluminum garnet (Nd:YAlG) laser through a crystal of KDP.

Phase matching in the crystal converted 80 per cent of the 10,600-A radiation to 5320-A visible light. When they then passed this visible light through the cadmium vapor-argon mixture, the frequency of the incoming light was tripled, producing coherent ultraviolet radiation at 1770 A. Only about 0.01 per cent of the light energy was converted to ultraviolet in this process.

The Stanford group then produced laser radiation at even shorter wave lengths. They mixed the same 5320-A radiation produced by the first KDP crystal with the residual 10,600-A radiation in a second KDP crystal. This produced radiation having a sum frequency at 3550 A.

This 3550-A radiation was then passed through the cadmium-argon cell, operated in this experiment at a Cd to Ar ratio of 1 to 2.5. But the overall efficiency of producing the output – 1180 A ultraviolet – was extremely low, with an energy conversion ratio of 10^{-5} per cent.

Xenon-argon systems. The Stanford physicists knew that the difficulty they had in producing a homogeneous mixture of the rare gas and metal vapor severely limited the conversion efficiency. To overcome this limitation, the group reported in early 1973 that they turned to a mixture of two of the so-called rare gases, xenon (Xe) and argon (Ar).

Their objective was to triple the frequency 3550-A radiation and thus produce a radiation at one-third its wave length, at 1180 A. Since xenon gas has a strong absorption line at 1190 A, its index of refraction on the higher frequency side of that line (at 1180 A) is smaller than its index in the near ultraviolet at 3550 A. Consequently, by mixing Xe with Ar, whose index is greater at 1180 A than at 3550 A, they equalized the indexes at the two wave lengths.

They achieved a satisfactory match with an Xe to Ar ratio of 1 to 430. They obtained the 3550-A radiation used as the starting point for this conversion by mixing 10,600-A with 5320-A radiation. An efficiency of 2.8 per cent was obtained for the overall conversion to 1180 A.

Physics

Three groups of researchers reported that lasers can generate far ultraviolet radiation without resorting to frequency multiplication. Two groups—A. Wayne Johnson and James B. Gerardo of Sandia Laboratories, Albuquerque, N. Mex., and a cooperative project involving Earl R. Ault and Mani L. Bhaumik from the Northrup Corporation Laboratories in Hawthorne, Calif.; William M. Hughes and Reed J. Jensen, Los Alamos (N. Mex.) Scientific Laboratory; and Alan Kolb and John Shamnon of Maxwell Laboratories, San Diego—reported the results of experiments in March, 1973. Charles Rhodes and Paul Hoff of the Lawrence Livermore Laboratories in Livermore, Calif., announced similar results in May. They all used Xe gas molecules as the active laser medium.

Molecular xenon (Xe_2) is not stable in its ground state, and quickly breaks apart into two stable Xe atoms. However, the molecule has metastable states, excited states that last long enough to permit efficient laser action.

In the reported experiments, a gas of Xe atoms is bombarded by a very intense beam of electrons. This beam ionizes and excites the Xe atoms, and many combine to form metastable xenon molecules. These molecules then give up their residual energy by emitting an ultraviolet photon and fall to the ground state, and the molecule immediately flies apart.

The ideal situation thus exists for laser action, with many molecules in an excited state and almost none in the ground state. The excited molecule emits radiation at a wave length of 1730 A. No operating lasers have been constructed, but these experiments all give strong evidence that laser action in the ultraviolet is taking place.

The Sandia and Northrup groups reported spectroscopic observation of a narrowing of the normally broad molecular ultraviolet spectrum. Such narrowing always happens in laser action, but by itself it is not proof. As further evidence of laser action, the Northrup group reported that the light intensity in the xenon cell melted a portion of the aluminum coating on the cavity mirrors. And the Lawrence report described not only line narrowing but also spacial coherence, thus confirming the existence of laser action.　　[Karl G. Kessler]

Elementary Particles.

In a field that has known several exciting years during its 25-year history, 1973 may go down as one of the most exciting of all. Evidence is mounting for a tiny new world within the so-called "elementary" particles. The new evidence comes mainly from studies of head-on collisions of protons in the intersecting storage rings (ISR) at the multinational CERN laboratory near Geneva, Switzerland.

These are the most violent collisions produced in any laboratory. But physicists studying the even more energetic cosmic-ray protons that come from outer space had seen hints of what was to come some time ago.

Electron-proton collisions at the Stanford Linear Accelerator (SLAC) first focused attention on the possibility of a new level of understanding elementary particles about three years ago. The high-speed electrons were deflected through large angles. They behaved as if they had recoiled from concentrations of electric charge much smaller than the diffuse cloud of lightweight, charged mesons thought to be the main constituent of the protons. Richard P. Feynman of the California Institute of Technology (Caltech) called these tiny concentrations partons, and developed ways of using the electron data to study their properties. But SLAC's electrons have too little energy to paint more than a fuzzy picture of the partons. James Bjorken of SLAC and a number of other theorists have been urging that similar studies be undertaken at the more powerful proton accelerators.

Small detail—high energy. Seeing finer details of structure requires higher energies. All fundamental forces in nature vary inversely as the square of the distance, at least at very short distances. Thus, it takes large amounts of energy to produce close encounters. There is also the dual wave-particle character of subatomic matter. A moving particle behaves as if it were a tiny bundle of waves: the higher its energy, the shorter its wave length.

In a giant accelerator, as with an ordinary microscope, it is impossible to see details much smaller than one wave length. When two particles collide, the amount of momentum transferred from one particle to another determines how fine a picture can be obtained from the collision data. When a fast-moving particle hits a stationary one, not all of its momentum can be transferred, because the momentum along its original flight path must be conserved.

The momentum of the colliding protons in the ISR is equal and opposite, so nearly all of it can be transferred, provided there is something inside the proton concentrated enough to absorb it. Because of this, a collision between the ISR's two 27-billion electron volt (GeV) beams is equivalent to a collision of a 1,500-GeV proton with a stationary one.

The most direct evidence for small concentrations of matter comes from 1972 collaboration among American physicists from Rockefeller and Columbia universities and an Italian group working at CERN. The team placed a detector sensitive to gamma rays, which are similar to X rays but more energetic, at right angles to the colliding beams. They expected this counter to be very "quiet" because meson-cloud collisions cannot send high-energy gamma rays off at right angles. To the scientists' surprise, they found quite a few gamma rays with energies as high as 7 or 8 GeV.

A number of theorists have interpreted this result as follows: A parton inside one of the protons has a near-miss encounter with a parton in the other proton, transferring a huge amount of energy and momentum. As the parton is bound to the other constituents in the proton by extremely strong forces, it must quickly brake to a half, shedding its excess energy.

One well-known way to do this is to radiate a gamma ray, a process that has been seen when electrons are stopped in matter. The theory permits a simple mathematical test – a graph of the number of gamma rays versus their momentum should be a straight line with a slight slope on logarithmic graph paper. Gamma rays produced in collisions of the meson clouds would produce a steeply falling curve. There are not enough events to be absolutely sure, but the logarithmic graph shows a straight line out to much higher momenta.

Like discovering the nucleus. This logarithmic behavior is remarkably similar to the scattering results that led

A 10 per cent increase in the total cross section, or size, of colliding protons in the newest, highest energy accelerators may mean that physicists are on the threshold of a completely new type of reaction between protons.

Proton-Proton Total Cross Section

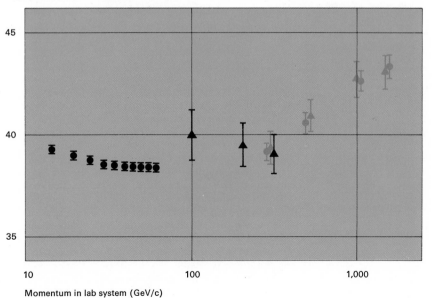

Serpukhov   ● ISR (Rome-CERN)
NAL   ▲ ISR (Pisa-Stony Brook)

Total cross section in millibarns

Momentum in lab system (GeV/c)

Physics

Continued

Parton-Parton Collision

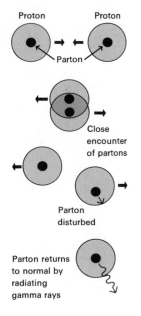

Proton   Proton

Parton

Close encounter of partons

Parton disturbed

Parton returns to normal by radiating gamma rays

Gamma-Ray Spectra

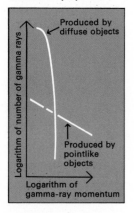

Logarithm of number of gamma rays

Produced by diffuse objects

Produced by pointlike objects

Logarithm of gamma-ray momentum

A slightly sloping line on a logarithmic plot of the results from the highest energy particle collisions indicates that pointlike objects, partons, exist within each proton and neutron.

the English physicist Ernest Rutherford to discover the atomic nucleus in 1911. Bombarding gold foil with alpha rays, Rutherford was startled to observe large numbers of them deflected through large angles. A plot of a number of alpha rays versus their momentum formed a straight line on a logarithmic graph, the characteristic "signature" of a pointlike center of force. He had expected a steeply falling curve, from the effects of the lightweight electrons, much like the curve expected from the meson cloud in the ISR experiment.

Large-angle scattering similar to that found in the ISR has long been common knowledge among researchers studying cosmic rays. The earth's atmosphere is continually bombarded by high-energy protons from space, and a few of these can transfer far greater momentum than protons in the ISR. Collisions of these energetic protons with nuclei in high-altitude detectors occasionally produce "jets" of mesons that travel nearly parallel to one another, and at a considerable angle to the proton's original line of flight. The momentum transfer is often as great as 30 GeV. Collisions between meson clouds, however, would produce a "splatter" of mesons centered on the original proton direction.

The parton model gives a natural explanation of these wide-angled jets. They are just an alternate way for a disrupted parton to shed excess energy, by converting it to matter in the form of a narrow jet of mesons along its original path.

Why were these cosmic-ray results, known since the mid-1960s, so long ignored? Cosmic-ray research is a very small corner of particle physics, cut off from the mainstream of research. The cosmic-ray researcher must haul his apparatus to a mountaintop or fly it in a balloon, because the atmosphere filters out most of the protons before they reach the ground. Data comes in at a painfully slow rate. It takes a daylong balloon flight or a summer at mountaintop altitude for a detector with an area of 10 square feet to see as many collisions as occur in one second in the ISR.

Now, the ISR results, combined with hard-to-get research money in particle physics, may well lead to a revival of interest in cosmic rays.

The proton grows. Another cosmic-ray result that adds to the case for a new

world within the particle has also been confirmed in the ISR. One of the most basic particle physics experiments is to measure the cross section, or "size," of the proton by seeing how often protons collide when passing one another. The resulting size is slightly smaller than that obtained by other methods, indicating that protons are somewhat "transparent" to one another. As energy increases, the transparency increases very gradually, and was expected to reach a steady value at energies as high as the ISR's.

Previous cosmic-ray measurements around 10,000 GeV showed the proton to be about 20 per cent "larger" than expected, but these results were not taken too seriously because the evidence was somewhat indirect. Now, two independent measurements made at CERN seem to confirm the cosmic-ray result. Over a range from 300 to 1,500 GeV single particle energy, the proton's size seems to "grow" by about 10 per cent. Measurements of this type were being analyzed in 1973 at the National Accelerator Laboratory (NAL) near Chicago, covering the energy range from 100 GeV to 400 GeV.

Similar growth occurs at much lower energies, as colliding protons reach the energy needed to dislodge a free meson. This new way of reacting makes the proton less transparent. The new, high-energy growth of the proton may be due to yet another new way of reacting—disrupting its parton "core."

Another Chinese box. We might reasonably ask, as has theorist Geoffrey Chew of the University of California, Berkeley, whether there is any point in forever extending the study of matter to smaller scales. Will this succession of new levels of reality, of "Chinese boxes" within boxes never end? John A. Wheeler of Princeton University has what he thinks is a plausible guess. Combining Einstein's general theory of relativity with the quantum theory, Wheeler suggests that on an extremely small scale, space itself ceases to be smooth and continuous, and is full of tiny holes like a "Swiss cheese." These holes are incredibly small, 10^{-33} centimeters, even by particle physics standards. Thus, there may well be many more Chinese boxes between the particle physics of 1973 and the ultimate structure of matter and space. [Robert H. March]

The End of Particle "Chemistry"?

In the light of the new discoveries—which indicate that pointlike objects, or partons, exist within the proton—the 1950s and 1960s may come to be regarded by historians of science as the era of "particle chemistry." During this period, the emphasis in research was on finding new particles and looking for order in their complicated interactions.

Chemistry went through a similar period in the 1800s, when most of the elements were discovered and their simplest reactions were studied. By the end of this period, enough regularity had been observed to permit the Russian chemist Dmitri I. Mendeleev to arrange the elements in the periodic table. But attempts to explain this table on the level of the atoms themselves, such as by assuming all atoms to be built up from hydrogen, were doomed to failure.

Chemical experiments involved too little energy to do more than scratch the surface of atoms; the nucleus remained unaffected. Only when researchers achieved higher colliding energies and discovered that incoming particles would rebound from a pointlike object within the atom—the nucleus—could the differences between atoms be explained.

The history of particle physics is strikingly similar. By the early 1960s, particle physicists had their own periodic table, developed by physicist Murray Gell-Mann of the California Institute of Technology. To explain the table, he assumed that all particles were made up of three kinds of subparticles, with peculiar electrical and nuclear properties, which he christened "quarks."

The quark model has been remarkably successful: There are few blank spots in the physicists' "periodic table," and all particles fit neatly into it. But all efforts to knock a quark loose from a particle or find a free one in nature have failed. Gell-Mann doubts that the quarks can exist as free particles, but there could be other explanations for their stubborn refusal to materialize. Perhaps they are bound together by strong forces that prevent their being knocked loose at presently available energies. In any event, the question for particle physics in the 1970s seems to be: "Are the partons quarks?"   [Robert H. March]

Reaching Ever-Smaller Sizes

Size of detail in centimeters

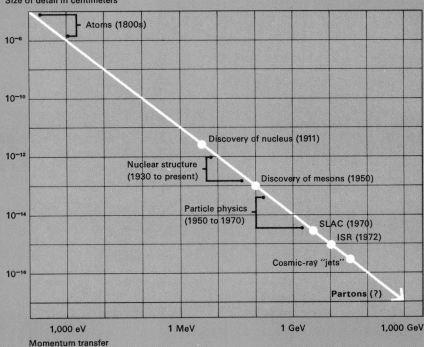

Nuclear Physics. Nuclear physicists in 1973 dramatically increased the accuracy of both their measurements and their predictions. During the preceding decade, they were understandably pleased at bringing their science to the point where it was possible, in almost all cases, to predict nuclear properties with from 20 to 50 per cent accuracy. But suddenly, through new discoveries, new calculating techniques, and including previously ignored phenomena, physicists have found ways to make nuclear predictions that match experimental results to within a few per cent in many instances. This marked improvement should have far-reaching consequences in both the fundamental and applied branches of nuclear science.

Much knowledge of nuclear structure has come from reactions in which physicists add or remove one or a few neutrons or protons from a target nucleus. The simplest of these reactions is the deuteron-stripping reaction, in which the incident deuteron, a single neutron and a single proton loosely bound together, interacts with the target. The neutron is captured in an orbit around the target to form a new nucleus one unit in mass heavier. The proton emerges from the reaction with a velocity that conserves energy and momentum. From the energy angular distribution of these protons, physicists can obtain the detailed characteristics of the orbit in which the neutron was captured.

In the simplest mathematical model of this reaction, the incoming deuteron and the outgoing proton are treated as completely free particles, and the neutron is simply transferred at the point of closest approach to the target.

However, the target nucleus affects the behavior of the incoming deuteron and the outgoing proton. As a result, Raymond Satchler of the Oak Ridge National Laboratory (ORNL) in Tennessee developed a more sophisticated computation, the so-called distorted-wave Born approximation (DWBA) in 1958. It has been used extensively throughout the world as a much more reliable method of finding nuclear structure from the reaction data, even though the best results were only 20 to 50 per cent accurate.

"The sky is falling! The sky is falling!"

This treatment ignored certain key aspects of the interaction. The first of these is recoil. When the target captures an additional nucleon it must recoil, and the DWBA approach simply ignores this fact. In 1973, Ralph DeVries of the University of Washington and Saclay, France, demonstrated that including this recoil effect produces a much more accurate description of the reaction.

The second ignored effect was that the target can be excited to a higher quantum state and that the neutron can be captured by the target while in this excited state. Moreover, the outgoing proton can interact with this system and lift the orbiting neutron into a still higher quantum state about the target. All three processes can occur simultaneously.

Charles King and Lance McVay of Yale University found striking evidence in 1973 for these processes in measurements on tungsten and on various rare-earth nuclei.

Robert Ascuitto and his colleagues, also at Yale, developed a new theoretical analysis that includes all these processes and found that they can reproduce experimental results to accuracies of within a few per cent and also get much more precise nuclear structure information. The study of these higher ordered effects is still in its infancy but it opens up a new dimension in nuclear reaction work.

New nuclear shapes. In the early days of nuclear physics, physicists assumed that nuclei were spherical. As better information became available, it was clear that they could be cigar-shaped or doorknob-shaped, and in some cases they could oscillate between the two.

Recent measurements in many laboratories have shown that even these models are much too simple, and that much more exotic shapes occur. The uranium nucleus, for example, apparently resembles a loaf of French bread, with a rather lumpy center, while some of the barium nuclei are reminiscent of the flying saucers that once were prominent in comic strips.

Having discovered these new shapes, physicists hope to use their knowledge of how the individual neutrons and protons move inside the nucleus, not only to fit the new shapes, but also to predict where in the table of the nuclei we should expect other shapes to occur. In 1973, for example, Cheuk-Yin Wong at ORNL predicted that doughnut-shaped nuclei would be stable, once formed, and experimenters are now examining the possibility of preparing such nuclear doughnuts by using very high-energy heavy ions to ream out the central hole from a heavy nuclear target.

Nuclear matter, consisting of protons and neutrons, is extremely dense—a cubic inch would weigh about a billion tons—and is virtually incompressible. This means that the volume of the nucleus must be directly proportional to the number of neutrons and protons present. For approximately spherical nuclei, the nuclear radius should be proportional to the cube root of the total number of neutrons and protons.

Since this simple prediction had long been considered correct, electron-scattering measurements produced a considerable surprise in 1971. Measurements by Robert Hofstader and his collaborators at Stanford University, by John Schiffer and his collaborators at the Argonne National Laboratory near Chicago, and by William Bertozzi and his Massachusetts Institute of Technology (M.I.T.) collaborators working at the National Bureau of Standards showed a quite different relationship for calcium isotopes. The experimenters moved systematically from Ca-40 targets to Ca-48 targets (these differ only in the addition of more neutrons). They found that the radius apparently decreased. Similar behavior was found in precise measurements on other families of isotopes.

This puzzle was resolved in 1973 by John Negele at M.I.T. Negele recognized that theorists could not consider the neutrons and protons as simple objects without an internal structure, as was traditional in nuclear physics, because electron scattering really measured the charge distribution in the isotopes under study.

Elementary particle physicists have found that the neutron has an internal meson structure—a tight, positively charged shell near its center and a much more diffuse cloud of negative charge outside that (the neutron as a whole is electrically neutral). By taking this internal structure into account in his analysis of electron scattering, Negele showed that the dependence of radius of the calcium isotopes on mass number remains valid. [D. Allan Bromley]

343

Plasma Physics. One of the most dramatic but least-heralded enterprises of our day is the struggle to harness the energy of nuclear fusion that powers the hydrogen bomb. Few people realize that the success or failure of this quest may literally determine the fate of civilization. All serious projections of mankind's needs into the 21st century seem to require this virtually unlimited, nearly pollution-free source of energy. The suddenness with which we perceived the specter of the energy crisis in 1973 and its intimate connection with the preservation of the environment attest to the physicists' sense of urgency in developing fusion reactors.

The working fluid of such reactors will be a very hot deuterium-tritium plasma, one attaining temperatures of 100 million°C. This gas of electrons and ions must be kept away from the walls of its container by a strong magnetic field, which also serves to confine it long enough to achieve a net energy gain. The criterion for adequate confinement is the Lawson criterion. It states that the product of the confinement time, measured in seconds, and the ion concentration of the hot plasma, measured in ions per cubic centimeter, should exceed 10^{14}. To date, no plasma device has been able to exceed 5×10^{12}.

The ATC experiment. The success of the Adiabatic Torus Compression (ATC) experiment at Princeton University's Plasma Physics Laboratory is the most significant advance in 1973. The experiment tested a suggestion by Princeton's Harold P. Furth and Shoichi Yoshikawa of how to heat a tokamak plasma more efficiently. Tokamaks are doughnut-shaped plasmas that have strong electric currents induced in them. This plasma current heats the plasma, because of its electrical resistance, and also generates a magnetic field that helps stabilize, or confine, the plasma.

One difficulty that had previously been encountered was the tokamak's inability to produce a high enough temperature, because the plasma's resistance decreases as the temperature increases. Furth and Yoshikawa suggested that a vertical magnetic field, decreasing in strength with increasing radius, be imposed perpendicular to the plane of the plasma doughnut. The vertical field drives the circular plasma current flowing in the plasma doughnut inward, into a stronger field, where it is compressed and heated.

The initial experiments, conducted under the direction of Robert A. Ellis, Jr., raised the electron temperature from 10 million°C. to 25 million°C., and the ion temperature from 2 million°C. to 6-million°C.

Recent experiments have achieved even greater heating. The Princeton group injected an energetic beam of electrically neutral deuterium atoms into the plasma prior to compression. Neutral particles, unlike charged particles, are not affected by magnetic fields. Consequently, they can penetrate the confining field, collide with the electrons and ions, and heat the plasma. They achieved a final ion temperature of 7.5 million°C. But this is still a long way from the 100 million°C. temperature needed in a fusion reactor.

Other tokamaks, notably the Oak Ridge National Laboratory's Ormak, and the Massachusetts Institute of Technology's Alcator, also went into operation during the past year. However, even if they are completely successful, none of them can demonstrate the "scientific feasibility" of controlled thermonuclear fusion, defined as satisfying the Lawson break-even criterion.

A new tokamak, the Princeton Large Torus (PLT), now under construction, is expected to reach to within 50 per cent of the break-even criterion. About three times as big as the largest existing tokamak, its confinement times should be from 0.3 to 0.5 second, typically 100 times those presently achieved. Ion densities will be about the same. The PLT is scheduled to go into operation in 1976, a year after the target date set for a comparable, but slightly smaller tokamak, T-10, being built at the Kurchatov Institute in Moscow.

Reactor design. The steady progress in understanding fusion plasmas has encouraged detailed theoretical design studies of a working reactor. The most extensive of these, by Arthur M. Fraas of the Oak Ridge National Laboratory in Tennessee, also attempts to study all the foreseeable problems of a tokamak reactor. Fraas's study envisages a huge, doughnut-shaped plasma container more than 30 feet high and more than 100 feet across, surrounded by a blanket of

A 7-ton superconducting magnet, shaped like the seams of a baseball, is gently lowered into its vacuum chamber at Lawrence Livermore Laboratory in California. This experiment, called Baseball II, is one of many attempts to create the plasma-containing magnetic bottle that will allow physicists to control thermonuclear fusion.

liquid lithium more than 3 feet thick. The lithium would be heated by the reaction to a temperature of about 1000° C. (1800° F.).

This thick blanket of lithium will serve three purposes. It will shield the superconducting magnet coils surrounding the outer chamber from the intense neutron radiation emerging from the plasma's fusion reaction. It will breed the fuel needed for the thermonuclear reaction; tritium is a recoverable reaction product when lithium absorbs the neutrons. In addition, it will provide a heat exchanger between the plasma and the conventional boiler that drives the electric generator. A single such plasma fusion reactor would be expected to generate up to 5,000 megawatts of electricity, which is about half the present power requirement of New York City.

The problems that must be overcome to build such a reactor are a little like trying to confine and extract energy from a portion of the interior of the sun without melting the container. The plasma must be kept away from the container walls, which must be effectively cooled. But the proposed coolant, molten lithium, is an excellent conductor of electricity and therefore can only be pumped along the direction of the magnetic field.

Furthermore, the molten metal must be prevented from coming in contact with the water in the boiler to avoid a destructive chemical explosion. The container walls, exposed both to the hot corrosive lithium and to the massive doses of energetic neutrons, must last from 10 to 20 years. None of the radioactive tritium that circulates through the system can be allowed to escape, despite a tendency that many materials have to become permeable to this gas at the temperature of the reactor. But Fraas's study indicates that these difficulties can probably be overcome.

Fusion reactors appear to have far fewer hazards than fission reactors. Fission reactors create much greater and longer-lasting radioactivity. This poses a difficult waste-disposal problem, as well as the danger of a catastrophic explosion. The production and transport of its fuel are also dangerous. Fusion reactors, though more complex than fission reactors, will not be beset by these hazards.　　[Ernest P. Gray]

Physics

Solid State Physics. Intense studies of magnetic materials that have extreme properties produced breakthroughs in 1973 that may soon affect such diverse fields as computer technology and air-pollution control. And physicists also found a new way to study crystal changes at high pressures.

Bubbles on the move. In 1960, C. Kooy and H. Enz of the Phillips Research Laboratories in the Netherlands demonstrated that they could produce a regular pattern of tiny cylindrical regions of reversed magnetization in certain carefully prepared materials. These regions are now called bubble domains. Then, in 1967, Andrew H. Bobeck of Bell Telephone Laboratories reported that he and several colleagues had discovered a new class of magnetic materials in which bubble domains could be generated. They demonstrated that these microscopic domains could then be moved about by applying electric current to miniature magnetic routing plates deposited on the magnetic material itself. Devices based on the motion of these bubbles, they reasoned, could carry out a variety of useful electronic switching and logic operations.

The materials they used are the so-called orthoferrites, which have the general formula $RFeO_3$. R stands for a rare earth atom, usually terbium, thulium, ytterbium, or in some cases mixtures of the rare earth elements.

Bobeck prepared a very thin platelet of the orthoferrite material for the experiment. A small magnetic field created a number of regular domains, or "stripe domains," in the sample that have alternating directions of magnetization. In this material the magnetization in each domain, however, is always perpendicular to the plane of the platelet.

When the suitable small external magnetic field is increased, the stripe domains collapse into bubble domains that appear as circles when viewed with polarized light. The bubbles, as small as 2 microns in diameter, remain stable as long as the external field exists.

Magnetic bubbles can be made to move quite rapidly in certain materials by applying a small, highly localized field near the domain. In order to construct a usable electronic device, however, the magnetic material must be highly homogeneous so that the density of the bubbles will be great, and each bubble will be stable in size and shape. Until now, the research effort in developing bubble-based devices had been limited by the very difficult job of preparing large single-crystal platelets of the several types of orthoferrite materials that seem best suited for a working device.

Praveen Chaudhari, Jerome J. Cuomo, and Richard J. Gambino of the I.B.M. Research Laboratory at Yorktown Heights, N.Y., reported in 1973, that they had discovered how to produce highly uniform arrays of small bubbles in several amorphous, or noncrystalline, alloys of gadolinium and cobalt. The researchers prepared films of these alloys by either sputtering—bombarding the surface with microscopic alloy particles—or evaporating a previously prepared ingot onto a suitable substrate. In these films, the local magnetic moments of the gadolinium atoms are directed opposite to those of the cobalt atoms.

The I.B.M. research team found that highly homogeneous and noncrystalline films could be prepared on a wide variety of substrates, even on a flexible heat-resistant plastic film. They varied the cobalt concentration between 65 and 90 per cent. Even though the atoms in the noncrystalline mixtures are not located at regular points, as they are in a crystal lattice, their random clustering about each other still has a high degree of homogeneity on the scale required for the bubble domains.

Such amorphous magnetic materials had been prepared previously by other investigators. The I.B.M. work has shown, however, that unique, controllable, and highly regular properties can be achieved without the expense of producing the perfect single crystals that researchers previously believed were necessary for these magnetic devices. See ELECTRONICS.

Shock-wave transformations. High-compression shock waves, such as those produced by explosions, should produce chaotic changes in a crystalline atomic lattice. Yet, physicists know little about what occurs at the atomic level when a shock wave passes through a solid.

Evidence over the past years indicated that shock pressures cause regular crystal transformations within very short times.

Physics

Continued

But it was not until 1972 that scientists directly observed a crystal changing its structure while a shock wave was passing through it.

Quintin Johnson and Arthur C. Mitchell of the Lawrence Livermore Laboratory in California explosively shocked boron nitride with a pressure wave equivalent to 245,000 times atmospheric pressure. Boron nitride is a solid with a crystal structure similar to that of graphite. Just before the explosion, they also flashed a very high intensity X-ray source through the sample to transmit a pulse of X rays that lasted only 50 billionths of a second.

The researchers then checked the X-ray lines that had formed on photographic film. Before the shock wave entered the sample, the crystal structure was characteristic of normal boron nitride. However, the film showed that new, regular lines formed during the compression. Johnson and Mitchell tentatively believe these short-lived lines were caused by a new crystal structure, similar to the structure of cadmium sulfide, CdS, a well-known compound.

However, the boron nitride that Johnson and Mitchell studied did not show this crystal change with lower-pressure shock waves. The researchers have not yet determined whether such structural transformations from one regular crystal to another will take place in all solids, especially if a long-range shift of atoms is required to form the new structure. But the properties of almost all materials depend drastically on pressure, and high-pressure techniques are currently used to synthesize numerous materials, such as artificial diamonds. Thus, these direct observations of the effects of compressional waves provide physicists with a new and important look at rapid atomic processes.

Catalytic materials. By using new ultrahigh-vacuum systems and high-resolution electron microscopes, scientists have increased their ability to prepare and study the character of surfaces. For instance, they can now examine the electronic bonding between small numbers of atoms and a metal surface to which they are attached. Such studies should help them better understand the

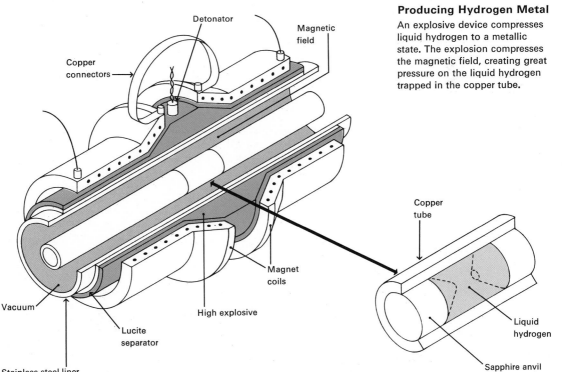

Producing Hydrogen Metal
An explosive device compresses liquid hydrogen to a metallic state. The explosion compresses the magnetic field, creating great pressure on the liquid hydrogen trapped in the copper tube.

Detonator

Magnetic field

Copper connectors

Magnet coils

High explosive

Vacuum

Lucite separator

Stainless steel liner

Copper tube

Liquid hydrogen

Sapphire anvil

Physics

gas-surface interactions that are believed to be most important in catalysis, where a metal such as platinum increases the rate of a chemical reaction.

In 1973, physicists made progress in understanding the microscopic processes in catalysis and synthesized additional catalytic materials. They stressed the relationship of a catalyst's electrical and magnetic properties to its activity.

Willard F. Libby of the University of California, Los Angeles, suggested in 1971 that a material called lanthanum cobaltite ($LaCoO_3$) might be a good and economical catalyst for oxidizing the carbon monoxide (CO) in automobile exhaust to carbon dioxide (CO_2). Libby's suggestion had been stimulated by the earlier work of D. B. Meadowcroft of the Central Electric Research Laboratory in Leatherhead, England. Meadowcroft discovered that the class of materials that included $LaCoO_3$ were good oxidizing agents.

While all of the oxides Meadowcroft studied were semiconducting, he noted that the molecules in some of the oxides would become magnetically ordered.

Meadowcroft mentioned a possible relationship of the catalytic activity to the magnetic properties of $LaCoO_3$.

Now, a team of scientists at Bell Labs – Rudolf J.H. Voor Hoeve, Joseph P. Remeika, Paul E. Freeland, and Berndt T. Matthias (also from the University of California) – prepared and tested several mixtures of compounds of the $LaCoO_3$ type. They used $RMnO_3$, where R may be the rare earth atom lanthanum, neodymium, or praseodymium. They also replaced some of the R atoms with lead. These oxides have a wide variety of physical and magnetic properties, depending on the atoms used.

The scientists then studied the conversion of CO to CO_2 by simultaneously flowing CO and oxygen gas (O_2) over a small quantity of these catalytic materials. They reported very favorable carbon dioxide conversion rates. Much study is underway on these and other new catalytic materials to determine their application to the chemical industry and pollution control and on how their activity is related to magnetic and electrical properties. [Joseph I. Budnick]

Psychology

Two approaches to the problem of learning disabilities in children received wide attention in 1973. One is typified by the work of Dr. Paul H. Wender, a pediatrician-researcher at the National Institute of Mental Health Laboratories in Washington, D.C. It is based on research in the neurochemistry of the brain and, more directly, on the effects of drugs that improve the behavior of such children.

The other approach is that of neuropsychologist Paul Satz and his colleagues at the University of Florida. They suggest that a child with learning disability suffers from delayed maturation of the neural systems of the brain that are related to the left hemisphere. The left hemisphere of the human brain is important for the use of language and for other symbolic activities, such as those used to solve arithmetic problems.

From 10 to 30 per cent of the children in public schools fail to do as well as their peers in schoolwork, even though their intelligence-test scores are well within the normal range. In most cases, difficulties with reading and arithmetic prevent them from realizing their academic potential. Specific symptoms may include disorders in listening, talking, writing, spelling, or thinking.

Several terms have been used to describe these difficulties. Among them are congenital word blindness, perceptual disorder, primary reading retardation, specific dyslexia, educational handicap, psychoneurological learning disability, and disabled learner. No single symptom identifies the child with such problems, and the symptoms overlap considerably.

Some scientists use the term minimal brain dysfunction (MBD) to describe a collection of behavioral symptoms that are presumed to represent a disturbance of brain activities. Scientists interested in a physiological or neurological approach to the problem usually use the term MBD, whereas people with a behavioral or educational approach use the term disabled learner.

Physical examinations sometimes reveal hearing and visual deficiencies, such as problems in hearing the difference between sounds like D and B or seeing the difference between letters like T and F. However, the behavior of these chil-

The mollusk, *top*, normally withdraws if touched with a sterile probe. But if trained to respond to a probe coated with food, and then to touch alone, *above*, it then follows, extends its tongue, and strikes. When its nervous system was isolated and stimulated, the effects of learning were detected at the level of single nerve cells for the first time in an animal.

dren cannot be explained on the basis of uncorrected hearing or visual disorders. Many of the children are hyperactive, finding it difficult to keep from getting into fights, throwing things, or dashing wildly about the schoolroom or house. They also show signs of withdrawal and are irritable, self-deprecating, anxious, and aggressive. Their parents and teachers consider them high-strung, or easily upset. Electroencephalographic (EEG) tests show that some of these children also have peculiar brain waves.

Family histories of such children indicate that their mothers tended to have more chronic illnesses than the mothers of normal learners. Also, the affected children are more likely to have been premature babies and to have been born to older parents.

At ages 5 to 6, many of these children seem to be as much as eight or nine months behind their peers in motor skills, or muscular coordination. They may often trip and fall, for example, or be unable to jump or climb a ladder or to get their eyes and hands to work in unison. Teachers and parents frequently ignore this delay in development until the children begin to have difficulty with the more formal aspects of education in the third and fourth grades. Almost all of these children have difficulty reading.

The overlap among the behavioral symptoms is substantial enough to raise the possibility that there is a single underlying factor that produces a particular child's behavior.

Too little pleasure? In his book *Minimal Brain Dysfunction in Children* (1971), Wender strongly advocates the view that there is an MBD syndrome stemming from physiological and neurological causes. It could have a basis in heredity, in accidents that occur to the fetus, or in accidents or unusual conditions at birth or shortly after. In any case, Wender believes that there is a deficiency in the brain mechanisms responsible for experiencing pleasure.

Much evidence suggests that the neurons, or nerve cells, essential to the experience of pleasure and responsive to rewards that modify behavior use the hormone noradrenalin to transmit nerve impulses across the synapses, or contact points, between the neurons.

Many cell bodies of the neurons in the noradrenergic systems lie in the brain

349

Psychology

Continued

With special equipment such as an abacus, a child with learning disabilities can take full advantage of his senses of sight and touch in order to overcome difficulty in conceptualizing numbers.

stem. Each of these cell bodies has a long fiber, called an axon, which projects forward, through, and along the bottom of the brain and forms part of a large collection of nerve fibers called the medial forebrain bundle, where the "pleasure center" is located. The axons end in various portions of the forebrain, or front section of the brain.

Injections of amphetamines such as Dexedrine or amphetaminelike drugs such as Ritalin stimulate the noradrenergic systems and usually make a normal person more active and enhance his pleasurable experiences. The odd fact about a child with signs of MBD is that if the child is hyperactive, these drugs can actually decrease his overactivity. Wender explains this by saying that the drug helps the child get more pleasure from his contacts with the environment because it increases the amount of noradrenalin in his brain. As a consequence, the child then exerts less effort to make contact with the environment and appears to be less active.

Maturation lag. Paul Satz and his colleagues present their theories in the book *The Disabled Learner: Early Detection and Intervention* (1972), which Satz edited with pediatrician John J. Ross of the University of Florida's J. Hillis Miller Health Center.

The effect of this delay varies according to the age at which the lag in maturation develops. In most cases the child's maturation is merely slowed down and a permanent deficit need not result. This optimistic prediction is tempered, however, by the possibility that a child may be especially ready to learn certain skills at a particular time. If the developmental lag is too great, however, the child may not be ready for new forms of behavior at the usual time. Thus, a "golden time," or critical period, for learning may slip by, never to be recovered.

The reasons for this neural and behavioral lag are as uncertain as those that may lead to the impaired noradrenergic systems that Wender postulates. Genetic factors are undoubtedly involved, because reading disorders are common in the families of some disabled learners, and six times as many boys as girls are affected. Some investigators believe that the reading difficulties found among a group of related people may be different from those found in the child without affected relatives. Environmental factors, such as trauma, disease, and undernourishment, may also slow a child's maturation rates.

Treatment possibilities. The most important· question that can be asked about the different views of Wender and Satz is, "What can be done for the child?" Treatments differ in one major regard: The value of using drugs. This is a most critical aspect of Wender's treatment, but it is not as important to Satz and those with his orientation. In part, this may result from the degree of functional abnormality that the two men usually encounter in their clinical work. Wender, a pediatrician at a research-oriented referral medical center may see more severely afflicted children than does Satz, who is a psychologist working within a medical complex.

Both would agree, however, that improving learning abilities depends on making adjustments that suit individual abilities and reactions. Both also think early detection is important because this allows the child, the parents, and teachers to work together to find new methods to deal with the problem and help the child before serious emotional problems begin to develop.

The disabled child is not lazy or dull; harsh disciplinary measures will not improve his learning. And individual psychotherapy will probably be ineffective. On the other hand, the consistent use of techniques that allow the disabled child to attend to his schoolwork and to be less distracted by external events at home and in the classroom may allow him to make remarkable progress in coping with his hyperactivity and his emotional problems.

It is especially significant that neither Wender nor Satz believes that the problem is based on inappropriate early learning or emotional development. Both believe that the basic defect has a neural origin. Consequently, the greatest hope for preventing or correcting learning disabilities lies in developing a better understanding of how the brain develops and functions.   [Robert L. Isaacson]

Science Support

The Administration of President Richard M. Nixon and the scientific community came into conflict on a variety of issues in 1973. The two most important points of contention involved dismantling the White House science advisory apparatus and reducing the funds available for scientific research and training.

The Nixon Administration described its actions as an economy drive and an effort to meet new national needs. But spokesmen for the scientific community charged that the Administration was manipulating the nation's research establishment for political reasons.

Science adviser resigns. The first hint of trouble came on Jan. 3, 1973. Edward E. David, Jr., the President's science adviser, announced to his staff that he was resigning immediately. David quit his government post to become executive vice-president of a Chicago-based manufacturing firm. There was no public explanation of why David had resigned so abruptly after 28 months on the job. However, it was reported that he was unhappy with Administration plans to revamp the advisory group he headed.

That advisory establishment consisted of four major parts: the science adviser; the Office of Science and Technology (OST), which includes about 50 technical specialists attached to the Executive Office of the President; the President's Science Advisory Committee (PSAC), made up of 18 experts who advised the President on technical matters; and the Federal Council for Science and Technology (FCST), composed of representatives from the major federal agencies concerned with science and technology.

On January 26, the White House announced a reorganization plan that would take effect July 1. It abolished both the OST and the post of science adviser. Their responsibilities were taken over by the director of the National Science Foundation (NSF), a federal agency whose chief purpose is to support scientific research. At the same time, the White House decided to let PSAC lapse into disuse and made plans to alter the functions of the FCST. The NSF director, in his new role as science adviser, was to report to an aide of Secretary of the Treasury George P. Shultz. Previously, the science adviser had reported directly to the President.

The reorganization was part of a White House effort to "streamline the executive branch of the federal government" by transferring responsibilities from the President's immediate staff to other agencies. President Nixon stated that he was abolishing the OST largely because it had completed its job of strengthening the scientific and technological capabilities of the government.

The abrupt changes caused consternation in the scientific community. Philip H. Abelson, president of the Carnegie Institution of Washington and editor of *Science* magazine, called David's resignation and the abolition of his office "a disappointing shock." Philip Handler, president of the National Academy of Sciences, the nation's most prestigious scientific organization, warned that the NSF director would find it difficult to carry out his science advisory chores. Major difficulties, he said, would be lack of support from his staff members; con-

flict of interest when, as science adviser, he must offer advice that affects his own agency; and communications problems in reporting to lower level White House aides rather than the President.

The budget. The Nixon Administration cut the budget for scientific projects and emphasized applied research, aimed at immediate, practical results, rather than basic research. The budget cuts were part of an Administration economy drive to curb inflation and avoid a tax increase by holding total federal spending below a limit of $250 billion in fiscal 1973 (July 1, 1972, to June 30, 1973) and $270 billion in fiscal 1974.

In order to stay below those levels, the Administration had to make cuts in the existing fiscal 1973 budget. In many cases, the Administration impounded, or refused to spend, funds that had been appropriated by Congress. When the proposed 1974 budget was released on Jan. 29, 1973, the Administration revealed that while it had originally planned to commit $17.8 billion for research and development in 1973, it had decided to cut this figure down to

Federal Spending for R & D

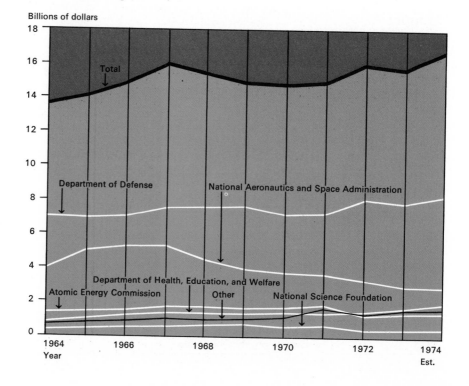

Billions of dollars

Science Support

Continued

Edward Gaensler, left, professor of physiology and surgery at Boston University School of Medicine, describes the medical facilities at the university hospital to a group of Chinese doctors who toured U.S. hospitals in October, 1972.

$17.1 billion, with only a slight boost to $17.4 billion in 1974.

Special areas of research were slated for budget increases. They were evenly divided between military and civilian research. The chief areas of civilian research that were scheduled to receive more funds involved energy, cancer and heart disease, transportation, prevention of natural disasters, drug abuse, and crime prevention.

Budget cuts fell particularly hard on certain areas of research and training in the basic sciences. Biomedical research and training supported by the National Institutes of Health (NIH) suffered most of all. The annual NIH budget was funded at $250 million less in fiscal 1974 than it had been in fiscal 1972.

One of the most drastic cuts affecting the NIH was the virtual elimination of the fellowship and traineeship programs used to support the advanced education of young scientists. The proposed 1974 budget provided money to complete the training of those who had previously won grants, but it contained no money for new awards. The Administration

suggested phasing out the training programs entirely, claiming that the nation had enough biomedical researchers. In July, after severe criticism, the Administration decided to continue the fellowship program, but at a reduced spending level. Another cut was imposed on the NIH competitive grants programs, the main vehicle for supporting basic biomedical research. Funds for new competitive grants were slated to drop more than 40 per cent over a two-year period.

The funding problems also affected other scientific disciplines. In a speech to the American Society for Microbiology on May 9, 1973, Handler asserted that nuclear structure physics was "in serious danger, with leadership now passing to other nations." He also stated that astronomy was hampered by lack of money for new instruments and operating funds and declared that the future of space science seemed dim, because the rising costs of the space shuttle would markedly reduce the money available for other areas of space science. Handler noted that, taking inflation into account, the proposed 1974 budget would actu-

Science Support

Continued

A group of students on the International Youth Science Tour react enthusiastically to the nighttime launch of Apollo 17 from Cape Kennedy, Fla.

ally provide less money for basic research than had the 1967 budget.

Education programs supported by the NSF also suffered heavy cuts, dropping from a peak of $135 million in fiscal 1968 to about $55 million in fiscal 1973. These and other federal budget cuts pushed the nation's research-oriented universities into a precarious financial condition, according to a survey report issued by the Carnegie Commission on Higher Education in April, 1973. The report stated that some institutions were demoralized because they would have to reduce the quality of their programs, while others had doubts about their future as research institutions.

Goal-oriented research. The Nixon Administration stressed research directed at specific goals, such as a cure for cancer or heart disease. Secretary of Health, Education, and Welfare Caspar W. Weinberger told a press luncheon on May 3, 1973, that the Administration was trying to move away from an overemphasis on research initiated solely by the scientists' interests and toward research determined by national needs.

The Administration believed this was the best way to make progress toward national goals. However, leading scientists argued that progress against major diseases could only be achieved by learning more about the fundamental biological processes.

The debate over the proper balance between goal-oriented and basic research was paralleled by disagreements between the scientific community and the Administration over the mechanisms by which the NIH awards research funds. In general, the scientific community favored the use of grants, which allow the scientist maximum freedom to design his own work. The Administration increasingly turned to the practice of awarding government contracts, which allow government officials to specify the type of work needed to be done. Moreover, the scientific community wanted committees of expert scientists outside of government to determine what research should be supported. However, the Office of Management and Budget challenged the desirability of such committees.

Woman heads AEC. On Feb. 6, 1973, Dixy Lee Ray became the first woman to head the Atomic Energy Commission (AEC). She replaced James R. Schlesinger, who resigned as AEC chairman to become director of the Central Intelligence Agency.

Ray, a marine biologist, had been a member of the commission since July, 1972. Besides specializing in marine invertebrates, Ray had previously been involved with environmental causes and with programs to promote public understanding of science and scientific issues. Ray reportedly was determined to remain neutral in the nuclear energy-environment controversy. But observers believed her experience in presenting science to the public would be a great asset in dealing with these problems.

Controversy over the safety of nuclear power plants became a major AEC issue in 1973. On February 16, the Environmental Protection Agency rejected the AEC's environmental impact statement on safety procedures and requested that an independent study be made of the possibility of accidents that might release catastrophic levels of radiation in populated areas. See ENERGY.

New scientific establishments. In October, 1972, Congress established the Office of Technological Assessment (OTA) to advise on pending legislation involving science and technology. The OTA will study new developments in technology and attempt to determine whether these developments will benefit or harm people and the environment.

On October 4, representatives from the official scientific organizations of 12 industrialized nations, meeting in London, signed a charter establishing the International Institute of Applied Systems Analysis, an East-West think tank that will consider the problems that beset advanced societies. In addition to studying such matters as population control, urban management, the energy crisis, and the environment, the new institute will provide an opportunity for scientists from both Communist and non-Communist nations to work together freely.

The institute is headquartered in the Laxenburg Palace near Vienna. Its

"We don't get many calls for your particular specialty. Have you ever operated a key punch machine?"

Science Support

members include scientists from Bulgaria, Canada, Czechoslovakia, East and West Germany, France, Great Britain, Italy, Japan, Poland, Russia, and the United States. Mathematician Howard Raiffa, a Harvard professor of managerial economics, was appointed director. A Russian management expert, Jermen M. Gvishiani, became chairman of the policy-making council.

The institute will operate on a budget of $3.5 million a year, with Russia and the United States contributing up to $1-million each. The other member countries will make up the balance. Eventually, the members hope to have 100 scientists working in the institute, applying the techniques of systems analysis and computer technology to the problems of the modern world. The work of the institute is pledged to peaceful purposes, and all of its research will be open to public scrutiny.

Youth science tour. A group of 78 students from 15 to 17 years old, each representing a different nation, toured various science centers in the United States in December. The tour was sponsored by the National Aeronautics and Space Administration and the U.S. Department of State.

The students watched the lift-off of Apollo 17 from Cape Kennedy, Fla., on December 7, then visited the Tennessee Valley Authority and the Oak Ridge National Laboratory in Tennessee; the Marshall Space Flight Center in Huntsville, Ala.; the National Oceanic and Atmospheric Administration's facilities in Boulder, Colo.; the Lyndon B. Johnson Spacecraft Center in Houston; the Jet Propulsion Laboratory and other space research centers in California; and U.S. Coast Guard facilities on Governor's Island, N.Y. The tour tried to give foreign students a better understanding of the state of science in the United States, with emphasis on advances being made through the space program.

Other foreign visitors included a delegation of 11 Chinese physicians who arrived in the United States in October. During their two-week stay, they toured medical facilities throughout the United States, including the NIH in Bethesda, Md. [Philip Boffey]

Space Exploration

The salvaging of Skylab, a large orbiting laboratory, was the outstanding achievement in space in 1973. Skylab was damaged during launch on May 14. A three-man astronaut crew rendezvoused with the crippled laboratory in orbit on May 25 and undertook a series of repair jobs that were unparalleled in the annals of space flight.

The first of three manned Skylab flights had originally been planned as a complicated mission for testing man's ability to live and work in space for a long period of time and to carry out an elaborate series of scientific and technical experiments. But from its very outset, the mission became a spectacular effort to save the Skylab program.

Trouble with Skylab developed soon after it was launched from Cape Kennedy, Fla. At first, the launch appeared to be perfect, with Skylab going into orbit 275 miles above the earth. But the orbital workshop suffered extensive damage at the moment, early in the flight, when the increasing speed of the rocket streaking up through the atmosphere greatly stressed the vehicle's surface.

The stress damaged two important components of Skylab. The first was two winglike arrays of solar cells that were to have automatically opened out from its side soon after reaching orbit. One of the solar panels was completely ripped off, and the other was jammed so that it could not unfold automatically.

The other component, a thin micrometeoroid shield, coated with heat-reflecting paint, was destroyed. It was designed to protect the walls of the orbital workshop from incoming particles of cosmic dust and reflect the sun's heat.

The damage to the solar panels drastically reduced the amount of electric power available to the workshop. An X-shaped group of solar panels for operating telescopic instruments on Skylab provided a limited amount of power, but not enough to keep all the workshop batteries charged. The loss of the micrometeoroid shield caused the workshop's internal temperature to rise sharply, reaching a high of 190°F.

Plans for the May 15 launching of the astronauts to link up with the Skylab were immediately abandoned, and the

The ignition blast from a Saturn 5 rocket lights up the midnight landscape around the Cape Kennedy, Fla., launching pad as Apollo 17 lifts off for the moon on Dec. 7, 1972. The dramatic nighttime launch marked the end of the Apollo moon-landing program.

flight was eventually rescheduled for May 25. This gave technicians time to assemble and test a repair kit and to train the astronauts to use the tools needed for emergency repairs. Skylab officials assigned top priority to designing a substitute sunshield that could be erected by the astronauts, and two sunshields were actually built. Officials believed that if the astronauts could set up one of these sunshields and if it reduced internal temperature, then all three manned missions could be flown, even with severe limitations on the use of electricity.

Beating the heat. On May 25, astronauts Charles Conrad, Jr., Paul J. Weitz, and Joseph P. Kerwin, who is also a physician, flew up to the crippled laboratory in their modified Apollo command module and assessed the damage. Of the two replacement sunshields, they favored one that could be erected from inside the pressurized workshop. It was built on the principle of a collapsible umbrella, consisting of a metal-coated plastic sheet about 20 feet wide and 24 feet long on a series of interlocking aluminum tubes. The shield was specially designed to fit in a sunward-side airlock, which had been intended for deploying scientific instruments.

The astronauts made a long pole by fitting aluminum tubes together and pushed the umbrella-type sunshield through the lock. Once outside in space, it unfurled to protect the central living quarters from the merciless heat raining from the sun. The shield quickly brought Skylab's internal temperatures down to livable levels.

Space power. By early June, temperatures inside Skylab stabilized at between 70° and 80°F., but a new problem developed. Two of 18 batteries drawing power from the solar cells of the telescope mount stopped functioning, and a third seemed to be weakening. At one point, the astronauts were getting only 600 watts of power beyond what they needed for survival. The only way to get more electric energy was to correct the problem that kept the jammed solar wing from opening out.

At the National Aeronautics and Space Administration's (NASA) facility in Huntsville, Ala., technicians immersed a full-scale model of the orbital workshop in a huge water tank to simulate as closely as possible the zero-gravity

On the way to the moon, the Apollo 17 astronauts took man's first look at the south polar icecap.

Space Exploration

Continued

conditions of earth orbit. Working in the tank, they found that the solar array could be freed, and on June 7, simulations on the ground were translated into action in space by Conrad and Kerwin.

Leaving Skylab through an airlock, the two astronauts used a pair of heavy-duty metal snipping shears rigged to a long pole to cut a thin aluminum strap that appeared to be holding back the solar-panel array. But after the strap had been cut, the panels opened only partway. Nevertheless, meters at Skylab control in Houston revealed that power was flowing into a set of batteries in the workshop for the first time.

Technicians on the ground had predicted that the hydraulic fluid needed to move the solar array into position was frozen because it was shaded from the sun's heat by the makeshift sunshield. Skylab was then angled so that the sun's rays fell on the hydraulic mechanism, thawing the fluid. Within a few hours, the solar array was completely deployed.

A final repair job was done by Conrad while he and Weitz were outside Skylab collecting film cassettes from the tele-scope mount. One of the battery-charging assemblies on the mount had given trouble early in the flight, and Conrad fixed it by hitting it with a hammer, thus freeing a stuck relay switch.

Accomplishments. Despite the problems that plagued the mission, the astronauts undertook impressive work and experimentation programs. They conducted a series of earth-resources experiments designed to measure ozone and airborne-particle concentrations in the atmosphere and to survey crops, mineral deposits, and ocean and weather conditions. They also carried out some of the 19 experiments proposed by high school students in astronomy, geology, meteorology, and in determining the effects of zero gravity on various biological processes. The telescope mounted on Skylab made it possible to photograph the sun without interference from the earth's atmosphere.

Kerwin, the first physician to fly in a U.S. space mission, was in charge of monitoring and testing the astronauts' physical responses to long periods of weightlessness. The Skylab crew showed

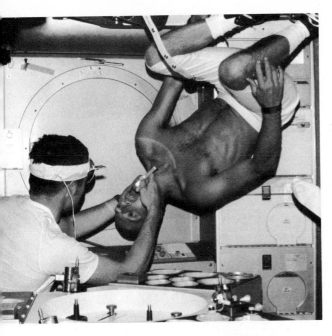

Conrad says "Ahh" for physician Joseph Kerwin
under weightless conditions in the Skylab, *above*.
Later, Conrad prepares to bathe in the zero-gravity,
vacuum-operated shower, *right*. Before the May,
1973, launch, engineers checked out equipment,
in a model of Skylab's spacious work area, *below*.

Space
Exploration

Continued

Huge antennas aboard new Russian tracking ship *Yuri Gagarin* can monitor spacecraft in earth or moon orbit and up to 50 million miles away from earth. The ship has more than 100 laboratories and carries a crew of 300 sailors and scientists.

a gradual decline in their physical stamina during the second half of their record 28-day mission, and at the end, all three showed some adverse effects from the long flight. Kerwin, the doctor, was in the poorest physical condition of the three after splashdown on June 22.

The second Skylab mission, a 59-day flight, was scheduled for launch on July 28. The third and final act in the Skylab drama was scheduled for launch sometime in November.

Beyond Skylab, only one manned mission was definitely scheduled in the U.S. calendar of space events. This was the Apollo-Soyuz Test Project, a joint flight with Russian cosmonauts, to be flown in 1975. The U.S. crew, Thomas P. Stafford, Donald K. Slayton, and Russell L. Schweickart, was designated on Jan. 30, 1973. The Soviet participants, Alexei A. Leonov and Valery N. Kubasov, were named in June. The mission will start with a Soviet launching, followed by the launching of an Apollo command-service module, which will rendezvous with the two-man Soyuz space capsule.

But after the destruction in orbit of a Salyut space station in April, doubts began to develop that this joint mission actually will occur. The Russian craft apparently had been launched on April 3 in preparation for a manned mission that was subsequently abandoned. Reportedly, the space station exploded on April 14, but the cause of the disaster was unknown. Western observers also speculated that an earlier Salyut launched in July, 1972, had failed.

Apollo 17, probably the last U.S. manned lunar-landing mission of the century, was launched on Dec. 7, 1972. While Ronald E. Evans piloted the command module in lunar orbit, mission commander Eugene A. Cernan and geologist Harrison H. Schmitt descended to the moon's Taurus-Littrow Valley.

Cernan and Schmitt spent a record 22 hours exploring outside their landing module. They traveled more than 20 miles on three separate trips in their lunar rover and collected 250 pounds of rock and soil samples, including a glassy, orange soil that indicated past volcanic activity. Scientists first thought the orange soil was relatively young, but when it was analyzed on earth they discovered it was about 3 billion years old. See GEOCHEMISTRY.

The astronauts also found a huge boulder that had broken in two after rolling down a slope. Schmitt called it Split Rock and described it as being of a crystalline nature. These and other finds made Apollo 17 the most scientifically rewarding moon mission. See APOLLO—THE END OF THE BEGINNING.

Unmanned flights. In January, Russian scientists sent an unmanned probe to the moon. It carried Lunokhod 2, a remote-controlled lunar rover. Lunokhod 2 explored areas of the Sea of Serenity about 140 miles north of the Apollo 17 landing site. The Russians reported that the rover had discovered an unusual, smooth slab of rock, which might be much younger than the micrometeorite-pocked rocks surrounding it.

In June, 1973, NASA launched an unmanned craft, Explorer 49, into moon orbit to gather scientific data. One of its prime objectives was to "listen" for radio signals from outer space during periods when the moon lies between the spacecraft and the earth. This blocks out interference created by noise on earth.

On July 23, 1972, NASA launched the first Earth Resources Technology Satellite (ERTS-1) into orbit 570 miles above the earth. ERTS-1 was designed to gather ecological information about the earth and its atmosphere. It takes photographs of features on earth and transmits data about air and water pollution. See AGRICULTURE.

Two other U.S. craft, Pioneer 10 and Pioneer 11, were on their way to Jupiter and beyond in 1973. Pioneer 10, launched in March, 1972, entered the asteroid belt between Mars and Jupiter in July, 1972, and passed through it unscathed about seven months later. This caused astronomers to lower their estimates of the amount and distribution of material in that region of the solar system. Pioneer 10 was scheduled to fly past Jupiter in December, 1973. Pioneer 11, launched on April 5, 1973, is scheduled to fly by the solar system's largest planet late in 1974.

Despite severe fund shortages in areas of basic scientific research, the government budget, released on Jan. 29, 1973, allowed funds for the Mariner 77 program, two interplanetary probes aimed at Jupiter and Saturn. Two Mariner spacecraft, scheduled for launch in 1977, will fly by Jupiter in 1979 and Saturn in 1981. They could also pass close to Titan, the largest of Saturn's 10 moons. Astronomer Carl Sagan of Cornell University believes that Titan has all the prerequisites for the existence of life.

Funds were also provided for a flight in 1973 and 1974 to the earth's two sunward planetary neighbors, Venus and Mercury. It will be the first flight ever aimed at Mercury. The Mariner-Venus-Mercury craft, carrying television equipment, will make one close approach to Venus followed by two Mercury fly-bys.

To Mars and Venus. On Oct. 27, 1972, NASA's Mariner 9 spacecraft completed its yearlong Mars mapping mission. The mission confirmed that Mars is geologically active and provided data for choosing landing sites for two unmanned U.S. Viking spacecraft. In May, NASA announced it will try to land the first Viking on Mars on July 4, 1976, the 200th anniversary of American independence.

The proposed site of this landing is in a region of Mars called Chryse, near the mouth of a vast supercanyon, first revealed by pictures from Mariner 9. The canyon, 3,000 miles long and 4 miles deep, looks like canyons created by the action of running water. The Chryse landing site is in a region marked by long channels resembling dried riverbeds.

The proposed landing site for the second Viking is far to the northwest in a region called Mare Acidalium, which also seems a likely place for detecting water. The detection of life is Viking's primary objective, and water is necessary for any imaginable form of life.

Russia's Venera 8 spacecraft made a soft landing on Venus in July, 1972. In spite of the searing heat, it was able to transmit information for 50 minutes (see ASTRONOMY, PLANETARY). During its brief lifetime, Venera 8 confirmed that the surface temperature of Venus is about 880°F. The chemical composition of the surface area around the landing site was similar to that of the earth. This indicated that Venus' history is similar to that of the earth—it was once a mass of molten material and this condition allowed heavy elements to sink, lighter elements to float to the surface. See GEOCHEMISTRY. [William Hines]

Transportation

In 1973, it became clear that the United States cannot solve its transportation crisis by technology alone. We must completely change the way we transport goods and people.

In the early 1960s, the foremost problems were congestion and improper land use in our urban centers. The rapid expansion of expressways had divided neighborhoods and lowered property values. Downtown parking became an increasing problem, and we slowly realized that building more highways and parking facilities only induced more travelers to forsake public transportation for automobiles. In the late 1960s, pollution became a major problem. We recognized that automotive exhausts contribute up to 60 per cent of the harmful atmospheric pollutants in urban areas.

There were some successful efforts to revive public transportation. San Francisco and Washington, D.C., began to build new rail rapid-transit systems, others were being planned or rebuilt, and bus fleets were streamlined and expanded (see GETTING AROUND THE RUSH HOUR). Despite these efforts, mass transit ridership continued to decrease. In 1970, automobile wholesale sales totaled $14.6 billion and motor vehicles traveled 1,120 billion miles, a 56 per cent increase over the 1960 figure. The airlines increased their passenger mileage 260 per cent over the 1960 figure.

Energy crisis. In 1973, a new factor became the driving force behind the need for urban transportation reform in the United States—the so-called energy crisis. The energy shortage is more serious for transportation than any other user, because more than 95 per cent of all transportation of freight or passengers requires petroleum, the most rapidly decreasing fuel resource.

The proportion of the nation's energy that is used for transportation has not increased; the transportation bite has always been about 25 per cent. However, the total petroleum consumed for transportation has been increasing. Americans also have been steadily switching from efficient rail and bus travel to transport systems that are much more wasteful of energy—automobiles and airplanes. Automobiles burn four times as much

Elastic plastic front end on an experimental car hits a post at 5 miles per hour. But it springs back to its original shape as soon as the car is backed away. Impact-absorbing material would reduce the weight of a car as well as repair costs.

Increasing Use of Petroleum

Year	Per cent of supply used in transportation	Per cent of transportation energy from petroleum
1950	50.3	77.8
1960	51.7	95.3
1970	53.2	95.5
1980 (estimate)	57.6	96.1
2000 (estimate)	72.3	97.1

Source: U.S. Bureau of Mines, 1961, 1971

Decreasing Use of Efficient Intercity Transportation

Year	Total passenger-miles (billions)	Per cent of total passenger-miles			
		Auto	Airplane	Bus	Train
1950	508	86.2	2.0	5.2	6.4
1955	716	89.0	3.2	3.6	4.0
1960	784	90.1	4.3	2.5	2.8
1965	920	88.8	6.3	2.6	1.9
1970	1,185	86.6	10.1	2.1	0.9

Source: U.S. Interstate Commerce Commission

Transportation

Continued

fuel per passenger-mile as buses, airplanes more than twice that.

Paradoxically, the earlier zeal to reduce air pollution helped bring on the energy crisis a few years before it would otherwise have arrived. Pollution-control devices on automobiles decreased their efficiency to the point that 1973 models consume 10 to 30 per cent more fuel per passenger-mile than earlier models.

An array of solutions were proposed in 1973 to meet this growing crisis – gas rationing, lower speed limits, increased rapid-transit subsidies, exclusive bus lanes on major streets, free bus transportation, increased taxes on automobiles by horsepower or weight, taxing automobile commuters, higher downtown parking fees, limited street parking, limited taxicab cruising, limited daylight truck deliveries, rescheduled airline flights to increase passenger loads, and revitalized intercity rail travel. Americans who once believed that technological breakthroughs would somehow dramatically improve transportation were beginning to realize that they had to do a better job with what they have.

Personal bus service. Typical of this new thinking is the increasing attention being paid to dial-a-bus. This demand-activated transit system combines energy-conserving bus service with the convenience of a taxi system. It uses computers to sift telephoned requests and reroute buses. A call for service is instantly processed and the caller informed of how many minutes he must wait. Such a system should give people door-to-door convenience at prices below regular taxi fares.

Haddonfield, N.J., is the site of Dial-a-Ride, the most ambitious demonstration of this system at work. Since February, 1972, a fleet of 18 minibuses seating 17 persons each have been manually dispatched using two-way radios, but computer controls were being tested in 1973 and will be added soon. Passengers usually wait less than 10 minutes for a bus and pay a fare of 60 cents per ride.

Variations of this service are also being considered by many other cities to provide transportation to health-care facilities for the elderly, handicapped, or others without access to transportation.

Catalysts for
Clean Air

American motorists added a new term to their vocabulary in 1973–catalytic converter. This device cleanses engine exhausts of nearly all carbon monoxide (CO) and hydrocarbons (HC).

The U.S. Environmental Protection Agency (EPA) set interim standards on April 11, 1973, limiting emissions of CO to 15 grams (gm) per mile and HC to 1.5 gm on 1975 model cars. Automobile makers had until 1976 to meet the standards provided by the Clean Air Act of 1970–no more than 3.4 gm of CO and 0.41 gm of HC a mile, or one-tenth of the amounts emitted by 1972 models.

However, special interim standards were set for California because of the unique smog problem in the Los Angeles area. The limits on 1975 California cars are 9 gm of CO and 0.9 gm HC. These interim standards are lower than can be achieved by fine-tuning the automobile engine. Thus, catalytic converters must be used on all cars sold in California, about 10 per cent of all U.S. car sales.

The converters will burn the HC and CO in the exhaust to form harmless carbon dioxide and water vapor. These reactions occur naturally when exhaust gases reach the atmosphere, but the change takes years to complete. Most converters will use a platinum catalyst to speed them up. Ideally, the catalyst will remain unchanged in the process.

The catalytic converter fits midway between the engine and the muffler and looks like a small muffler. It has either a ceramic honeycomb structure or a stack of ceramic pellets as its core. The ceramic is coated with alumina (Al_2O_3) and dipped in a salt or acid solution of platinum and palladium. Alumina's rough, porous surface gives the core an enormous surface area per unit volume. The converter will work in conjunction with an air pump to bring in fresh air. Exhaust gases and oxygen come in contact with the metal catalyst as they pass through the converter.

The EPA calculates that the converter will cut the CO level from 87 gm a mile on a car without exhaust controls to no more than 9.0 gm, and the HC level from 8.7 gm a mile to no more than 0.9. This meets the new 1975 California limits. Car buyers in the rest of the coun-

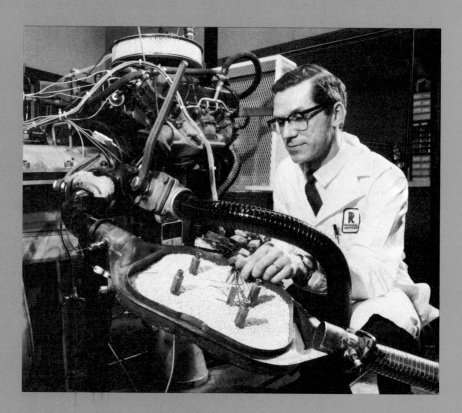

A General Motors technician installs an experimental catalytic converter for a test.

try, the EPA estimated, could meet their less-stringent standards by fine-tuning.

However, Ford Motor Company President Lee A. Iacocca said in March that 25 per cent of Ford's 1975 models would need catalytic converters to meet the federal standards and maintain acceptable performance. In May, General Motors Chairman Richard C. Gerstenberg said that converters might be needed on all of its 1975 cars.

The cost per converter "will not be exorbitant," according to GM President Edward N. Cole. Gerstenberg claimed the system would cost about $150 a car, while Ford put a $300 price tag on it.

The great advantage of the catalytic converter is that it treats the exhaust after combustion. Therefore, engines can be modified and tuned to achieve optimum fuel economy and performance.

The catalysts can be poisoned beyond repair, however, by lead, sulfur, and phosphorus, all of which may be present in gasoline. Thus the EPA also required that unleaded gasoline be available throughout the nation by July 1, 1974. Catalysts also can be destroyed by overheating; for instance, when the engine backfires or gets too hot, the ceramic may melt.

Despite these potential problems, the catalysts were expected to be on several million 1975 model cars. Still unsettled, however, was whether a second type of catalyst might be needed on 1976 models to meet EPA standards on the third major auto pollutant—oxides of nitrogen (NO_x), the most difficult to control.

In 1973, automakers began to limit NO_x emissions by fine-tuning to 3 gm per mile. But drivers were paying for emission controls. Ford said the fuel consumption of a typical 1973 model was 21 per cent higher than that of a 1965 model and acceleration was down 12 per cent. On a 1975 model, the economy would be off 26 per cent and performance down 18 per cent from 1965.

In the 1976 models, NO_x must be reduced to 0.4 gm per mile. A NO_x catalyst would break NO_x down to nitrogen and water vapor. However, in June, 1973, auto manufacturers said there was no suitable catalyst ready for use. They and the EPA were seeking changes in the limits on NO_x emissions that would make a second catalyst system unnecessary.   [Robert W. Irvin]

For instance, health-care delivery on a shared-cab basis is a simplified version of dial-a-bus. Computers would sort out requests for service and dispatch a cab to pick up four or five persons per trip.

Taxi-car pools. In 1973, Carnegie-Mellon University in Pittsburgh acquired a bankrupt taxi company and its dispatchers, radio communications, cabs, and repair facilities. It will begin using the company in 1974 as a nonprofit laboratory to test new concepts in door-to-door transportation offered at mass-transit prices.

Another plan Carnegie-Mellon will try is to use the dispatching facilities to convert private vehicles into part-time taxi operation. Automobile commuters who now find car pools inconvenient would serve as the drivers. With licenses and proper insurance, they would be put in touch daily with others who make the same work trip and, for a fee, provide a taxi-car pool at a profit to the driver. By thus increasing the number of riders in cars going into urban areas, these part-time taxis would reduce congestion, pollution, and fuel consumption.

Sails for cargo ships. Perhaps the most innovative of the suggestions that have been put forth, however, has been the one least expected: The return to sail for ocean freighters.

Engineers at the University of Hamburg's Schiffbau Institute have drawn up plans for a 400-foot, 17,000-ton freighter whose square sails are set, reefed, and adjusted by computers, completely without the aid of sailors in the rigging. In fact, the ship's five masts are free-standing—no rigging supports the 200-foot masts. The sails fit in tracks on aerodynamically curved, stainless steel yards and are set by rotating the entire mast using hydraulic winches.

Continuous weather information by satellite would enable the crew, smaller than that of a conventional freighter, to seek the best winds. An auxiliary engine used for lighting, refrigeration, and associated power could move the ship at 8 knots in calm winds. But when winds are favorable—85 per cent of the time—the large vessel would attain speeds of 12 to 20 knots. This compares well with conventional freighter speeds of 10 to 15 knots. As of mid-1973, there were no firm plans to build this remarkable ocean-going ship. [James P. Romualdi]

Zoology

Physiologists M. Colleen McNamara and Marvin L. Riedesel of the Department of Biology at the University of New Mexico in Albuquerque investigated the memories of golden mantled ground squirrels. These animals normally hibernate soon after exposure to the cold.

The scientists put the squirrels into a tank of water and taught them to escape by swimming down either a black or a white channel, only one of which led to an escape ramp. Then the colors of the channels were reversed and the squirrels had to learn to swim the channel of the opposite color. Since these squirrels hate the water, they soon learned to select the channel that led to dry land.

The squirrels were then placed in a hibernaculum, a secluded chamber in which animals can be exposed to cold temperatures of 7° to 10°C. (44.6° to 50° F.). These temperatures are borderline for hibernation, causing only some of the ground squirrels to hibernate. Eleven days later, both hibernators and nonhibernators were tested again to find out how well they remembered what they had learned. Those that had hiber-

nated remembered the learned behavior significantly better than those that did not hibernate. Moreover, the nonhibernators forgot the most recently learned experiences, the reversal tasks.

These experiments suggested that memory loss may be due to a displacement of what was initially learned by new material rather than the amount of time that had elapsed from learning to remembering. Evidently hibernation enhances memory by blocking interfering experiences.

Fish antifreeze. Fish that live in the supercooled waters of the Antarctic Ocean need some mechanism to keep their blood from freezing at the prevailing temperatures of 29°F. Biochemist Robert E. Feeney of the University of California, Davis, has isolated the antifreeze from the blood of polar fish collected on several trips to the Antarctic. The antifreeze is a glycoprotein, a pure-white material never before found in any other living thing. "Hopefully, the discovery will add to our fundamental knowledge of how water changes to ice," said Feeney in March, 1973. "Somehow

Some 6-day-old mouse embryos grown in a collagen tissue culture at the Johns Hopkins University will develop normally for eight more days. The technique allows scientists to study for the first time many aspects of cell differentiation that occur after implantation on the uterine wall.

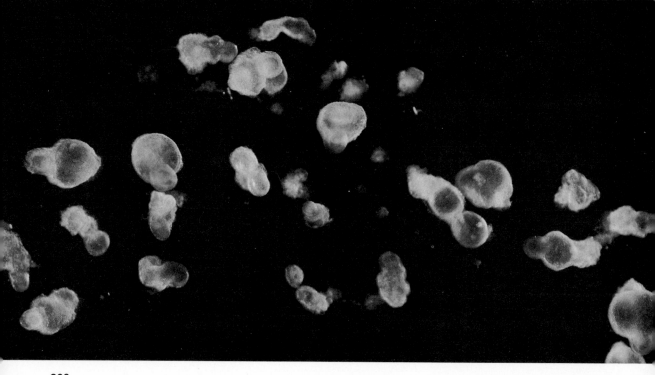

Zoology

Continued

this antifreeze in polar fish lowers the temperature at which this normally happens. Yet it does not do it the same way salt or alcohol or Prestone does it."

Lobster cannibals. The rising price of lobsters and their increased scarcity in offshore waters have raised hopes that sooner or later it may be possible to grow lobsters in captivity the way we raise domestic farm animals. To this end, zoologist J. Stanley Cobb and biological oceanographer Akella N. Sastry of the Narragansett Marine Laboratory of the University of Rhode Island, Kingston, have found out why it is difficult to keep lobsters alive and well in the laboratory. "The main problem is that lobsters fight," said Cobb in March, 1973. The youngsters are little cannibals that seem to enjoy eating other lobsters as much as people do. When the growing lobsters shed their protective shells, which they must do from time to time in order to grow, they are vulnerable to attacks by other lobsters.

Cobb and Sastry are exploring two possible ways of avoiding this difficulty. One is to provide the young lobsters with individual shelters they can hide in to protect themselves from their more belligerent brothers and sisters. The other is to find out enough about molting in lobsters to synchronize shedding within an entire lobster population. In this way, they will all be at the same disadvantage at the same time.

Saving the wolves. At McGill University in Montreal, wildlife biologist Roger J. Bider was working with a group of students to help save the Canadian wolf from impending extinction by making the public aware of the truth about wolves. As many as 50,000 wolves may still live in the Canadian wilderness. They are ruthlessly hunted by people who believe that wolves are vicious marauders of game animals and given to unprovoked attacks on human beings. Experts insist that wolves kill only what they need for food, and rarely attack any but sick and senile prey.

To learn more about the biology of wolves, and to help perpetuate the species, Bider was establishing a semi-captive wolf herd on several acres of woods and fields about 20 miles west of

"Brother, what I mean, we sure got the dirty end of the stick when natural habitats were handed out."

Drawing by John Corcoran © 1973 The New Yorker Magazine, Inc

367

Zoology

Continued

A female cricket will "walk" toward the sound of mating songs of male crickets played over loudspeakers to her left and right. The mazelike pathway rotates under her feet, permitting biologists to measure her responses.

Montreal. The nucleus of this wolf pack is Macaza, a healthy female recently trapped in the wild. She will be joined by a male, and, hopefully, will reproduce to establish a captive pack that zoologists can study. What they learn about the biology of wolves and their social organization may yield valuable information to help preserve these animals from extinction.

Compensating kidneys. If one kidney is removed from an adult mammal, the other one will enlarge. Called compensatory hypertrophy, this reaction is achieved by cell divisions in the remaining organ. The remaining kidney enlarges by increasing the size rather than the number of its nephrons – those tiny tubules upon which the function of the organ depends.

This supports a long-held theory that the number of nephrons that develops very early in life is genetically determined and is all the nephrons that an animal will ever have. New evidence indicates that this may not necessarily be the case. Dr. Jean-Pierre Bonvalet of the Hôpital L. Bernard at Limeil-Brevanes, France, reported experiments on young rats which indicate that the kidneys may be able to make more than the normal complement of nephrons.

The rat kidney normally contains about 32,000 nephrons (the human kidney contains about 1 million). But Bonvalet found that if one kidney is removed before the animal is 50 days old, the opposite kidney not only grows larger, but also develops more nephrons. By the time the animal is mature, the remaining kidney will have about 47,000 nephrons – almost a 50 per cent increase.

The opposite of compensatory hypertrophy is the response of an animal to more than enough of a particular organ. Each animal normally has two kidneys, but what would happen if a third kidney were transplanted into a rat? Surgeons Thomas W. Klein and Ruben F. Gittes of the University of California, San Diego, grafted third kidneys into rats. They used animals from an inbred strain to avoid immunological rejections. The extra kidneys survived and functioned normally, but underwent varying degrees of atrophy, or wasting away, during the first few weeks. However, if one of the rat's own kidneys were removed at the time of transplantation,

the graft atrophied, but to a lesser extent, and the rat's own remaining kidney enlarged. If both of the rat's kidneys were removed at the time of grafting, however, the transplant enlarged.

This proves that the grafted kidney is capable of compensatory growth, but only if it is forced to fully take over the excretory burdens of the animal. Otherwise, the original kidneys do most of the work, thus rendering any extra kidney tissue superfluous and thereby promoting its atrophy.

Compensating lungs. The lungs are also capable of compensatory growth when lung tissue is removed by pneumectomy, an operation that lung cancer often makes necessary. Biologist Paul I. Tartter of Brown University, Providence, R.I., investigated the physiological factors responsible for enlarging a remaining lung in the rat.

After confirming that a single lung enlarges to do the work of the original two, Tartter tied off the bronchus, a branch of the windpipe, leading to one lung instead of removing the organ altogether. The opposite lung still enlarged as much as after a pneumectomy. This showed that physiological deficiency, not the physical absence of the lung, causes compensatory growth.

When Tartter tied the right pulmonary artery, which carries blood from the heart to the lung, the opposite lung did not grow as much as when he had tied off the bronchus. In still another experiment, he caused one lung to collapse by injecting silicone plastic into the chest cavity. Under these conditions, the opposite lung underwent even less growth, but nevertheless it enlarged moderately in an apparent attempt to make up for the nonfunctioning lung.

These experiments prove that an overworked intact lung can stimulate its own growth, but they leave unanswered the question of which aspect of the complicated process of respiration is directly responsible for triggering compensatory lung growth. The possibilities include: A change in the rate of breathing, a lack of oxygen, and a shift in blood flow.

Space spiders. Dr. Peter N. Witt of the North Carolina Department of Mental Health in Raleigh, an authority on web-spinning by spiders, was involved in a study of how spiders will spin webs in the weightless environment of

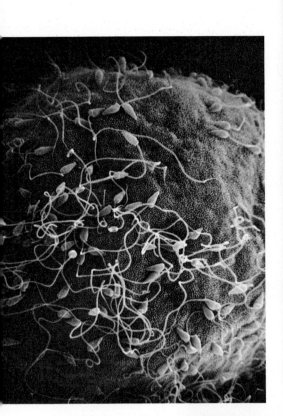

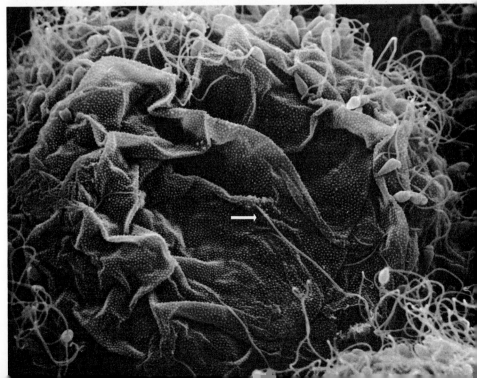

Photomicrographs show
how sperm attempt
to fertilize sea urchin
eggs. Between 5 and
15 seconds after the
insemination, *above*,
an increasing number
of sperm attach their
heads to the egg. At 30
seconds, *right*, a sperm
has penetrated the egg's
outer envelope. Note
its tail trailing from
hole in envelope, *arrow*.
The remaining sperm
have left the area.

Zoology

Continued

This tiny pig, developed in West Germany, weighs only 10 ounces. When full grown, it will tip the scales at a mere 75 pounds. It has a big future in medical research because its size makes it convenient for laboratory studies.

space. The experiments will be conducted as part of the second manned Skylab mission, which was launched in late July, 1973.

Using female cross spiders, a species that spins a new web almost every day in a highly predictable pattern, researchers hope to learn how the webs' geometry may be affected by zero gravity. If the spiders spin, pictures of their webs will be televised to the earth, for measurement and comparison. Any variation from the normal oval-shaped web structure will yield valuable information about how this creature's instincts express themselves. A movie Witt made on web-building and the behavior of spiders was used to train Skylab astronauts to care for the world's first "space spiders."

Diving gorillas. Curator Robert R. Golding of the University of Ibadan Zoo in Nigeria added a new dimension to the lives of a pair of lowland gorillas. Not content for these powerful but gentle animals to sit around all day in naked cages, Golding provided his gorillas with an outdoor enclosure sur-

rounded by a water-filled moat. To prevent them from wandering into the deepest water of the moat, three electrically charged wires were strung down the middle of the moat.

When first released into the new compound, the gorillas cautiously inspected the moat and soon climbed into it and grasped the wires. Upon receiving a few shocks this way, however, they learned that the wire was to be scrupulously avoided. The water, on the other hand, fascinated them. They were soon frolicking about, sitting on the bottom with the water up to their chins, but never going so far as to duck their faces. With much splashing and chest-beating, their confidence in the water soon increased to the point where they learned to propel themselves along by sitting and pushing their hands and feet against the bottom of the moat. In about a week they were actually swimming with something resembling a breast stroke. At last report, the gorillas were diving into the water with tremendous bellyflops and backward twists in midair, but never submerging their faces.   [Richard J. Goss]

Men and Women Of Science

Few scientists have the energy, the talent, and the imagination required to plan, build, and direct a research laboratory. This section, which recognizes outstanding scientists, features two whose laboratories reflect their unique abilities and beliefs.

372 Robert Wilson by Robert H. March
A man of incredible energy and single-mindedness, this physicist, sometimes-sculptor overcame a host of obstacles to build and get underway the world's largest particle accelerator.

388 Jonas Salk by John Barbour
Doing things his own way, this biologist created a polio vaccine that won him fame and the opportunity to create a unique institute where science and the arts blend in an attempt to help all mankind.

404 Awards and Prizes
A list of the winners of major scientific awards over the past year and summaries of the work that earned them special recognition.

413 Deaths
Brief biographical notes on world-famous scientists who died between June 1, 1972, and June 1, 1973.

Robert Wilson

By Robert H. March

Scientist and artist, a loner directing a team effort, this physicist guides a giant machine that forges the tiniest of subnuclear particles

Back in 1961, students in the sculpture class at the Academy of Fine Arts in Rome, most of them Italian youths still in their teens, puzzled over a lively, hard-working, 47-year-old American classmate. They were even more puzzled when they found that he already enjoyed a considerable reputation in a field quite remote from the fine arts. The classmate was Robert Rathbun Wilson, on a year's leave from his post as professor of physics and director of the Laboratory of Nuclear Studies at Cornell University.

The sculpture course was Wilson's investment in his creative future. "A sculptor can get better and better as he gets older," he points out, "while in physics it generally goes the other way." He had no way of knowing then that some of his most noteworthy achievements were still ahead of him, in physics as well as in sculpture.

Today, Bob Wilson is director of the National Accelerator Laboratory (NAL), a vast $200-million installation of the Atomic Energy Commission (AEC), which sprawls across 6,800 acres of Illinois prairie 35 miles west of Chicago. The NAL was built to continue the search

NAL director Robert Wilson admires unique roof made of discarded beer and pop cans. An associate designed roof for an NAL building.

A map of the 4-mile, ringlike tunnel housing the NAL accelerator frames Wilson as he sits in his office. From here, his personal touch extends to every phase of the operation.

The author:
Robert H. March is a professor of physics at the University of Wisconsin and works on a project at NAL. He wrote the article "CERN: Experiments in Physics and People" in the 1973 edition of *Science Year.*

for new insights into the nature of matter by probing the internal workings of subnuclear particles. This laboratory is clearly the product of thousands of hands and minds, yet everywhere it bears two unmistakable Wilson imprints: One of Wilson the physicist, the other of Wilson the artist. For Robert Wilson is a most contradictory figure – a romantic individualist leading a great scientific team effort.

A visitor to the NAL site would have to look closely for clues of a major research laboratory. The heart of the project, a 4-mile ring of machinery designed to push protons up to 99.99 per cent of the speed of light, lies buried in a tunnel beneath untilled farmland. Were it not for one architecturally striking high-rise office building, strange and aloof in this rural setting, it would be easy to believe that the state of Illinois had taken over the land for a wildlife refuge.

The hundred or so farm buildings and low-cost tract houses that the NAL inherited when the site was purchased in 1967 still stand, although many were moved into a compact "village" that serves as the workplace for most of the laboratory's 1,000 employees. At the edge of the village stands a group of farmhouses. Their lonely dignity evokes the turn-of-the-century Midwestern mood of a Grant Wood painting. They provide housing for the hundreds of visiting scientists who come from throughout the world to use the NAL. The taste, the mood, and the practicality of the village are all pure Wilson.

"I couldn't bear the thought of just knocking them down," he explains, "especially the old farm buildings. Besides, this way we could

At the edge of the sprawling NAL complex, a circular 4-mile access road outlines the underground ring tunnel. Inside the tunnel, *below,* the particle accelerator sends protons whizzing around the ring at ever-increasing speeds until they reach an energy of 200 GeV, or even higher. Iron electromagnets hold the protons in their path.

move right in without waiting for construction.'' The fun of playing city planner in the layout of the village was one of the aspects of his job that Wilson likes to refer to as a ''bonus.''

Wilson took a hand in the design of nearly every building at NAL. He takes particular pride in the high-rise and in the 1,000-seat auditorium adjoining it. He also got the chance to apply his sculptor's talents to avoid a traditional eyesore: The metal towers that carry high-tension power lines. Electric power comes into NAL on whitewashed wooden towers designed by Wilson.

This creative spirit spread to other members of the staff, one of whom got the idea of making translucent building panels from used beer and soft-drink cans pressed between plastic sheets. The panels were incorporated into a dome atop a building that serves as the control center for a number of experiments. Collecting enough cans for the project kept several Boy Scout troops from nearby communities busy for months.

The man who shaped this vast collective effort largely from his own ideas and energy chose scientific research because he believed it was the

The moment of triumph came on March 1, 1972, as Wilson led the NAL staff in a Chianti toast to celebrate the first successful acceleration of a proton beam to 200 GeV. This energy has since been doubled.

one really important thing a man could still do all alone. Long days spent riding the range as a boy in the West gave him an early lesson in the joys of solitude. He has spent most of his working life not as a loner, however, but as the director of one scientific project after another. Confronted with this contradiction, Wilson can only laugh at himself: "I guess it shows a basic defect of character," he says.

Wilson's career is by no means unique, however, for his generation of American physicists had a "rendezvous with destiny." This was the generation that came out of the universities just prior to World War II. Until that time, physics in America had been carried on by a small group of scientific tinkerers with little funding and no central direction. The frantic wartime effort, followed by a steady postwar expansion in physics research, was to thrust many Robert Wilsons into administrative roles they had never imagined for themselves when they chose science as a career.

Wilson was born in 1914 in Frontier, a mining town in southwest Wyoming. His father had come to Wyoming as surveyor for a railroad, liked what he saw there, married a rancher's daughter, and settled down. In the isolated life of the still-Wild West, young Bob Wilson found plenty of time to read. He particularly liked the works of the Greek philosopher Plato. "When I arrived at Berkeley [the University of California] as a freshman," he says, "I really believed in Plato's idea that you could figure out everything really important by pure reason, without having to learn a lot of messy facts. I tried to study on that basis in my first semester, and it was a total disaster. I suppose by today's standards I would have been thrown out on my ear. So I backed off. Physics looked like a small and reasonably self-contained corner of human knowledge, one in which Plato's dream might actually come true. Again, it was too much to handle, but, by the time I found that out, I was hooked on physics."

In 1935, Wilson went to work in the laboratory of Nobel prize-winning physicist Ernest O. Lawrence, who had just invented the cyclotron, a relatively tiny ancestor of NAL's giant proton accelerator. "It didn't seem so small to me then. I suppose it was one of the most impressive scientific instruments that had been built up to that time. There was a big switch that turned on the motor-generator unit that powered the cyclotron magnet. We had to be careful how we hit it, because one wrong move might turn off the whole town of Berkeley. That was something to think about, a foretaste of the kind of big science that was to come in a few years."

While he completed his doctoral research on collisions between protons, Wilson courted Jane Scheyer, an English literature major who came from San Francisco, across the bay from Berkeley. They were married in 1940, just before he left for his first job, at the new cyclotron laboratory at Princeton University, where he hoped to follow up his thesis research.

But fate had different plans in store for a nuclear physicist at that time. Wilson was one of a handful of specialists in this field on the East Coast, and his skills were commandeered by Enrico Fermi. Fermi had just fled Fascist Italy and was working on uranium fission at Columbia University. The Princeton cyclotron was an essential tool in the informal "uranium project" launched by Fermi and other physicists before the federal government created the Manhattan Project. The Princeton group was assigned to study neutron capture in uranium.

As a junior member of the group, and also a confirmed pacifist, Wilson found himself in a serious moral dilemma. Was he being asked to do war research? At the time, the physicists' best guess was that a bomb would probably not be feasible, but there was no way to be sure. The Battle of Britain was in full swing, and Wilson had little time left in which to make a decision. He concluded that Hitler must not be allowed to rule the world through nuclear blackmail. Wilson became committed to the project without further hesitation. He was soon named to head a research team to follow up an idea he had for separating uranium 235 from the more abundant U-238. In 1943, when J. Robert Oppenheimer opened the Los Alamos, N. Mex., laboratory where the first atomic bombs were designed and built, Wilson took his team there. He became the head of nuclear physics research though he was not yet 30 years old.

Los Alamos was a strange community, isolated from the world atop a desert mesa. Because the very existence of the laboratory was supposed to be a secret, most of its staff members were rarely allowed to

Driving through the administrative area, Wilson passes a tank sporting the official NAL symbol and some of the original houses, converted into offices when the laboratory took over the site.

leave the site. Jane Wilson worked as a schoolteacher for the many children who were trapped there by their parents' work.

After the war, Wilson joined the scientists' fight to keep control of atomic energy out of military hands. Out of this political battle grew the Federation of American Scientists (FAS), a liberal organization often referred to as "the conscience of the physicists." Wilson served as the group's first chairman, and is still a member of the FAS advisory board. Jane Wilson is an editor on the staff of *The Bulletin of the Atomic Scientists,* a monthly journal on science and politics.

Returning to civilian research, Wilson spent the year 1946 designing a new cyclotron at Harvard University, and then accepted the post of director of nuclear studies at Cornell University in 1947. At last, he thought, he had found the kind of life he had dreamed of when he went into physics. The Cornell laboratory in Ithaca, N.Y., was eminent, but small. He could teach classes in the morning, handle administrative chores in the afternoon, and have evenings free to pursue research projects on his own. There was even time for sculpturing, and Wilson's works of this period can be found on a number of university campuses and in private homes around the world. Here the Wilsons raised three sons who are now of college age.

During Wilson's 20 years at Cornell, team research gradually became necessary in particle physics. Characteristically, Wilson bucked the trend, working long into the night to carry on experiments he could handle himself. During the early 1960s, he began his last round of solitary experiments, using electron beams to probe the internal structure of protons and neutrons. When it finally became clear that the results of his research were too important to await the slow pace of a solo effort, Bob Wilson reluctantly threw in the towel and asked his colleagues for their help. He began to make an uneasy peace

A sense of history led Wilson to combine old farmhouses for visitors' housing, and provide a haven for buffalo on the NAL fields.

with team research. But Wilson's stubborn individualism still continues unabated. He may not be able to handle the entire job himself, but he makes it clear that he will be personally involved in everything important that happens.

While Wilson was at Cornell, he built a succession of increasingly powerful electron accelerators, but the laboratory was kept small. "It was an easy group to work with," Wilson recalls. "We had been together so long that we understood each other's minds, and there were rarely any personal problems."

But in the early 1960s, a project was being launched at Wilson's alma mater that would eventually reach out and tear him away from the idyllic Cornell scene. A group of University of California physicists

Perched precariously on a high ladder, Wilson points out unusual
features of the meson lab roof, one of his own architectural designs.

had designed a 200-billion electron volt (GeV) proton synchrotron, seven times more powerful than the biggest accelerator then in existence. It soon became clear, however, that it would be politically impossible to build this large accelerator laboratory in the San Francisco Bay area, which already had giant machines at Berkeley and at Stanford University. The AEC turned the project over to a nationwide consortium of universities. The site was chosen in an open contest, with over 100 communities bidding for it. With great fanfare, the AEC announced the choice of Batavia, Ill., as the site for the new accelerator in 1966.

By the time this plum was awarded, however, it was no longer quite so sweet. The post-Sputnik science boom was over, and Washington ordered $50-million cut from the project's $300-million budget estimate. The new site, with its severe winter climate, could only add to construction costs. To the Berkeley physicists, this was the last straw. Already upset at having the labor of years taken from their control, most of them refused to move to Illinois to join the project. The consortium had to find someone else to build the machine.

In the new mood of austerity, Bob Wilson seemed a logical choice. He had just completed an electron accelerator at Cornell University for nearly 15 per cent less than the original cost estimate. Although a much smaller machine, it was a cousin of the NAL giant. Using an excavating "mole," Wilson had built it under the beautiful Cornell campus without disturbing the surface. In an ecology-conscious era, this was an important plus. But the most important factor leading to his selection was Wilson's criticism of the Berkeley group for "overdesigning" the 200-GeV machine. In effect, Wilson was offered the job on a "put-up or shut-up" basis.

Building a major new particle accelerator is always a step into the unknown. Such a machine costs less than even one of the rockets that took men to the moon, but the expenditure is an enormous one for pure science. Only about a dozen have been built. Each has been unique, justifiable only if it could perform experiments that had hitherto been impossible. To cover the unknown factors that may develop in a one-of-a-kind device, accelerator builders have usually left a large margin of error in the design of each component.

City planner Wilson had a part in designing nearly every building at NAL, but he takes special pride in the modern office building.

The working principle of the NAL machine is simple. Protons, produced by a smaller accelerator, are steered into a circular race course where they are held in their path by 954 electromagnets. Once each time around the ring, they are given a "kick" by radio waves. As they speed up, the magnetic field must be raised to keep them "on path" at the higher energy. Because the rise in speed is automatically synchronized with the rise in magnetic field, a machine of this type is called a "synchrotron." There is a final limit to the field strength that can be reached with iron electromagnets; this is why higher speeds require a bigger diameter machine.

Wilson's career has spanned nearly the entire history of particle accelerators and their phenomenal growth in size. He was a freshman at Berkeley when Lawrence turned on the first cyclotron, in which protons spiraled out between the poles of an ordinary laboratory electromagnet. Their final orbit was 11 inches in diameter. The cyclotron Wilson used in his thesis research had a diameter of 27 inches. The first electron synchrotron he built at Cornell was just under 7 feet in diameter; the last one was exactly 100 times larger. The NAL ring has a diameter of 1.3 miles.

To build the NAL machine on a stringent budget, Wilson cut out all the margin of error that had been allowed in construction design. The tunnel holding the pipe in which the protons travel was originally designed as a massive, air-conditioned structure anchored in bedrock. It became a simple cut-and-fill concrete sewer pipe. The magnets were redesigned so they could be produced on an assembly line. They were to prove so inexpensive to build that Wilson found his budget would cover enough for a machine twice as big as the original Berkeley design. This meant that it could ultimately reach 400 GeV simply by drawing more electric power.

To Wilson's delight, his design for the magnet supports satisfied economics and aesthetics. The artistic stands were cheaper to cast than conventional frames.

In NAL control area, Wilson and the author check a computer display screen that monitors the machine. The entire accelerator complex is controlled from a single room.

Even this economy drive felt the impact of Wilson the sculptor. He was delighted to find that his art-nouveau design for the stands on which the accelerator magnets rest would be cheaper to cast than a simple rectangular frame.

Of course, this approach was a gamble. In a device so complex, with every component designed to just barely do its job, something was bound to fail. Wilson hoped that correcting the inevitable failures would cost less than building in the safety factors. The risks were not his alone. Given the stringent economic mood in Washington, any serious mistake he made might spell the doom of particle physics research in America.

Wilson saved still more money by deliberately understaffing the lab. "People are happier and work better when they can see there's plenty to be done," he says. Just as at Cornell, chains of command were kept informal with Wilson intervening directly on all levels. Many staff members grumbled at the lack of help and chafed at having the director looking over their shoulder. Yet the pace of work grew.

Then, in July, 1971, the weak link in the machine exposed itself. The ring tunnel had been completed the previous winter, leaving it buried in frozen mud. When hot, humid July air hit its cold walls, the tunnel was drenched in condensation. The insulation on the electrical windings of the mass-produced magnets could not stand the miniature rainstorm. One by one, the magnets shorted out. Each magnet, weighing several tons, needed to be removed, disassembled, dried out, rebuilt, and put back into place. It was a herculean task. The only way that it could be done was to draft nearly every physicist and engineer at NAL, most of whom had been working on other aspects of the project, to spend extra shifts at menial tasks in the damp tunnel.

Under this huge influx of new people, the informal command structure of the lab broke down. Although the "draftees" were willing enough to dirty their hands, there was often nobody there who knew what needed to be done. The effort became known as the "Great Leap Forward." For a few frustrating months, the laboratory was plunged in gloom. Wilson admits that he never fully anticipated how difficult repairs would be on a machine when the faulty component could be miles away from the control center.

Wild rumors circulated about the condition of the machine. Even wilder ones circulated about Robert Wilson:

"We thought he was done for," recalls one staff member. "Either he would quit, or have a heart attack, or one of us would go berserk and shoot him." Jane Wilson recalls this period as one in which she rarely saw her husband, and during which he lost his deceptively youthful appearance and began to look his age.

Finally winter came again, bringing crisp, dry air. The rebuilt magnets began to hold fast, and the heat they generated slowly warmed

At home, Wilson pours wine and enjoys a moment of relaxation before dinner. The sculpture on the table is one of his creations. Such periods of calm were rare during the days when the NAL team was trying to get the machine into operation.

the tunnel walls, ensuring there would be no repeat of the previous summer's disaster when warm weather returned. New problems arose, but by this time the conscripts in the tunnel had become expert in solving them. On March 1, 1972, the machine finally managed to hold a beam of protons in line for the 300,000 turns required to reach an energy of 200 GeV. A rising line on an oscilloscope screen told the tale. In minutes, word spread around the lab. The control room became pandemonium as most of the scientists crowded in to savor the nearly missed triumph. Bob Wilson led his exhausted staff in a toast with Italian Chianti wine, a tradition started among physicists when Fermi turned on the world's first nuclear reactor at the University of Chicago in December, 1942.

Wilson and his team had much to be proud of. Against many obstacles, the NAL staff had built a machine that was capable of reaching twice the proposed energy, four months ahead of the original target date. And when they finished, there was still money left in the budget. In an era when cost overruns, performance cutbacks, and production delays have become the rule in government projects, Wilson's triumph at NAL ranks as a minor miracle.

After dinner, Bob and Jane Wilson take coffee in the living room of their home in Batavia.

At the end of the day, Wilson the scientist heads for his studio in an old barn behind his house. There, he becomes Wilson the sculptor as he turns his hand to creating images in wood and stone.

Despite this triumph, the new accelerator still faces a lot of growing pains before it reaches its full potential. It has reached the 300-GeV and 400-GeV milestones and has solved the ticklish problem of ejecting the beam from the giant ring and sending it to a specified target area. Eventually, the giant synchrotron must deliver protons on a round-the-clock basis to experiments scattered about the NAL site. At full capacity, it must be able to serve from 10 to 20 experiments at a time. Work on these aspects of the project lagged during the push to finish the main ring.

So the pressure is still on Wilson and his staff. Much of it comes from the experimenters who are installing their elaborate particle detectors at the target ends of the beam lines. Many of the researchers have spent years tooling up to use the NAL, and they are eager to get to work. At most other laboratories, the users run the show through committees that approve the experimental program. Although there are such committees at NAL, Wilson insists on having the final say. This brings him into frequent controversies with experimenters, who are often proud men with strong scientific reputations, and accustomed to having their own way.

The first experimenters to get a crack at the machine were members of a Soviet-American team that proposed to study proton-proton collisions inside the main ring by squirting a thin, controlled jet of hydrogen in the path of the beam. This technique had been developed at the 70-GeV accelerator at Serpukhov, Russia, the previous world leader in proton energy. It was a natural choice for the experimental kickoff at NAL, since it could be performed before the beam had been extracted from the ring.

Wilson is particularly pleased at this milestone in international science because it symbolizes a thaw in the Cold War hysteria he bitterly opposed while head of FAS. Jane Wilson also had a role in this venture, helping settle the Soviet physicists and their wives in rural Illinois. The visitors moved into NAL farmhouses and planted vegetable gardens, showing that they intended to stay for a while. Jane and other NAL wives taught English classes for the visitors; before long, the Soviet wives were teaching Russian to the NAL staff.

For NAL, its controversial director, and the scientific community they serve, the real challenge still lies ahead. That the machine will run is no longer in doubt; what it will contribute to man's understanding of the age-old riddle of matter remains to be seen. There are hopeful signs; recent preliminary experiments indicate that the NAL machine will be capable of exploring a new level of reality far smaller than the so-called elementary particles themselves. But for now, exhausted NAL staff members can find time for long-postponed vacations. And Robert Rathbun Wilson can again experience, for a few hours each week, the solitary, deep creative satisfaction that comes from driving a chisel into a block of wood or marble. Which may well be the one really important thing a man can still do all alone.

Jonas Salk

By John Barbour

**His antipolio vaccine made him a folk hero and the
fame enabled him to create his own research institute**

The sun burnishes the hardwood floors in Jonas Salk's office at the
Salk Institute for Biological Studies. Surrounded by the paintings of
his artist wife, Salk spends the afternoon preparing reports and writing
letters, meeting with the young scientists who work with him on immunological research, talking with other scientists on the institute
staff, and receiving visitors—scientists from other institutions, public
officials, even delegations from foreign countries.

One hundred yards away, reminders of an earlier part of his life
gather dust in an empty area that will someday be converted into a
laboratory. Scattered about are pieces of old lab equipment. Filing
cabinets and cardboard boxes contain the records of children immu-
nized a generation ago. Drawers are filled with legislative resolutions
from dozens of states and scrolls and certificates from civic clubs and
foreign governments. They all seem to begin: "Whereas, the dreaded
disease of polio has year after year struck fear and anguish into the
hearts of mothers and fathers everywhere...."

These are the mementos of the scientist who became a folk hero
after developing the first vaccine against poliomyelitis, or infantile
paralysis. Before Salk's discovery, parents, dreading the summer and
its annual polio epidemics, lived in fear that their children might be
attacked by this crippling or fatal disease.

Until the spring of 1953, Salk had been working in the relative
seclusion of his laboratory at the University of Pittsburgh, doing immunological research on both influenza and polio. Then, in March, he
reported in the *Journal of the American Medical Association* (*JAMA*) his

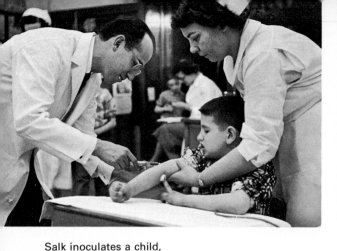

Salk inoculates a child, *above,* during the 1954 testing of his vaccine. Flanked by National Foundation president Basil O'Connor and HEW Secretary Oveta Culp Hobby, Salk is honored by President Dwight D. Eisenhower, who presented him with a presidential citation in April, 1955.

The author:
John Barbour is a science writer for the Associated Press. His article, "Probing the Pawnee Grassland," in the 1972 edition of *Science Year,* won the AAAS-Westinghouse science writing award.

findings that a killed-virus vaccine had produced antibodies to polio in human beings. This indicated the possibility of a polio vaccine and Salk was swept into the headlines. For many years thereafter he would be beset by public demands and pressures.

Twenty years later, Salk continues his immunological work in the institute he created in 1963 at La Jolla, Calif. High on the seaside cliffs in this northern section of San Diego, the concrete walls of the institute rise like a piece of sculpture—some say a fortress; others, a cathedral. Salk stands at a window in his office at the west end of the building, looking out at the Pacific Ocean.

"My desire was not to have the institute named after me," he says. "All I wanted was to create the institute and then work there as an individual and not as its administrator. I wanted to create a place, an atmosphere, in which I could feel comfortable in expressing my many interests, while at the same time continuing my laboratory work."

The main activity of the institute is experimental biology. But Salk has always intended that his institute be a place in which the economist, the sociologist, the philosopher, the urban developer, and even the novelist, poet, and painter would be able to work and draw upon the knowledge and insights that come from advanced biology. "What we've tried to do," Salk explains, "is create an institute that would be concerned with all aspects of man."

The projects going on in the institute reflect this philosophy. They range from studies of genetic markers in the immune system of the mouse to symposia on the Renaissance. Among the scientists probing scientific mysteries is Nobel prizewinning geneticist Robert W. Holley, who examines the chemical factors in the blood that affect cancer cells differently from normal cells. Biologist Melvin Cohn studies the mechanisms of immune responses. Immunologist Edwin Lennox examines how the immune system deals with tumors. Physiologist Roger Guillemin isolates hormones from the hypothalamus that control reproduction and growth. And chemist Leslie Orgel studies the chemical and physical events that led to the beginning of life on earth. Salk and his research team are looking for ways to stimulate or suppress specific responses of the immune system. They would, for example, have the immune system reject cancer cells, but accept transplanted organs.

In another area, mathematician and philosopher Jacob Bronowski searches into the nature of human language. The broad span of the research that is in progress at the institute in many ways mirrors Salk himself, and in a sense, it is an expression of his personality and of the various forces that molded it.

Jonas Edward Salk was born on Oct. 28, 1914, in New York City, the first of three boys. He received his early education in the city's public schools, but he also attended Hebrew school and remembers being "much more religious than anyone in my family."

Salk's mother, Dora, had a major influence on his life. "She was a formidable person," Salk says, "very dynamic, very energetic." She would tell her family how she escaped from anti-Semitic persecution in Russia at the age of 12, came to the United States, and worked as a forelady for a dress manufacturer during the sweatshop era.

By contrast, Salk remembers his father, Daniel, as a kind, gentle person. He designed the detachable lace collars and cuffs that were fashionable for women during the 1920s. "He liked to paint and draw," Salk recalls. "My mother never appreciated that. She was very materialistic. Struggle—she knew what struggle meant."

Dora Salk invested everything in her children. "I was the first born, the preferred one," Salk says, "and I had to suffer her overattention." He escaped this overattention by retreating into himself. "I learned to listen and then do as I pleased. I was a very good boy, of course, and my mother looked on me as the apple of her eye. But I learned to proceed on my own, to live within myself and by myself. And that experience has stood me in good stead."

Inspired by the ocean view, Salk often sits at dawn in this corner of the living room in his southern California home, dictating into a small tape recorder.

The Salk Institute for Biological Studies rises from the oceanside cliffs of La Jolla, Calif. The $15-million concrete building, designed by Louis I. Kahn, consists of two wings divided by a courtyard, with a fountain and informal sitting area at one end. Stairways are built in service columns outside the main building.

Early in his life, Jonas developed a deep interest in law. When he was about to enter college, he intended to study law, but his mother was against it. "She thought that a lawyer should be very articulate, and with her, I wasn't very articulate. In fact, I tended to stutter and stammer." His mother thought he should be a teacher, and not wanting to go completely against her wishes, Salk began to consider a third alternative, medicine. His home did not seem to have the kind of atmosphere that would bend a young man toward science. Yet, Salk reasons, the same thirst for the intellectual problems involved in the laws of man began leading him toward the laws of nature.

Once Salk entered college, he committed himself whole-heartedly to medicine. But when he decided on medicine as a career, it was not to practice as a physician. Instead his goal was to do research. Salk single-mindedly pursued his studies. He earned his B.S. degree in 1934 at the age of 19. Five years later, he had his M.D. from New York University, having taken a year off from his studies to work as a research fellow in chemistry.

While he was a medical student, he met two people who would play important roles in his life: Donna Lindsay, a student at the New York School of Social Work, and Thomas Francis, Jr., head of the bacteriology department at New York University's medical school. Jonas and Donna were married on June 8, 1939, the day after he graduated from medical school. In the years ahead, they had three sons, Peter, Darrell, and Jonathan.

While in his senior year of medical school, Salk elected to study with Francis, who was doing research on influenza viruses. By this time, Salk had already developed an interest in immunology, and his early association with Francis stimulated this growing interest. In 1941, Francis became head of the epidemiology department at the University of Michigan in Ann Arbor. The following year, after completing his internship at Mt. Sinai Hospital in New York City, Salk received a fellowship that made it possible for him to work with Francis. Salk and his wife moved to Ann Arbor and rented an old farmhouse, complete with a wood-burning stove and a World War II victory garden.

The U.S. Army, swelling with new recruits, needed a way to control flu epidemics among troops. So research on influenza viruses became a top-priority item. Francis and Salk were studying influenza viruses killed with ultraviolet light or a formaldehyde mixture called formalin. They wanted to know if, used properly, this material would kill the viruses without destroying their ability to stimulate the production of antibodies, substances made by the body to fight off infectious agents (see New Insights into Immunology). Francis and Salk developed a killed-virus flu vaccine which was field-tested in 1943 and proved effective against the Type A flu virus. In 1945, they also found it was effective against the Type B flu virus. However, in 1947, this vaccine proved ineffective against a strain, or variety, of Type A virus. This illustrated the importance of creating multivirus vaccines, for

only they would be effective against the numerous, changing strains of viruses that cause influenza epidemics.

As Salk gained more experience in epidemic diseases and immunology, the scope of his research broadened. He was now interested not only in the viruses themselves and the antibody response to them, but also in their epidemic effects on populations. In 1947, he took a job as associate research professor of bacteriology and director of the Virus Research Laboratory at the University of Pittsburgh's medical school. Even though it was a research-poor institution, he believed that by heading his own laboratory there he could satisfy his deeply ingrained need for independence. He also saw this as an opportunity for expanded research and learning.

At that time, the National Foundation for Infantile Paralysis was gearing up for a massive attack on polio. The disease was a public horror and a summertime scourge. Although at epidemic levels polio

Salk begins an afternoon at the institute working in his office. Later, he leaves to visit a laboratory, and on the way, greets a staff member, *above.* In a laboratory, Salk and Roger Guillemin talk about research on the hypothalamus, *above left.*

struck only 20 out of every 100,000 persons, most of its victims were children. Parents kept their youngsters at home, particularly shunning swimming pools, drinking fountains, and movie theaters. They knew and dreaded the first signs of polio—sore throat, headache, and stiff neck. Furthermore, it was a disease that struck without regard to class or economic status. Even President Franklin D. Roosevelt had been crippled by polio. He had founded the National Foundation in 1938 and put his influence behind its March of Dimes fund drives.

In the late 1940s, the National Foundation outlined a program to classify all polio viruses. Scientists knew there were at least three types of polio virus. But each type included many strains, and there were strains that had not even been classified. It was possible that more than three types of polio virus existed, and scientists needed to know this before they could develop a vaccine that would provide immunity to all types of polio. Salk was one of four scientists who collaborated to classify the untyped virus strains.

Although he was a relative youngster and a newcomer to the polio-research scene, Salk proposed an innovative classification method that was eventually accepted by the National Foundation's advisory committee. He injected a monkey with an unknown virus strain, then tested the antibodies it produced in the monkey's blood serum against the three known types of polio virus. If the antibodies reacted with one of the known virus types, Salk would know that the unknown strain belonged to that type. If there was no reaction with any of the three types, the unknown strain would have to represent a fourth. However, Salk never found a fourth type, and by 1951, he and the other three scientists were satisfied that only three types of polio existed.

While the virus classification work was going on, other important advances were made. In 1949, a researcher at Johns Hopkins University produced antibodies in monkeys with a formalin-killed polio-virus vaccine made from monkey brain tissue. That same year, at Harvard University, John F. Enders, Thomas H. Weller, and Frederick C. Robbins grew polio viruses in human tissues that were not from the nervous system. Scientists had previously thought that polio virus multiplied only in nerve tissue. A vaccine from virus grown in nerve tissues could create an allergic reaction that might cause paralysis.

Salk began thinking about a polio vaccine while he was working on virus classification. He knew that he would have to master the Enders tissue-culture technique, which was a valuable tool both for typing viruses in tissue culture rather than monkeys and for making a safe vaccine. But the National Foundation, not yet ready for Salk to take up a new project, offered him neither funds nor encouragement. So he raised money locally and in 1950 set up his own tissue-culture laboratory. "I had a way of doing things anyway," Salk says. Later that year, when the National Foundation's research director visited Salk's laboratory and saw the tissue-culture work underway, he offered foundation funds with which to support it.

Research goes on at the Salk Institute not only in traditional laboratory settings, but also in the courtyard where Salk and his co-workers in various fields meet to exchange information.

Salk was working toward a killed-virus vaccine. But other researchers favored a live-virus vaccine. With a live-virus vaccine, the viruses would continue to multiply in the body and, therefore, produce an immunity similar to that created by a natural infection. But Salk, influenced by his work with Francis, favored a killed-virus vaccine, even though it might not be as potent as one made with weakened live-virus. Salk believed there were too many uncertainties about a live-virus vaccine. For instance, would the weakened virus revert to a virulent form? As the polio research progressed, scientists disputed the principle of a killed- rather than a live-virus vaccine.

Salk believed he could produce a safe killed-virus vaccine that would be effective if a sufficient amount of the killed virus was injected into the human body. But he also knew that he would have to find a sure way of killing the viruses without impairing their ability to stimulate antibody production.

Using formalin to kill the polio viruses, Salk experimented until he learned which concentrations and temperature were best for the process. He also developed a technique for growing and reaping large amounts of polio virus from both monkey kidney tissue and a synthetic chemical medium. He knew from tests he made on monkeys that his vaccines were safe. Although they were sure these vaccines could be improved, Salk and his research staff began to think that the time had come to try them on people instead of monkeys.

Meanwhile, work was also progressing on a live-virus vaccine. In March, 1951, Hilary Koprowski, a researcher at Lederle Laboratories in Pearl River, N.Y., reported that he had tested a weakened live-virus polio vaccine on 20 institutionalized children and 2 adults with no signs of physical illness. It produced antibodies to polio.

On July 2, 1952, Salk began inoculating children and adults at the D. T. Watson Home for Crippled Children near Pittsburgh with a killed-virus vaccine. They had already been stricken with polio, so their blood contained polio antibodies. Salk checked the antibodies in each person's blood to determine the type of polio virus with which he had been infected. Then, he injected each with a vaccine containing that type of virus. The antibodies already in their blood ensured they would not get polio again from any virus that might still be alive in the vaccine. Beginning two weeks later, Salk checked polio antibody levels in each person's blood and found that the vaccines were effective. They substantially increased polio antibody levels. Next, he injected them with a vaccine made from a polio virus type for which they had no antibodies. Again, the tests were successful.

Salk was now ready for a crucial test. He chose a group of children whose blood contained no antibodies to any type of polio and injected them with what became known as the Salk vaccine, a vaccine containing all three types of killed polio viruses. The children formed antibodies to polio, and none of them became ill. In January, 1953, Salk reported the results of his tests at a meeting of the immunization committee of the National Foundation, and in March, he published his findings in *JAMA*.

The National Foundation began to discuss preparations for large-scale field trials, but scientists had differing opinions about how to conduct them. How many children would have to be inoculated to give statistically reliable results? How many uninoculated children should be observed as a control group? Virologist Albert Sabin, who was working toward a live-virus vaccine, frequently criticized the whole idea of a killed-virus vaccine.

Finally, Francis took over the direction of the field trials and settled these questions. Francis was also chosen to evaluate the results of the field trials, which began on April 26, 1954, and were completed in June. In that time, some 440,000 children received the vaccine, about 210,000 were injected with a placebo (a substance containing no vaccine), and an additional 1,180,000 children were put under observation as a sample of the uninoculated population. Coded information about the inoculations and booster shots was collected by local public health officials and sent to Francis at the University of Michigan. Then there was nothing to do but sit back and wait.

However, the burning interest of millions of American parents, fearful for their children's safety during the annual summer polio epidemics produced a steady flow of headlines speculating about the outcome of the field tests during the year that Francis worked on the vaccine data. By the time the verdict was announced on April 12, 1955, public interest had reached fever pitch.

Television and newsreel cameras ground away at the rear of the University of Michigan's Rackham Auditorium as Francis stood up behind the lectern and told a group of 500 scientists that the Salk vaccine was from 80 to 90 per cent effective against all three types of polio. However, he said it was least effective against Type I, the most common form. Then Salk stepped to the podium and reported that the use of merthiolate as a preservative in the vaccine had impaired its effectiveness against the Type I virus. He also said that lengthening the time between the original inoculations and the booster shot would increase the antibody response.

The nation was overjoyed at the news. Polio had been conquered, and Salk was the conquering hero. Hollywood wanted to make a movie of his life, starring Marlon Brando. He was besieged with endless requests for his autograph and dozens of invitations to speak at banquets and lecture before civic clubs and state groups. President Dwight D. Eisenhower presented Salk with a presidential citation in 1955, and he was also awarded a congressional gold medal. In addition, he won the 1956 Albert Lasker Award of the American Public Health Association and the 1958 Bruce Memorial Award of the American College of Physicians.

Salk conducts a private tour for opera singer Beverly Sills, who made a public-service film about the Salk Institute.

But two weeks after the Francis announcement, tragedy struck. Word came from California and Idaho that children inoculated with the vaccine had contracted polio. The casualty list grew. The problem was eventually traced to live virus found in a batch of vaccine from one pharmaceutical laboratory. On May 7, 1955, the surgeon general halted all inoculations with the Salk vaccine and did not allow them to resume until May 26.

The number of children vaccinated during that first summer was far below the number predicted. However, public confidence in the Salk vaccine was gradually restored and by 1958, half of the population below the age of 40 had been vaccinated. The incidence of polio dropped 86 per cent below the prevaccine level. Part of the drop was due to the so-called herd effect, in which the reduction of infectious sources through vaccination helps protect even unvaccinated individuals. However, in succeeding years, the disease pattern changed. Polio cases among white suburban children dropped dramatically, but epidemics still erupted among black inner-city children.

Meanwhile, Sabin decided he had developed sufficiently weakened live polio viruses to use them in a vaccine. A live-virus vaccine could be taken orally, on a sugar cube or in a spoon. Many scientists and public health officials believed this made the vaccine easy to administer to large segments of the population, because a doctor and a needle were not necessary. By 1961, the Sabin vaccine was licensed in the United States, and medical authorities urged that everyone who had been inoculated with the Salk vaccine be revaccinated.

However, Salk does not believe that his killed-virus vaccine is inferior to or more difficult to administer than the live vaccine. "I feel the changeover to a live vaccine was unnecessary," he says. "By 1961, the incidence of polio in the United States had dropped to 7.2 cases per million persons. From 1950 through 1954, there had been an annual average of about 255.5 per million.

"Convenience was not the question. All other vaccines are administered by injection or by scarification, scratches on the surface of the skin. What really was at issue was a principle: Could a killed-virus vaccine eradicate polio, produce the herd effect, and provide long-lasting immunity? The evidence accumulated by 1961 indicated that the answers were yes. What was presumed to be possible only with a live vaccine could be accomplished with a killed vaccine, without the risk that is always present when a live vaccine is used. There is evidence that the live vaccine has been responsible for recent cases of paralytic polio, both in those who have taken the vaccine and in persons with whom they have been in contact."

Nevertheless, six years after it was introduced, the Salk vaccine was considered to be almost obsolete in the United States. However, it is still widely used in other countries.

As the public clamor over his vaccine died down, Salk began to reflect on the events surrounding its introduction. He wondered what

At home, Jonas and Francoise Salk enjoy a walk in their garden, *opposite page.* Sitting in their living room before one of Francoise's paintings, they discuss a book on her work.

In the doorway to his study at home, Jonas Salk ponders the role that his institute will play in the future of biological research.

directions biology would take in the future and how it would influence human affairs. He also began to think about what he wanted to do with the rest of his life. As early as 1956, he began to dream of an institute that would bring together people of various disciplines and specialties to work on the increasingly complex problems of the modern world. He envisioned a research institute where biological advances would be used to benefit all aspects of man.

The founding of the Salk Institute for Biological Studies in 1963 made this dream a reality. The city of San Diego donated 27 acres of coastal land, and the National Foundation pledged the funds for the building and equipment. The institute's building, completed in 1965, was designed by architect Louis I. Kahn.

At the time the institute was taking shape, there were great changes in Salk's personal life as well. He and his wife Donna separated in 1967 after 28 years of marriage and were divorced in 1968. In 1970, at the age of 55, Salk married Francoise Gilot, 48, an artist and author of the book, *Life with Picasso*.

Salk's interests also began to broaden, reaching out to include art and philosophy. He believes that the seeds of his new interests have always been with him. But in his goal-oriented early years he had little time or energy to pursue them. "I found out there were aspects of myself that I wasn't previously aware of, probably more philosophical and artistic aspects than I ever realized," he says.

Now, Salk devotes almost as much time to his other interests as he does to his immunological research. He spends several months each year in Paris, where his wife maintains a studio. While in Paris, he devotes all of his time to writing.

Although Salk is still the director of the institute, he has entrusted much of the administration to others. Eight resident fellows and six nonresident fellows now help shape policy and provide leadership. Nonresident fellows are those who work elsewhere, but are associated with the institute. Among those who have served as nonresident fellows are three Nobel prizewinners—Francis H. C. Crick, Salvador E. Luria, and Jacques Monod.

The long-term policy of the institute is guided by a board of trustees, which includes such prominent people as Chief Judge David Bazelon of the U.S. Court of Appeals for the District of Columbia and U.S. Senator Jacob K. Javits (R., N.Y.). The immediate business of the Salk Institute falls to its president, Frederic de Hoffmann; its director, Salk; and an executive committee made up of trustees and one of the institute's fellows.

The institute has more plans than money at present. Three huge areas that could be converted into laboratories stand idle—including the one used to store Salk's mementos—because there is not enough money to equip them or pay the scientists who would work in them. But Salk takes an optimistic view. "The space we have is one of our greatest assets," he says. "When the money does become available, we will have room to expand and continue our growth and development in new areas and directions."

However, there is not yet enough money to pay for all the projects Salk would like the institute to undertake. The institute operates on an annual budget of $6 million, with the National Foundation contributing $1 million a year. But hard-won private donations and federal grants must make up the balance.

In spite of these problems with the institute, Salk seems satisfied and comfortable with life. He has time now for art and literature and philosophy. He retires early and rises early, spending part of each day writing in his study at home. He has completed two books, both with a philosophical flavor, *Man Unfolding* (1972) and *Survival of the Wisest* (1973). The books he is writing now are autobiographical in tone.

As the afternoon sun sinks beyond the ocean cliffs and the rugged eminence of the institute, Salk's past—Pittsburgh, Ann Arbor, and his work on a polio vaccine—seems far away. He leans back and looks about. "The institute is growing," he says, "maturing and solidifying. It has a life of its own.

"As for myself," Salk continues, "I have reached a point where I can freely do the things that I want to do most. I can deal with problems concerning research in my own laboratory, and I can pursue my own writing. I was never sure that I would ever arrive at this point. But now I can say that I have."

Awards And Prizes

A listing and description of major science
awards and prizes, the men and women
who won them, and their accomplishments

Earth and Physical Sciences

Chemistry. Major awards in the field of chemistry included:

Nobel Prize. Three biochemists shared the 1972 Nobel prize in chemistry for their pioneering studies in enzymes, one of the key substances of life. They are Christian B. Anfinsen of the National Institutes of Health (NIH) in Bethesda, Md., and professors Stanford Moore and William H. Stein of Rockefeller University in New York City. They shared a cash award of $100,000.

They studied ribonuclease, the enzyme that breaks down ribonucleic acid (RNA) in a cell after the RNA has transcribed the genetic material in deoxyribonucleic acid (DNA). Anfinsen discovered how the ribonuclease molecule develops its three-dimensional structure. Moore and Stein determined the sequence of the 124 amino acids in the ribonuclease molecule and developed new analytical methods to speed their work. The work of all three laid the groundwork for unraveling the structure of other enzymes.

Anfinsen received his Ph.D. degree in biochemistry from Harvard University in 1943. He has been at NIH since 1950, and has been chief of the laboratory of chemical biology at the National Institute of Arthritis, Metabolism, and Digestive Diseases (formerly Arthritis and Metabolic Diseases) since 1963. His specialty is the relationship between protein structure and function, and the genetic basis of the protein structure.

Moore received a Ph.D. degree in organic chemistry from the University of Wisconsin in 1938 and joined Rockefeller University in 1939. His work emphasizes the importance of enzyme and protein research in the prevention of disease.

Stein has been studying proteins, peptides, and amino acids for more than 30 years. He received his Ph.D. degree from Columbia University in 1935 and came to Rockefeller University in 1938.

Cope Award. Roald Hoffmann of Cornell University and Robert B. Woodward of Harvard University were joint recipients of the first Arthur C. Cope Award for contributions to organic chemistry. The award, presented in December, 1972, by the American Chemical Society (ACS), carries an honorar-

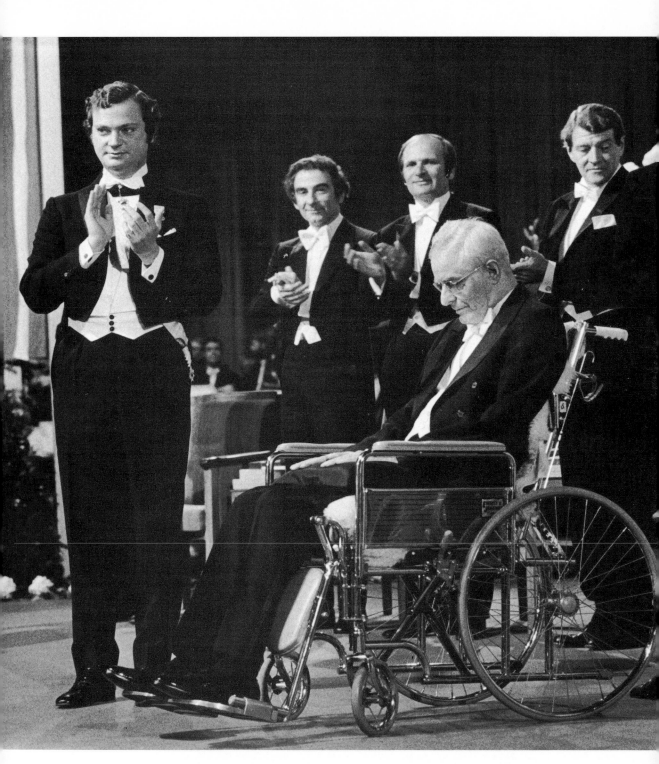

William H. Stein, in wheel chair, receives applause after accepting the Nobel prize in chemistry from Sweden's Crown Prince Carl Gustaf, left. Joining in the applause are fellow prizewinners, from left, physicists Leon N. Cooper and J. R. Schrieffer, and chemist Christian B. Anfinsen.

Earth and Physical Sciences

Continued

Harold C. Urey

John Bardeen

Manson Benedict

ium of $40,000 to be used primarily for continuing research.

The scientists were honored for developing the "Woodward-Hoffmann rules" that allow organic chemists to predict if either heat or exposure to light can initiate a reaction between organic substances. The rules, introduced in 1965, have been called the most significant theoretical advance made in organic chemistry in 30 years.

Woodward received the 1965 Nobel prize in chemistry for his syntheses of natural substances, especially chlorophyll. He is an international authority on the chemistry and synthesis of natural substances. After receiving his Ph.D. from the Massachusetts Institute of Technology (M.I.T.) in 1937 at the age of 20, he was awarded a postdoctoral fellowship to Harvard and has been there ever since.

Hoffmann was born in Poland and earned his Ph.D. degree in chemical physics at Harvard in 1962. He spent the next three years in a research position there, during which time he began his collaboration with Woodward.

Perkin Medal. Theodore L. Cairns won the 1973 Perkin Medal from the American Section of the Society of Chemical Industry. The award is the highest honor given in the United States for applied chemistry.

Cairns is director of the central research department of the Du Pont Company in Wilmington, Del. He is credited with opening the field of cyanocarbon chemistry through his work in synthesizing tetracyanoethylene. He was born in Edmonton, Canada, and received his Ph.D. degree in organic chemistry from the University of Illinois in 1939.

Priestley Medal. Harold C. Urey, professor emeritus at the University of California, San Diego, won the 1973 Priestley Medal, the highest award given by the ACS.

A physical chemist and cosmochemist, Urey has made fundamental contributions to knowledge of the solar system and its history. He won the 1934 Nobel prize in chemistry for his discovery of deuterium. He also took part in experiments in which amino acids, necessary for life, were formed by passing electric discharges, simulating electrical storms, through the gases believed to compose the atmosphere of the primitive earth.

He received his Ph.D. degree in physical chemistry from the University of California, Berkeley, in 1923. In 1945, he became professor at the University of Chicago's Institute of Nuclear Studies, and became interested in the chemical history of the earth and solar system. He joined the faculty of the University of California in 1958.

Physics. Awards recognizing major work in physics included:

Nobel Prize. Three scientists who originated a theory explaining superconductivity shared the $100,000 1972 Nobel prize for physics. They are John Bardeen of the University of Illinois, Leon N. Cooper of Brown University, and J. Robert Schrieffer of the University of Pennsylvania.

The BCS theory, named after them, explains why certain metals lose all resistance to an electric current when they are cooled to temperatures near absolute zero. Thus, the current in a superconducting magnet will flow forever, once started. The scientists developed the theory between 1955 and 1957 while working together at the University of Illinois.

Bardeen received his Ph.D. degree in mathematical physics from Princeton in 1936. He shared the 1956 Nobel prize in physics for the discovery of the transistor effect, developed while he was at Bell Telephone Laboratories. He thus becomes the first person to win two Nobel prizes in the same field.

Cooper received his Ph.D. degree in 1954 from Columbia University, and has been on the Brown University faculty since 1958. Schrieffer earned his Ph.D. degree in 1957 at the University of Illinois, where he was a graduate student when he collaborated in the superconductivity research.

Buckley Prize. Gen Shirane, a senior physicist at Brookhaven National Laboratory in New York City, won the 1973 Oliver E. Buckley Prize for solid state physics, given by the American Institute of Physics. He was honored for his "broad contributions to the understanding of structural phase transitions by means of inelastic neutron scattering." The Buckley award carries an honorarium of $2,000.

Shirane was born in Japan and received his doctorate in physics from the University of Tokyo in 1953.

Earth and Physical Sciences

Continued

Gen Shirane

William A. Fowler

Comstock Prize. Robert H. Dicke, an experimental physicist, won the 1973 National Academy of Sciences (NAS) Comstock Prize. The $4,000 prize is given every five years.

Dicke is Cyrus Fogg Brackett professor of physics at Princeton University. He was recognized for his development and use of high-precision instruments to test the general theory of relativity and the "big bang" theory concerning the origin of the universe.

Fermi Award. Manson Benedict, professor of nuclear engineering at M.I.T., received the $25,000 Enrico Fermi Award from the Atomic Energy Commission (AEC) in 1972.

Benedict was honored for being one of the principal designers of the nation's first gaseous diffusion plant in Oak Ridge, Tenn., which produced the first uranium-235 for wartime use. He also made major contributions to the development of the nuclear reactor for commercial power uses and established a school of nuclear engineering at M.I.T.

He received his Ph.D. degree in chemistry from M.I.T. in 1935. After working on the Oak Ridge project, he worked for the AEC for one year before joining the M.I.T. faculty in 1952.

Franklin Medal. George B. Kistiakowsky, physical chemist, won the 1972 Franklin Medal, the highest award given by the Franklin Institute in Philadelphia. He received the award for his research in chemical kinetics and the thermodynamics of organic molecules.

Kistiakowsky was born in Russia and received his Ph.D. from the University of Berlin in 1925. He came to the United States in 1926 and joined the Harvard University faculty in 1930. He served as President Dwight D. Eisenhower's special assistant for science and technology from 1959 to 1961.

Geosciences. Awards for important work in the geosciences included:

Day Medal. Frank Press, head of the department of earth and planetary sciences at M.I.T., received the 1973 Arthur L. Day Medal of the Geological Society of America (GSA).

Press is one of the world's leading seismologists. He received his Ph.D. degree from Columbia University in 1949 and taught there until 1955. From 1955 to 1965, he was professor of geophysics and

director of the seismological laboratory at the California Institute of Technology (Caltech). He came to M.I.T. in 1965.

Penrose Medal. Wilmot Hyde Bradley, retired member of the U.S. Geological Survey, won the 1973 GSA Penrose Medal. His main area of research is the paleoecology and sedimentation of the Green River Formation in the Western United States. He also pioneered in studies of the early North Atlantic deep-sea cores that laid the basis for current studies of ocean sedimentation.

Bradley received his Ph.D. degree from Yale University in 1927 and has spent his entire career with the U.S. Geological Survey. He founded the military geology unit and was chief geologist of the geologic division for 15 years.

Thompson Medal. Hollis D. Hedberg, professor emeritus of geology at Princeton University, won the NAS Mary Clark Thompson Medal for 1973 and an honorarium of $1,000.

Hedberg was cited for combining industrial research with basic research in geology and for bringing the effects to bear on public policy. In 30 years as an exploratory geologist for the petroleum industry, he established an international reputation in theoretical geology. He also contributed to public discussion of the role of national governments in offshore exploration for mineral resources.

Vetlesen Prize. William A. Fowler, professor of physics at Caltech, won Columbia University's 1973 Vetlesen Prize for achievements in the earth sciences. He was cited for his work in nuclear physics and its applications to astrophysics and geophysics. The award includes a gold medal and $25,000.

In awarding the prize, the jury noted that almost all quantitative information about the basic nuclear processes that enter into the generation of stellar energy and element synthesis is due to Fowler or to work instigated by him.

One consequence of his work is a chronologic account of major events in the history of the universe. In a paper in 1960, Fowler showed that observed uranium and thorium isotope ratios are consistent with a synthesis process extending over several billion years before the earth was formed. His calculations encouraged serious consideration for the first time of ages as great as 20 billion years for the universe.

Life Sciences

Biology. Among the awards presented in biology were the following:

Horwitz Prize. Dr. Stephen W. Kuffler, a neurophysiologist at Harvard Medical School, won Columbia University's 1972 Louisa Gross Horwitz Prize for his research on the nervous system.

The award cited Kuffler for "outstanding experiments" that have provided "information of fundamental importance to the understanding of the nature of neuromuscular and synaptic transmission, the mechanisms responsible for inhibition in the nervous system, the functional organization of the retina and visual system, and the role of the neuroglia in the central nervous system." The prize includes a $25,000 honorarium.

Lilly Award. Leland H. Hartwell, associate professor of genetics at the University of Washington in Seattle, received the 1973 Eli Lilly and Company Award in Microbiology and Immunology. The award, presented by the American Society for Microbiology, consists of a medal and $1,000. It honors a scientist under 35 for outstanding research in microbiology or immunology.

Hartwell was recognized for his pioneering work on the regulation of the cell division cycle in yeast. His research applied the tools of genetics, cell biology, and biochemistry to identify the critical points of control and provide the first substantial details of the molecular events in the division cycle. His collection of yeast mutants is providing starting points and tools for other investigators interested in related problems.

U.S. Steel Foundation Award. Donald D. Brown, an embryologist at the Carnegie Institution of Washington, won the 1973 U.S. Steel Foundation Award in molecular biology. The NAS award includes a $5,000 honorarium.

Brown was recognized for his investigations into the structure, regulation, and evolution of genes in animals. His studies have contributed to the understanding of ribonucleic acid and its involvement in the synthesis of proteins and the process of heredity.

Medicine. Major awards in medical science included the following:

Nobel Prize. The 1972 Nobel prize in physiology and medicine was awarded jointly to Dr. Gerald M. Edelman, professor of biochemistry at Rockefeller University in New York City, and Rodney R. Porter, professor of biochemistry at Oxford University in England. They will share a $100,000 cash prize.

Working independently, Edelman and Porter unraveled the chemical structure of a gamma globulin, the commonest of the antibodies that circulate in the blood in order to provide resistance to disease.

Porter became interested in antibodies while a Ph.D. student at Cambridge University. In 1950, while at the National Institute for Medical Research in England, he showed that antibodies are not fundamentally different from other gamma globulins, and that they could be partially broken down without losing their antigen-combining activity. In 1959, he found that the enzyme papain would split the antibody molecule into large fragments. This work eventually led to his proposing a four-chain model to depict the structure of the molecule.

Edelman also tackled the problem of subdividing antibody molecules, and broke the molecule into two kinds of polypeptide chains. The climax of his work came in 1969 when he and his colleagues announced that they had obtained the complete amino acid sequence of a human immunoglobulin molecule. Subsequently, they established the chemical structure of the complete molecule, consisting of 19,996 atoms, a structure far larger than any protein successfully analyzed up to that time.

Gairdner Awards. Five medical scientists received awards in 1972 from the Gairdner Foundation of Willowdale, Canada, for their research.

Karl Sune Detloff Bergstrom of Stockholm, Sweden, received a $7,500 award for contributions to the identification and chemical characterization of prostaglandins, a group of fatty acids. He is chairman of the World Health Organization task force that is exploring the potential of prostaglandins in regulating human reproduction.

Another $7,500 award went to Britton Chance, professor of biophysics at the University of Pennsylvania, for developing biophysical techniques for observing molecular events in living tissues.

Oleh Hornykiewicz, Canadian pharmacologist, received a $7,500 award for his contribution to understanding the physiology of the brain, especially in

Donald D. Brown

Gerald M. Edelman

Life
Sciences

Continued

Jean Piaget, right, winner of the Kittay International Award for psychiatry, accepts check for $25,000 from Sol Kittay, president and founder of Kittay Scientific Foundation, as Mrs. Kittay watches.

regard to Parkinson's disease. Hornykiewicz is a professor of pharmacology at the University of Toronto and head of psychopharmacology at the Clarke Institute of Psychiatry in Toronto.

Robert Russell Race and Ruth Sanger, an English husband-and-wife team at the Lister Institute in London, received $5,000 awards for the study of human blood groups.

Kittay Award. Jean Piaget, Swiss child psychologist, won the first Kittay International Award for psychiatry, a $25,000 prize, for his analysis of the mental development of children.

Piaget, considered the father of modern child psychology, has described four basic phases in the development of the child's ability to reason abstractly. He showed that young children were self-centered and often unresponsive to logical persuasion because they reasoned entirely differently than adults. His work has implications for psychiatry by providing clues to emotional responses that may result in adult neuroses.

The award, presented by the Kittay Scientific Foundation of New York City,

will be given annually to a researcher who has done outstanding work in the field of mental health.

Stouffer Prize. Four scientists shared the $50,000 Stouffer Prize in 1972 for their pioneering research on how cholesterol and other fats are carried in the blood. They are Dr. Vincent P. Dole of the Rockefeller Institute, New York City; Dr. John W. Gofman of the University of California, Berkeley; Dr. Robert S. Gordon, Jr., of NIH; and John L. Oncley of the University of Michigan in Ann Arbor.

Dole and Gordon were recognized for their discovery of the importance of the free fatty acids in blood. Working separately, they developed methods of measuring the free fatty acids and reported discoveries that laid a foundation for a better understanding of how these fats are used as a source of energy.

Gofman and Oncley were honored for their pioneering studies of lipoproteins. Working independently, they developed new techniques that can separate lipoproteins from other substances in the blood.

Space Sciences

Aerospace. The highest awards in the aerospace sciences included:

Goddard Award. Edward S. Taylor, professor emeritus at M.I.T., won the 1972 Goddard Award for his contributions to the advancement of air-breathing propulsion. The award, given by the American Institute of Aeronautics and Astronautics (AIAA) includes a gold medal and an honorarium of $10,000.

Taylor was cited for 45 years "as designer, inventor, researcher, teacher, adviser, and as founder and leader of a major educational and research center of aircraft engine activity." His research has included extensive work on internal-combustion engines, gas turbines, and jet engines.

Guggenheim International Astronautics Award. German physicist Reimar Lust won the 1972 Daniel and Florence Guggenheim International Astronautics Award given by the International Academy of Astronautics. The $1,000 award recognizes his investigation of the movement of the extraterrestrial plasmas in the earth's ionosphere and magnetosphere, and his contribution to determining the shape of the earth's magnetic and electric fields.

Lust is president of the Max Planck Society for the Advancement of Science and former director of the Institute for Extraterrestrial Physics of the Max Planck Institute for Physics and Astrophysics in Munich. He was scientific director of the European Space Research Organization in 1962 and 1963 and was one of its vice-presidents from 1968 to 1970. He received his Ph.D. degree in 1951 from the University of Göttingen.

Hill Space Transportation Award. The AIAA's 1972 Hill Space Transportation Award was shared by Richard H. Battin and David G. Hoag of M.I.T.'s Charles Stark Draper Laboratory. Battin is director of Apollo mission development, and Hoag is director of Apollo guidance and navigation.

They were honored for developing the hardware and software design of the Apollo spacecraft primary control, guidance, and navigation system. Their work demonstrated the feasibility of on-board autonomous space navigation during the flight of Apollo 8.

The award included a joint honorarium of $5,000 donated by the Louis W. and Maud Hill Family Foundation.

Astronomy. Important research in astronomy won the following awards:

Bruce Medal. Lyman Spitzer, astrophysicist at Princeton University, won the Catherine Wolfe Bruce Medal for "astronomical work of the highest quality." The gold medal is awarded by the Astronomical Society of the Pacific.

Spitzer has made fundamental contributions to many areas of astrophysical research. He explored the physical processes governing temperature in the interstellar medium and studied the occurrence of interstellar sodium gas clouds. He has made continuing contributions to plasma physics and the effort to achieve controlled thermonuclear fusion. In stellar dynamics, he has proposed a theory relating quasars to violent activity in galactic nuclei. His crusade for an orbiting astronomical observatory finally came to fruition with the launching of the Copernicus satellite in 1972.

Gould Prize. Kenneth I. Kellermann, radio astronomer at the National Radio Astronomy Observatory in Charlottesville, Va., received the NAS Benjamin Apthorp Gould Award in 1973. The award includes a $5,000 honorarium.

Kellermann was honored for his contributions to radio astronomy and long baseline interferometry, the use of two radio telescopes to pinpoint radio sources in space. In 1969, he was involved in conducting such observations with telescopes in Virginia and the Crimea, the first United States-Russian effort of its kind. He earned his Ph.D. degree at Caltech in 1963.

Michelson Medal. Herbert Friedman of the U.S. Naval Research Laboratory in Washington, D.C., won the Franklin Institute's Michelson Medal in 1972 for his pioneering work in solar and X-ray astronomy.

Friedman is superintendent of the atmosphere and astrophysics division of the Naval Research Laboratory. His research in rocket astronomy goes back to 1949, when he studied V-2 rockets and pioneered in developing rocket and satellite astronomy. Other achievements include tracing the solar cycle variations of X rays and ultraviolet radiation from the sun, producing the first astronomical photographs made in X-ray wave lengths, discovering the hydrogen corona around the earth, and measuring the ultraviolet fluxes of early-type stars.

Edward S. Taylor

Lyman Spitzer

General Awards

Science and Man. Awards for outstanding contributions to science and mankind included the following:

Founders Medal. Warren K. Lewis, professor emeritus of chemical engineering at M.I.T., won the 1973 Founders Medal given by the National Academy of Engineering (NAE) for contributions to engineering and society. The accompanying scroll cited Lewis for his contributions to chemical engineering education, the technical development of important industrial chemical processes, and his government service.

Lewis received his Ph.D. in chemistry from the University of Breslau in Germany in 1908. He has specialized in research on the distillation and thermal properties of liquids, and on colloids.

Kalinga Prize. Pierre V. Auger, French physicist, won the Kalinga Prize in 1972 from the United Nations Educational, Scientific, and Cultural Organization (UNESCO) for his contributions to the popularization of science.

Auger writes for the popular press on scientific subjects and has lectured to the general public in many countries. He is also active in broadcasting and is vice-chairman of the Radio Programs Committee of the French Radio and Television Service. Among his publications are a popular work on cosmic radiation and *L'Homme Microscopique* (1952).

From 1948 to 1959, Auger was director of UNESCO's Science Department and helped to create the European Organization for Nuclear Research (CERN) in Geneva. In 1959, he edited the UNESCO report: "Current Trends in Scientific Research."

The Kalinga Prize includes 1,000 British pounds sterling (about $2,500). The prize was founded by an Indian industrialist, Bijoyanand Patnaik, and named for an ancient Indian empire.

Oersted Medal. Arnold B. Arons, professor of physics at the University of Washington in Seattle, received the Oersted Medal in January, 1973. The medal is the highest honor awarded by the American Association of Physics Teachers (AAPT). Arons, who has an active interest in training elementary and high school science teachers, was cited for his teaching contributions.

Arons received his Ph.D. in physics at Harvard in 1943. He was a professor of physics at Amherst College in Massa-

Warren K. Lewis

Arnold Arons

chusetts from 1952 until he moved to the University of Washington in 1968. His writings include *Development of Concepts of Physics* (1965). Arons is a former president of the AAPT.

In addition to being a physicist, Arons is also an oceanographer. In 1943 he joined the Woods Hole Oceanographic Institution in Massachusetts as a research group leader, and he was a research associate in physical oceanography when he left in 1950.

Sperry Award. Leonard S. Hobbs and Perry W. Pratt, both retired executives of the Pratt & Whitney Aircraft Division of United Aircraft Corporation, shared the 1972 Elmer A. Sperry Award. The award medals honored them for their roles in directing the design and development of the JT3 turbojet engine, the first twin-spool jet engine to go into commercial production.

The Sperry Award recognizes an engineering contribution that has been applied in actual service to advance the art of transportation. It is sponsored by five engineering groups: American Institute of Aeronautics and Astronautics, American Society of Mechanical Engineers, Institute of Electrical and Electronics Engineers, Society of Automotive Engineers, and Society of Naval Architects and Marine Engineers.

Zworykin Award. Donald L. Bitzer, professor of electrical engineering at the University of Illinois in Urbana, received the 1973 Vladimir K. Zworykin Award from the NAE.

Bitzer is director of the university's Computer-based Education Research Laboratory. He received the award for inventing the PLATO computer-based education system, which is designed to assist in routine teaching, to teach concepts that may be difficult to present without the computer, and to free teachers to give more personal attention to students. Under Bitzer, the laboratory has developed PLATO equipment, techniques for controlling the computer, and teaching materials.

Bitzer's early work was in radar and radio direction-finding. He turned to developing large-scale computer-based education systems in 1960. He is co-inventor of the plasma display panel, an electronic method of presenting visual information, and he holds other patents in electronics.   [Kathryn Sederberg]

Major Awards and Prizes

Award winners treated more fully in the first portion of this section are indicated by an asterisk (*)

Adams Award (organic chemistry): Georg Wittig
American Chemical Society Award in Enzyme
  Chemistry: Howard M. Temin
American Institute of Physics—U.S. Steel Foundation
  Award (science writing): Edward Edelson
American Physical Society High-Polymer Physics Prize:
  H. Douglas Keith and Frank J. Padden, Jr.
Bertner Foundation Award (cancer research):
  Dr. George Klein
Bonner Prize (nuclear physics): Herman Feshbach
*Bruce Medal (astronomy): Lyman Spitzer
*Buckley Solid State Physics Prize: Gen Shirane
 Chemical Industry Medal: Ralph Landau
*Comstock Prize (physics): Robert H. Dicke
*Cope Award (chemistry): Roald Hoffmann and
  Robert B. Woodward
*Day Medal (geology): Frank Press
 Debye Award (physical chemistry): William N.
  Lipscomb, Jr.
*Fermi Award (physics): Manson Benedict
*Founders Medal (engineering): Warren K. Lewis
*Franklin Medal (physics): George B. Kistiakowsky
*Gairdner Awards (medicine): Dr. Karl Sune Detloff
  Bergstrom, Dr. Britton Chance, Dr. Oleh
  Hornykiewicz, Dr. Robert Russell Race, and
  Dr. Ruth Sanger
 Garvan Medal (chemistry): Mary L. Good
*Goddard Award (aerospace): Edward S. Taylor
*Gould Prize (astronomy): Kenneth Kellermann
*Guggenheim Award (aerospace): Reimar Lust
 Haley Astronautics Award: James Irwin, David
  Scott, and David Worden
 Heineman Prize (mathematical physics): Kenneth G.
  Wilson
*Hill Space Transportation Award (astronautics):
  Richard H. Battin and David G. Hoag
*Horwitz Prize (biology): Dr. Stephen W. Kuffler
*Kalinga Prize (science writing): Pierre V. Auger
*Kittay Award (psychiatry): Jean Piaget
 Kovalenko Gold Medal (medicine): Seymour S. Kety
 Langmuir Prize (chemical physics): Peter M. Rentzepis
 Lasker Awards (medical research): Dr. Joseph H.
  Burchenal, Dr. Denis P. Burkitt, Dr. Paul P. Carbone,

  Dr. Vincent T. DeVita, Jr., Dr. Isaac Djerassi,
  Dr. Emil Frei, Dr. Emil J. Freireich, Dr. Roy Hertz,
  Dr. James F. Holland, Dr. Edmund Klein, Dr. Min
  Chiu Li, Dr. V. Anomah Ngu, Dr. Donald Finkel,
  Dr. Eugene J. Van Scott, and Dr. John L. Ziegler
*Lilly Award (microbiology): Leland Hartwell
 Meggers Award (optics): Charlotte Moore Sitterly
*Michelson Medal (astronomy): Herbert Friedman
 National Academy of Sciences (NAS) Award in
  Aeronautical Engineering: Donald Wills Douglas
 NAS Award in Applied Mathematics: Samuel Karlin
 NAS Award for Environmental Quality: W. Thomas
  Edmondson
*Nobel Prize: chemistry, Christian Anfinsen, Stanford
  Moore, and William H. Stein; physics, John Bardeen,
  Leon N. Cooper, and John Robert Schrieffer;
  physiology and medicine, Dr. Gerald Maurice
  Edelman and Rodney R. Porter
*Oersted Medal (teaching): Arnold B. Arons
 Oppenheimer Memorial Prize (physics): Steven
  Weinberg
 Pendray Award (aerospace): Marcus F. Heidman and
  Richard Priem
*Penrose Medal (geology): Wilmot Hyde Bradley
*Perkin Medal (chemistry): Theodore Cairns
*Priestley Medal (chemistry): Harold Urey
 Reed Award (aerospace): I. Edward Garrick
 Smith Medal (astronomy): Clair C. Patterson
 Soviet Academy of Sciences Lomonosov Gold Medals:
  Nikolai Ivanovich Muskhelishvili and Max Steenbeck
*Sperry Award (engineering): Leonard S. Hobbs and
  Perry W. Pratt
*Stouffer Prize (medicine): Dr. Vincent P. Dole,
  Dr. John W. Gofman, Dr. Robert S. Gordon, Jr., and
  John L. Oncley
*Thompson Gold Medal (geology): Hollis D. Hedberg
 Tillyer Award (optics): Robert M. Boynton
*U.S. Steel Foundation Award (molecular biology):
  Donald D. Brown
*Vetlesen Prize (geophysics): William A. Fowler
 Warner Prize (astronomy): George R. Carruthers
 Warren Prize (geology): Luna B. Leopold
*Zworykin Award (engineering): Donald L. Bitzer

Deaths of Notable Scientists

Notable scientists who died between June 1, 1972, and June 1, 1973, include those listed below. An asterisk (*) indicates that the person has a biography in *The World Book Encyclopedia*.

Aiken, Howard H. (1900-March 14, 1973), mathematician, designed and built the first digital computer, the Mark I, in 1944. He was professor emeritus of applied mathematics at Harvard University in Boston.

Albert, Abraham Adrian (1905-June 6, 1972), mathematician, was called the father of modern algebra. He was dean of the University of Chicago's division of physical sciences from 1962 to 1971.

Artsimovich, Lev A. (1909-March 1, 1973), Russian nuclear physicist, was a pioneer of Russia's atomic research program. A member of the Soviet Academy of Sciences, he led efforts to control thermonuclear power for peaceful uses.

Birch, Herbert G. (1918-Feb. 4, 1973), pediatrician, studied the relationship of malnutrition and other social problems to mental retardation and investigated child development and brain injury.

Bohler, Lorenz (1885-Jan. 20, 1973), Austrian surgeon, developed a technique for healing bone fractures and lacerated muscles. His three-volume *Technique of Bone Fracture Treatment* was translated into 14 languages.

Bowen, Ira Sprague (1898-Feb. 6, 1973), astronomer, presented evidence to prove that the universe is composed of the same elements found on earth, and that nebulae are made up largely of hydrogen and helium. He was director of combined operations for the Mount Palomar and Mount Wilson observatories in California from 1946 to 1964.

Dreyfuss, Henry (1904-Oct. 5, 1972), industrial designer, worked on products ranging from clocks, telephones, and cameras to trains and ocean liners. He contributed to the design of the *20th Century Limited* passenger train and the luxury liner S.S. *Constitution*. He also designed the interior of the Boeing 707 for American Airlines.

Eddy, Nathan B. (1890-March 28, 1973), pharmacologist at the National Institutes of Health, developed a new class of pain-relieving drugs, the bensomorphans, to substitute for morphine and codeine.

Foelling, Ivar A. (1888-Jan. 25, 1973), Norwegian physician, discovered phenylketonuria and established that this form of infant mental retardation resulted from the body's inability to use certain amino acids.

Gast, Paul W. (1930-May 16, 1973), geologist, directed the scientists studying lunar rocks at the Manned Spacecraft Center in Houston. He led in developing isotopic methods of dating rocks and applying these techniques to studies of the moon's age. He also studied the rare earth elements to understand the earth's interior and its origins.

Gibbon, John H., Jr. (1903-Feb. 5, 1973), surgeon, performed the first successful open-heart operation in 1953 using a heart-lung machine that he designed. It by-passed the heart while maintaining normal blood circulation.

Hamilton, Walter C. (1931-Jan. 23, 1973), chemist, developed computer techniques for determining chemical crystal structures. He worked on a systematic study of the precise location of hydrogen atoms in amino acids.

Hitchcock, Lauren B. (1900-Oct. 15, 1972), chemical engineer, was a consultant on air-pollution control. He submitted a report in 1954 pointing to automobiles and back-yard incinerators as prime causes of smog in Los Angeles.

Hume, David M. (1917-May 19, 1973), surgeon, helped develop the technique of human organ transplants and pioneered in kidney transplants. He assisted in developing a technique connecting the livers of baboons to the blood streams of patients to cleanse human blood. He also designed a portable feeding apparatus for persons who have had most of their intestines removed.

Hutchinson, John (1884-Sept. 2, 1972), British botanist, developed a new plant classification system that spurred considerable discussion and new research. He was adept at popularizing science, and wrote such books as *Families of Flowering Plants*.

Jones, Chester Morse (1891-July 26, 1972), gastroenterologist, specialized in ulcer research. He was a member of the Harvard Medical School group that developed the technique of cutting vagus nerves to heal stomach ulcers in 1947.

*Leakey, Louis S. B. (1903-Oct. 1, 1972), British anthropologist and paleontologist, discovered skulls and bone remnants in the Olduvai Gorge in Tanzania that proved early man lived in Africa at

Lev A. Artsimovich

Louis S. B. Leakey

Harlow Shapley

Deaths of Notable Scientists

Continued

Igor I. Sikorsky

Max Theiler

Georg von Békésy

least 2 million years ago. Previously, the earliest known man had been placed in Asia about 500,000 years ago.

Lefshetz, Solomon (1884-Oct. 5, 1972), Russian-born mathematician, pioneered in developing topology, the study of constant properties in changing geometric forms.

Lehrman, Daniel S. (1919-Aug. 29, 1972), animal psychologist, studied the effect of hormones on behavior in animals and how social stimuli affect hormone production. He also analyzed the effect of learning on the development of instinctive behavior. He was director of Rutgers University's Institute of Animal Behavior since 1959.

MacArthur, Robert H. (1930-Nov. 1, 1972), Canadian-born biologist, studied the interaction of plant and animal life and formed mathematical theories to predict the effects of making environmental changes.

Millionshchikov, Mikhail Dmitrievich (1913-May 27, 1973), Russian physicist, promoted international cooperation in science. His main work was in turbulence, filtration, and applied gas dynamics. He was a vice-president of the Soviet Academy of Sciences.

Murphy, Robert Cushman (1887-March 19, 1973), ornithologist, was an authority on marine birds. His research demonstrated the close relationship between sea birds and their environment. He found the bones of the extinct New Zealand moa, and in 1951 discovered a colony of cahows, a bird thought to be extinct, in Bermuda.

Pardee, Harold E. B. (1886-Feb. 28, 1973), cardiologist, was the first to describe the changes in electrocardiograms that indicate coronary disease. Two electrocardiographic terms were named for him, Pardee's sign and Pardee's T wave.

Pecora, William T. (1913-July 19, 1972), geologist, directed the U.S. Geological Survey from 1965 to 1971, when he was named undersecretary of the interior. He was associated with Project Eros, an attempt to use space technology to assess the earth's resources, and with the National Center for Earthquake Research in California.

Rabinowitch, Eugene (1901-May 15, 1973), Russian-born chemist, was a section chief on the Manhattan Project, which developed the first atomic bomb. After World War II, he devoted his life to the search for an international framework to control atomic energy. In 1945, he helped found the *Bulletin of the Atomic Scientists.*

Seegal, David (1899-July 24, 1972), physician, was a leading researcher in cancer and other chronic diseases, including those affecting the kidney, liver, and the circulatory and respiratory systems. He was professor emeritus of medicine at Columbia University's College of Physicians and Surgeons and co-editor of *The Journal of Chronic Diseases.*

*Shapley, Harlow (1885-Oct. 20, 1972), astronomer, demonstrated that the Milky Way was much larger than previously thought and that the solar system is on its fringe, rather than in its center. He directed the Harvard Observatory from 1921 to 1952.

Sheps, Mindel (1913-Jan. 15, 1973), Canadian-born biostatistician, used mathematical and computerized methods to study the dynamics of population growth. She was an authority on applying statistical study to the biological factors affecting fertility and reproduction.

*Sikorsky, Igor I. (1889-Oct. 26, 1972), Russian-born aviation engineer, developed the world's first practical helicopter. He also built the first multiengine aircraft, in Russia, and invented the flying boat, a large, commercial plane that could land on water.

Stack, John (1906-June 17, 1972), aeronautical engineer, helped develop the first high-speed wind tunnel. From 1961 to 1962, he directed aeronautical research for the National Aeronautics and Space Administration.

Stearns, Carl L. (1892-Nov. 28, 1972), astronomer, spent his life measuring stellar parallaxes (apparent changes in stars' positions due to changes in the observer's position) to determine the position, motion, and distance of stars.

*Theiler, Max (1899-Aug. 11, 1972), South African bacteriologist, developed a vaccine against yellow fever after proving the disease was caused by a virus, not a bacterium. His work won the 1951 Nobel prize in physiology and medicine.

*Von Békésy, Georg (1899-June 13, 1972), Hungarian-born physicist, won the 1961 Nobel prize in physiology and medicine for his research into the mechanism of hearing. His work led to major advances in diagnosing and correcting damaged hearing. [Kathryn Sederberg]

Index

This index covers the contents of the 1972, 1973, and 1974 editions of *Science Year,* The World Book Science Annual.

Each index entry is followed by the edition year in *italics* and the page numbers:

Copernicus, Nicolaus, *74*-262, *73*-266

This means that information about Copernicus begins on the page indicated for each of the editions.

An index entry that is the title of an article appearing in *Science Year* is printed in boldface italic letters: ***Archaeology.*** An entry which is not an article title, but a subject discussed in an article of some other title, is printed: **Behaviorism.**

The various "See" and "See also" cross references in the index are to other entries within the index. Clue words or phrases are used when the entry needs further definition or when two or more references to the same subject appear in *Science Year*. These make it easy to locate the material on the page.

Continental drift theory: geology, *74*-305; geophysics, *73*-320, *72*-322; ocean drilling: *Special Report, 72*-128; oceanography, *74*-334; plume theory, *Special Report, 73*-163; primitive mammals, *73*-318. See also **Plate tectonics; Sea floor spreading.**

The indication *"il."* means that the reference is to an illustration only, as:

Artificial insemination: bees, *il., 74*-158

Index

A

Abouna, George, 73-328

Accelerator, particle: European Center for Nuclear Research (CERN), 73-238, 72-347; Los Alamos, 73-351, *il.,* 350; National Accelerator Laboratory (NAL), 73-349, *il.,* 348, 72-205, 347; Stanford Linear (SLAC), 74-338, 73-346, 72-348

Acetylcholine: muscle cell, 73-275; nerve cell, 72-274

Acne: diet and, 72-159, 174; *Special Report,* 72-152; *Trans-Vision®,* 72-162

Actin: role in muscle building, 73-155, *ils.,* 156, 158; role in muscle contraction, 73-275

Actinomycin D: interaction with DNA, 72-289

Actomyosin, 73-157, 72-395

Acupuncture: anesthesia, 73-192, 330, *ils.,* 194, 328; medical illustration, 72-11, *il.,* 13; medicine, 74-312; psychosomatic health, *il.,* 74-146

Adams, Roger, 73-423

Additives, food, 73-274, 295

Adenosine (A), 72-289

Adenosine diphosphate: role in muscle power, 73-157

Adenosine monophosphate: gene expression, 72-338

Adenosine triphosphate, 73-157, 275, 72-395

Adipose tissue: fat production, 73-122; glucose level and food intake, 73-122

Aerospace: awards and prizes, 74-410; industry unemployment, 73-360, 72-361, 364

Afar Triangle, 73-170, *il.,* 173

Africa: archaeology, 74-249

Aggression: population and crowding, 72-305

Aging: immunology, 74-119; *Special Report,* 74-120

Agranoff, Bernard, 72-145

Agriculture, 74-244, 73-254, 72-254; archaeology, 72-259, 262; asphalt moisture barrier, 72-256; *Books of Science,* 73-277; crop damage, 74-297; drugs, 74-282; effect of farming methods on climate, 72-101, *il.,* 103; in China, 73-197, *ils.,* 198, 199; irrigation, *Special Report,* 74-213; plant cell fusion, 72-203, *il.,* 202; plant disease, 74-326; Triticale, synthetic grain species, 72-255. See also **Aquaculture;** *Botany; Chemical Technology;* **Climate;** **Food; Pesticide.**

Agriculture, U.S. Department of: bees, 74-153

Aiken, Howard H., 74-413

Air-circulation patterns, 72-104

Air conditioning: energy, 74-292

Air cushion vehicle, 73-373, *il.,* 371, 72-375

Air Delivered Anti-Pollution Transfer System, 72-344

Air pollution: acidic rainfall, 73-297; atmospheric monitoring stations, 73-331; automobile exhaust, 73-311; catalytic converters, *Close-Up,* 74-364; chemical technology, 72-284; *Close-Up,* 74-364; electric transit system, 74-192; energy, 74-292; environment, 74-293; Environmental Protection Agency, 72-34; heart disease, 74-312; increased levels of CO and CO_2, 73-331; lead poisoning, 72-302; meteorology, 72-333; natural gas, 73-306; ozone absorption by plants, 72-282; particulate matter in atmosphere, 73-331; public health, 73-325; transportation, 74-362, 72-374. See also **Automobile; Climate; Pollution.**

Airlines: electronics, 74-287

Alaska: archaeology, 74-251

Alaska pipeline, 72-34, 372

Albert, Abraham Adrian, 74-413

Albright, William F., 73-423

Alcoa Seaprobe, il., 73-342

Aleutian Islands: archaeology, 74-251

Alfven, Hannes O.G., 73-416, 72-416

Algae: cell, *Special Report,* 73-109; source of protein, 72-284

Algal bloom, 74-323

Alikhanov, Abram I., 72-423

Allelopathy, 73-300

Alpha-bungaro: acetylcholine receptor isolation, 72-274

Alpha waves, *il.,* 73-338

Alvin: submersible, 74-333

American Association for the Advancement of Science (AAAS): awards and prizes, 73-422; *Close-Up,* 73-302; *Special Report,* 72-240

American Association for Thoracic Surgery, 72-189

American Chemical Society Awards: Analytical Chemistry, 72-422; Biological Chemistry, 72-422; Creative Invention, 73-422; Enzyme Chemistry, 73-422; Inorganic Chemistry, 72-422; Nuclear Applications in Chemistry, 73-422; Pure Chemistry, 72-422

American Medical Association: awards and prizes, 72-422; drug evaluation, 72-298

American Sign Language: chimps, *Special Report,* 74-37

American Telephone and Telegraph Company (A.T.&T.), 73-291, 72-293

Amino acids: aging, *Special Report,* 74-128; Al protein, 72-274; botany, 74-269; chemical synthesis, 74-278; lunar, 74-301; Murchison stone meteorite, 72-315, *il.,* 316; tryptophan, 72-276

Ammonia: interstellar molecules, *Special Report,* 72-67; plant absorption, 73-255

Amniotic fluid: prenatal diagnosis, 72-315

Amphetamines, 72-299

Amtrak, 72-374

Anabolic steroids, 73-161

Analogues: pesticide, 74-246

Anatomy: medical illustration, 72-11

Ancobon: drugs, 74-283

Andrews, E. Wyllys IV, 73-423

Anesthesia: by acupuncture, 73-192, *il.,* 194

Anfinsen, Christian B., 74-404

Angina pectoris: *Special Report,* 72-176

Angiography, 72-178, 187

Angiosperms, 74-309, 72-281

Anorthosite: in lunar rocks, 73-313

Antarctic Ocean, 72-118

Anthropology, 74-247, 73-256, 72-256; *Books of Science,* 74-266, 73-277, 72-278; *Special Report,* 73-224. See also *Archaeology.*

Antibiotics: acne therapy, 72-172; actinomycin D, interaction with DNA, 72-289; drugs, 74-283; in combination, 72-298; protein synthesis inhibition, 73-114, 72-145; tetracycline, 72-172; trobicin, 73-297

Antibodies: immunology, *Special Report,* 74-112

Antiferromagnets, 73-353

Antigen: immunology, *Special Report,* 74-113

Ants, 74-286

Aphasia, 72-151

Apollo Program: Apollo 8, *il.,* 74-13; Apollo 10, *il.,* 74-13; Apollo 11, *ils.,* 74-14, 15; Apollo 12, *il.,* 74-16; Apollo 14, 72-43, 366, *ils.,* 74-10, 17, 72-39, 41; Apollo 15, 73-363, 72-367, *il.,* 74-18; Apollo 16, 73-366, *il.,* 74-20; Apollo 17, 74-360, 73-366, *ils.,* 74-22, 23, 354, 357, 358; communications systems, 73-292; geochemistry, 74-299; Kraft, Christopher, 73-398; science support, 73-360; space exploration, 73-362, 72-366; *Special Report,* Photo Essay, 74-11

Apollo-Soyuz Test Project: Russian science, *Special Report,* 74-239; space exploration, 74-360

Appetite, 73-117

Applied chemistry: Perkin Medal, 74-406

Applied Mathematics Award, 73-421

Aquaculture: raft culture, 72-114; *Special Report,* 72-107; zones of upwelling, marine life concentrations, 72-118

Archaeology, 74-249, 73-259, 72-259; *Books of Science,* 74-266, 73-277, 72-278; interdisciplinary, *Special Report,* 74-93; looting, 74-249,

Index

73-260; megalithic archaeology,
Special Report, 72-228. See also
Anthropology; Geoscience.
Architectural engineering: *Books of
Science,* 73-277.
**Arctic Ice Dynamics Joint
Experiment,** 73-339
Arctic ice floes, 73-339
Arctic tundra: Alaska pipeline, 72-34
Arctowski Medal, 73-419
Arm movements: infant, 72-359
Armstrong, Neil A., 72-367
Arons, Arnold B.: Oersted Medal,
74-411
ARPANET: computer network, 73-295
Arroyo Hondo, 73-262
Arsenic: used in insecticides, 72-291;
water pollutant, *Close-Up,* 72-303
Arteriosclerosis: and diet, 73-322;
heart, *Close-Up,* 72-181
Arteriovenous fistula, 72-332
Artificial insemination: bees, *il.,*
74-158
Artificial upwelling, 72-343
Artsimovich, Lev, 74-413, 72-215, 353
Ascorbic acid: and iron availability,
73-145; identification, 72-387
Asphalt moisture barrier: agriculture,
72-256
Asthenosphere, 72-322
Astronautics: *Books of Science,*
73-277. See also *Space Exploration.*
Astronomy: awards and prizes, 74-410;
Books of Science, 74-266, 73-277,
72-278; cosmology, 74-260, 73-272,
72-271; heliocentric system, 73-266;
high energy, 74-257, 73-270, 72-269;
planetary, 74-253, 73-264, 72-263;
quasars, *Close-Up,* 74-258; stellar,
74-255, 73-268, 72-265; Very Long
Baseline (VLB), 74-259; X-ray, 74-65,
73-270, 72-269. See also *Space
Exploration.*
Asymptopia: particle physics, 72-347
Atherosclerosis: medicine, 74-313
Athletics: *Special Report,* 73-146
Atmosphere. See **Climate;**
Meteorology; **Weather.**
Atomic absorption spectroscopy,
73-141
Atomic and molecular physics. See
Physics (atomic and molecular).
Atomic Energy Commission (AEC),
74-354, 296, 73-309, 360, 72-32, 205,
352, 400
Atomic fission, 72-214
Atomic fusion, 72-214
Atomic safety: environment, 74-296
Auditory feedback, *ils.,* 74-139, 140
Auger, Pierre V.: Kalinga Prize, 74-411
Australia: irrigation, *Special Report,*
74-213

Australopithecus: anthropology,
74-247, 72-257; anthropology, *Special
Report,* 73-228; *ils.,* 74-248, 73-225,
234, 235
Auto-immunity: immunology, *Special
Report,* 74-117
Auto-Train, 73-373
Autogenic feedback training:
Special Report, 74-141
Automobile: *Close-Up,* 74-364;
electronics, 74-287; emission standards,
73-311, 72-373; environment, 74-293;
impact on society, 72-28; medicine,
74-312; tire safety test, *il.,* 73-374;
transportation, 74-362; Wankel engine,
Close-Up, 73-372. See also **Air
pollution;** *Transportation.*
Autonomic nervous system:
psychosomatic health, *Special Report,*
74-138
Aves Swell: submarine ridge, 72-135
Awards and Prizes, 74-404, 73-415,
72-414
Axons, 73-336, *il.,* 337
Aztec period, 72-262

B

Babbles and coos, 72-359
Babies: test-tube, 72-362
Bacteria: genes incorporated into other
cells, 73-333; microbiology, 74-323;
protein, 72-340; used to control oil
pollution, 72-344
Bacteriology. See *Microbiology.*
Bagasse: source of protein, 72-284
Barabashov, Nikolai P., 72-423
Bardeen, John: Nobel prize, 74-406
Barnacle glue: dentistry, 72-325
Bastnasite, 73-349
Bats: fog avoidance, 72-380; hiber-
nation, 73-376
Batten's disease, 74-128
Battin, Richard H.: Hill Space
Transportation Award, 74-410
Baynard's Castle, 73-259
BCS theory: physics, 74-406
Beal, George D., 73-423
Bear seamount, 73-339
Bears, grizzly: *Special Report,* 73-35
Beata Ridge: submarine ridge, 72-135
Beberman, Max, 72-423
Bees: *Special Report,* 74-148
Beetles: heat experiments, 72-300
Behavior: animal, 74-160; nutrition,
74-330; psychology, 74-348
Behaviorism, 73-356, 72-150
Bell, Charles, 72-12, *il.,* 16
Benedict, Manson: Fermi Award,
74-407
Benzene: chemical synthesis, 74-279
Bergstrom, Karl Sune Detloff:
Gairdner Award, 74-408
Bernal, John Desmond, 73-423
Berson, Solomon A., 73-423
Bezymianny volcano, 72-318

Big bang theory, 74-253, 73-272
Bile: formation of gallstones, 73-327
Binary stars: radio and X-ray radiation
from, 74-73, 73-268
Bioavailability: of drugs, 73-296
Biochemistry, 74-262, 73-274, 72-273;
Books of Science, 73-277; neural,
74-327, 72-137; trace elements, 73-133.
See also **Amino acids; Biology;**
Chemistry; **DNA; Enzymes;
Hormones; Plant; RNA.**
Biofeedback: aging, *Special Report,*
74-124; psychosomatic health, *Special
Report,* 74-138
Biological magnification: pollutant
concentration in food chain, 73-52
Biologics, Bureau of, 74-283
Biology: aging, *Special Report,* 74-120;
awards and prizes, 74-408;
Books of Science, 74-266, 72-278.
See also **Anatomy;** *Biochemistry;*
Botany; Cell; *Ecology;* **Enzymes;
Genetics;** **Marine life;** *Micro-
biology;* **Pharmacogenetics; Virus;
Zoology.**
Biomedical research: science support,
74-353
Biomes: grassland, 72-81
Biophysics: communication, 72-137
Birch, Herbert G., 74-413
Birds: endangered species, 73-50;
grassland study, 72-85, *ils.,* 86
Birth control: animal, 74-167
Birth defects: prenatal diagnosis,
72-315
Bitzer, Donald L.: Zworykin Award,
74-411
Black holes: *Special Report,* 73-77;
X rays, 74-75, 73-271
Black lung disease, 73-324
Blackheads, 72-158
Bladderworts, *il.,* 73-13
Blastoids, 73-96
Blegen, Carl W., 73-423
Blight watch, 72-254
Blood: erythropoietin, 72-330;
immunology, *Special Report,* 74-108;
macrophages, 74-265; phonoangiog-
raphy, 72-329, *il.,* 326; red blood cells,
73-155, 378, *il.,* 72-327; red blood cell
stimulant, 72-330; self-generated flow
technique, 74-140; sickle cell anemia,
73-323, 72-328, *il.,* 327; surgery,
74-317; T-globulin, cancer test, 72-326;
2,3-DPG, 73-155
Blood pressure: drugs, 74-283
Blood sugar: insulin deficiency, 73-140
Blue-green algae, 73-109
Body temperature: aging, *Special
Report,* 74-124, *il.,* 125
Bohler, Lorenz, 74-413
Bones: electric fracture treatment,
73-330; repair with ceramics, 73-284
Books of Science, 74-266, 73-277,
72-278
Bootstrap hypothesis, 73-348

Index

Borlaug, Norman E., 72-254, 421, *il.*, 415
Boss, Benjamin, 72-423
Botany, 74-269, 73-280, 72-281; aging, *Special Report,* 74-130, *il.*, 129; *Books of Science,* 74-266, 73-277, 72-278; ecology, 74-286; endangered plants, *Essay,* 73-11; paleontology, 74-309; plant cell fusion, 72-203, *il.*, 202; plant growth regulators, 73-255. See also *Agriculture;* **Plant.**
Bowen, Ira Sprague, 74-413
Boyd Orr, Lord, 73-423
Brachiosaurus: paleontology, 74-308
Bradley, Wilmot Hyde: Penrose Medal, 74-407
Bragg, Sir William Lawrence, 73-423
Brain: neurology, 74-327; psychology, 74-348; *Special Report,* 72-137; surgery, acupuncture anesthesia, 73-192. See also *Neurology; Psychology.*
Brazil: communications satellite systems, 73-292
Breccias: Apollo 14 rocks, 72-43
British soldier lichen, *il.*, 73-12
Brödel, Elizabeth, *il.*, 72-21
Brödel, Max, 72-13, *ils.*, 18, 19
Bronze casting, 72-259
Brown, Donald D.: U.S. Steel Foundation Award, 74-408
Brownouts: energy crisis, *Special Report,* 72-213; peak demand, 73-308
Bruce Medal, 74-410
Brucellosis, 73-256
Bubble technology: electronics, 74-289
Bucha, Vaclav, 72-231
Buckley Prize, 74-406
Bumblebees, 73-375
Bursa equivalent: immunology, 74-110
Buses: transportation, 74-363, 73-374

C

Cairns, Theodore L.: Perkin Medal, 74-406
Calcium metabolism, 72-276
Calcitonin, 72-276
Calculators: electronic, 72-308; portable, 73-293, 72-308
Caldera, 73-62, *il.*, 63
California: archaeology, 74-251
Calories: utilization of by exercise, 73-125
Canada: communications satellite systems, 74-280, 73-292
Canada geese, 72-210
Canals: on Mars, 73-62
Cancer: abnormal properties of cells, 73-25; betel nuts, mouth cancer, 72-258; biochemistry, 74-265; bladder, 74-311; blood test, 72-326; bowel,

74-312; breast, 72-326; carcinoembryonic antigen, 74-312, 73-327; carcinogens, 73-25; caused by virus, 73-26, 90, 72-299, 326; cell fusion, 72-198; chemotherapy and cancer research, 73-27; chicken vaccine, 72-255; clusters of incidence, 73-95; detection, 73-327, *il.*, 72-328; diagnosis, 74-311; diet, 74-312; drugs, 74-282; esophagus, 74-312; immunology, 74-117, 73-26, 72-327; kaunaoa worm, anticancer ingredient in, 72-377; L-asparaginase, 72-298; laser therapy, *il.*, 74-204; National Cancer Act, 73-27; National Cancer Institute, 73-28; National Cancer Plan, 73-30; nuclear magnetic resonance, 72-327; occupational, 73-324; research, *Essay,* 73-24; research, monetary considerations, 73-29, 72-363; RNA-directed DNA polymerase, 72-275; Russian science, *Special Report,* 74-234; stomach, 74-312; surgery, 74-315; SV-40, cancer-producing virus, 72-201
Cannibalism: anthropology, 74-249; zoology, 74-367
Cannon, Walter B., 73-118
Capillaries: increase by exercise, 73-155
Carbohydrates: and glycogen content in muscles, 73-158
Carbon, activated: used in waste control, 73-284
Carbon dioxide (CO_2): increased level in atmosphere, 73-331, 72-98; use in heart surgery, 72-189, 330
Carbon dioxide laser, 74-206, *il.*, 209
Carbon-14 dating, 72-230
Carbon monoxide (CO): catalytic converters, *Close-Up,* 74-364; environment, 74-293; increased level in atmosphere, 73-331; medicine, 74-312
Carcinogens. See **Cancer.**
Cardiovascular . . . . See headings beginning **Heart** . . . .
Caribbean Sea floor, 72-123
Carnap, Rudolf P., 72-423
Carp, 72-108
Cartilage: regeneration, 72-378
Carty Medal, 72-417
Casman, Ezra P., 72-423
Castellani, Marquis Aldo, 73-423
Cat: visual cortex experiments, 72-139, *ils.*, 142, 143
Catalytic converters: *Close-Up,* 74-364; environment, 74-293
Catfish, 72-109, *il.*, 110
Catheterization, 72-331
Cattle: crossbreeding of, 72-255; drugs, 74-282; new breeds, 74-245
Cedarwood extract: insect repellent, 73-280
Cell: aging, *Special Report,* 74-122, *il.*, 123; biochemistry, 74-262; cancer (detection), *il.*, 72-328, (abnormal properties), 73-25; evolution, 73-105,

333; fusion, *Special Report,* 72-191; hybrid cell formation, 72-193, *il.*, 195; immunology, *Special Report,* 74-108; multidisciplinary study, 73-25. See also **Amino acids; Blood; Cancer; DNA; Enzymes;** *Genetics;* **RNA; Virus.**
Cellulose: and protein-forming microbes, 72-284
Centaurus X-3, 74-73, 73-270
Central Arid Zone Research Institution, 72-101
Central nervous system, 73-336
Centromeres, 72-274
Ceramic: used in bone repair, 73-284
Cerebral hemispheres: neurology, 74-328
CERN (European Center for Nuclear Research): physics, 74-338; proposed accelerator, 72-350; proton collisions, 72-347; *Special Report,* 73-239
Cernan, Eugene A., 74-360, 73-366
Cesium 137, 72-300
Chance, Britton: Gairdner Award, 74-408
Chaney, Ralph W., 72-423
Channelization of streams: *Close-Up,* 74-294
Chao, Edward, 72-402
Chapman, Sydney, 72-423
Charge-coupled device, 73-304
Charney, Jesse, 73-103
Charpak, Georges, 73-239, *il.*, 242
Chemical and biological warfare, 72-247
Chemical disease, 73-215
Chemical Industry Medal, 73-414
Chemical insecticides, 74-246
Chemical reaction: angular distribution of product molecules, 73-286; chemical dynamics, 74-274
Chemical Technology, 74-271, 73-282, 72-283
Chemistry: Books of Science, 74-266, 72-278; dynamics, 74-274, 73-285, 72-287; organic, Cope Award, 74-404; structural, 74-276, 73-287, 72-289; synthesis, 74-278, 73-289, 72-291; synthetic scotophobin, 74-328. See also **Biochemistry; Chemical Technology.**
Chenodeoxycholic acid, 73-327
Chicken leukosis, 72-255
Child development, 72-357
Childhood disease: aging, *Special Report,* 74-128; catch-up growth, 74-330
Chimps: anthropology, 74-249; *Special Report,* 74-34
China: archaeology, 74-251; social structure and scientific progress, *Special Report,* 73-188; space program, 72-371
Chiral chemicals: chemical synthesis, 74-278
Chiral recognition: chemical synthesis, 74-279

Index

Chlorinated hydrocarbon insecticides, 72-291
Chloroplasts, 73-106, *il.,* 111
Cholesterol gallstones, 73-327
Choline: botany, 74-269
Chromium: human requirement, 73-139
Chromosomes: cell fusion, *Special Report,* 72-192; gene location, 74-297; stain technique, 73-311, *ils.,* 74-298, 73-312, 72-314. See also **Cell;** *Genetics.*
Circulatory system, 73-155. See also **Blood; Heart; Revascularization.**
Clairaudience, 74-145
Clairvoyance, 74-145
Classroom, open, 72-307
Claytonia virginica: chromosomal drift, 72-281
Clean Air Act, 74-293, 72-374
Cleocin: drugs, 74-283
Cleveland Clinic, 72-182
Climate: animal adaptation, 73-375; Caucasus Mountains, dust-snow studies, 72-95; dust-climate relationship, 72-96; effect of cities on, 73-332; effects of atmospheric pollution, 73-331; farming methods, effect on climate, 72-101, *il.,* 103; India, dust concentrations, 72-99, 100; postglacial vegetation and, 73-298; *Special Report,* 72-95; sun, 72-96. See also *Meteorology; Weather.*
Cloud seeding, 74-220, 73-333, 72-334, *ils.,* 334, 335
Coal: energy, 74-290, 72-30, 214; energy, *Special Report,* 74-52; gasification of, 73-283, 307, 72-31; natural gas substitute, 73-306
Coastal Zone Management Act, 74-332
Coaxial cable system, 72-293
Cobb seamount, 73-339
Cocking, Edward C., 72-203
Coesite: discovery of, 72-402
Coherin: hormone, 74-313
Cold-water eddies, 74-336, 73-340
Colds: vitamin C, 74-329
Colleges and universities: discriminatory admissions practices, 73-302
Communications, 74-279, 73-291, 72-293; *Books of Science,* 72-278; extraterrestrial civilizations, 73-364; satellites, 72-293; transit system, *Special Report,* 74-194. See also *Computers; Electronics.*
Communications Satellite Corporation (Comsat), 73-291, 72-293
Compensatory hypertrophy: zoology, 74-368

Computers, 73-293, 72-295; *Books of Science,* 74-266; CCD shift registers, 73-304; communications, 74-279; data transmission services, 72-293; electronics, 74-290; ferrite core memory, 72-297; grassland study, 72-82; semiconductor memory, 72-297; transit system, *Special Report,* 74-194; use in supertankers, 73-216
Comstock Prize, 74-407
Condors, *il.,* 73-375
Conrad, Charles, Jr., 74-357, 73-366
Conservation: *Books of Science,* 72-278; legislation, 73-13; new strategy to save birds, 73-50. See also *Agriculture; Air pollution; Birds; Ecology; Environment; Fish; Food; Medicine; Natural resources; Pesticide; Petroleum; Pollution; Public health; Zoology.*
Consumer affairs: food labeling, 74-331
Continental drift theory: geology, 74-305; geophysics, 73-320, 72-322; ocean drilling, *Special Report,* 72-128; oceanography, 74-334; plume theory, *Special Report,* 73-163; primitive mammals, 73-318. See also **Plate tectonics; Sea floor spreading.**
Cooper, Leon N.: Nobel prize, 74-406
Cope Award, 74-404
Copernicus, Nicolaus, 74-262, 73-266
Copper: geology, 74-304
Corn: blight-resistant seed, 73-254; experimental hybrid, 73-254; leaf blight, 72-254
Coronary . . . . See headings beginning **Heart . . . .**
Corpus callosum, 72-150
Corynebacteria acnes, 72-158, 172
Cosmic rays, 74-256, 340, 73-271
Cosmodom (home in space) project, 72-365
Cosmological red shift, 74-258
Cosmology, 74-260, 73-272, 72-271
Cosmonauts: deaths, 72-365
Cotrell Award, 73-421
Council on Environmental Quality, 72-34, 361, 373
Coupling constants, 73-347
Crab Nebula: astronomy, 74-260; X-ray emission, 74-69, 73-271
Cretaceous Period, 74-309, 72-281
Croneis, Carey, 73-423
Crop damage: environment, 74-297. See also *Agriculture.*
Cross-Florida Barge Canal, 72-361, 373
Crown ether: chemical synthesis, 74-278
Crown gall: plant tumor, 73-334
Cucumber, 73-254
Cultural Revolution (China): effect on science, 73-191

Cumulus clouds, 74-222
Currents: measurement of, 72-343
Cyanoacetylene, 72-73
Cyanocobalamin, 74-262
Cycladic Islands: archaeological investigations, 72-229
Cyclic AMP, 72-338
Cyclohexaamylose molecule: enzyme synthesis, 72-292
Cyclones, 72-333
Cygnus X-1, 74-75, 73-271
Cygnus X-3, 74-256
Cytidine (C), 72-289

D

Data communications, 74-279
Data processing. See *Computers.*
David, Edward E., Jr., 74-351, 72-362
Davis, Raymond, Jr., 74-79, *il.,* 81
Day Medal (geology), 74-407, 73-417
DDE: breakdown product of DDT, 73-53
DDT: agriculture, 74-244; birds, *Special Report,* 73-53; chemical synthesis, 72-291; environment, 73-310; ocean pollution, 73-215, 341
Death: aging, *Special Report,* 74-120
Death rate: United States, 73-322
Deaths of Notable Scientists, 74-413, 73-423, 72-423
Decathlon, 73-148
Deep Sea Drilling Project, 74-333, 72-122, 323
Deer, 73-380
Deimos, 73-65, *il.,* 66
Dendrochronology, 72-231
Dental glue, 72-325
Dentistry, 74-310, 73-321, 72-325; implants, 73-321; plaque inhibition, 73-321; preventive, 73-305, 72-325; tooth impressions, age determination, 73-39, *il.,* 38
Deoxyguanosine groups, 72-289
Deoxyribonucleic acid. See **DNA.**
Department of Agriculture, U.S., 74-244
Dermis, 72-154, *il.,* 155
DES. See **Diethylstilbestrol.**
Descartes Highlands, 73-366
Detergent. See **Phosphate detergent.**
Deuterium: energy crisis, *Special Report,* 72-218; on Jupiter, 74-253
Deuteron: high-energy neutron transfer, 73-349; neutrinos, *Special Report,* 74-81; stripping reaction, 74-342
De Vaux, Roland, 73-423
Diabase, 72-134
Diabetes mellitus, 73-140, 72-298
Dial-a-bus: transportation, 74-363
Dialysis, 74-314, 72-332
Diamond knife, 73-382
Diamonds: geology, 74-304
Diesel engines: environment, 74-295

Index

Diet: aging, *Special Report*, 74-126. See also *Nutrition;* Obesity.
Diethylstilbestrol: agriculture, 74-244, 73-255; drugs, 74-282
Differential solubility, 73-212
Diffusionism, 72-227
Digoxin, 73-296
Dilatancy: geophysics, 74-306
Dinosaurs: paleontology, 74-308
Discrete system: hi-fi channels, 73-306
Disease: aging, *Special Report*, 74-134; animal, 73-256; chemical, 73-215; genetic, 72-315; psychosomatic health, *Special Report*, 74-138; viroids, 74-326. See also **Childhood disease;** *Medicine;* **Public health;** also names of specific diseases.
Dispersants, 72-344
Dissection, human, 72-11
Distorted-wave Born approximation, 74-342
Dmochowski, Leon, 73-90
DNA (deoxyribonucleic acid): aging, *Special Report*, 74-128; bacterial, 73-333; biochemistry, 72-273; cell evolution, 73-110; *Close-Up*, 72-275; denaturation of in stain technique, 73-311; gene synthesis, 73-313; medicine, 74-311; microbiology, 72-337; procaryotic and eucaryotic cells, 73-106; satellite DNA, 72-273; structural chemistry, 74-276, 73-287, 72-289; viral, 73-94; viroids, 74-326. See also **Cell;** *Genetics;* **RNA;** **Virus.**
Dole, Vincent P.: Stouffer Prize, 74-409
Dogs: *Special Report*, 74-160
Domestication of animals, 74-162
Doppler effect, 73-343
Doppler radar, 74-230
Dorsal Column Stimulator, 74-319
Double-clutching, 73-56
Douglas fir, *il.,* 73-14
Dreikurs, Rudolf, 73-423
Dreyfuss, Henry, 74-413
Drones, 74-230
Drought, 72-334
Drug-abuse center, *il.,* 73-296
Drug control legislation, 72-299
Drugs, 74-282, 73-295, 72-298; bioavailability, 73-296; *Books of Science*, 74-267, 73-277, 72-279; fixed-ratio combinations, 72-298; in athletics, 73-160; over-the-counter, 73-295; pharmacogenetics, 73-325; psychology, 74-351. See also **Antibiotics; Cancer; Food and Drug Administration;** *Medicine;* **Psychiatry.**
Duke, Charles M., Jr., 73-366, *il.,* 369

Dust: effect on climate, 72-95; storms on Mars, 73-65, 264
Dutch elm disease, 73-255
Dwarf galaxies, 72-267
Dwarf grains, 73-254
Dyer, Rolla E., 73-423
Dynamic positioning, 72-122
Dyslexia, *il.,* 72-140

E

Earth: age of crust, 72-122; magnetic field, 72-231
Earth Day, 72-308
Earth Resources Technology Satellite, 74-247, 332, 361
Earth sciences: *Books of Science*, 73-278, 72-279. See also *Agriculture; Geoscience; Meteorology; Oceanography.*
Earth Week, 72-308
Earthquakes: building codes, 72-321; geophysics, 74-306; man-made, 73-320, 72-323; ocean ridge system, 72-322; polar wobble, *il.,* 72-320; scientific monitoring, 72-321; surface faulting, 72-321; zones, *Trans-Vision®,* 73-177
East Pakistan, 72-333, *il.,* 336
Echo sounder, 72-124
Ecological Society of America, 72-300
Ecology, 74-284, 73-297, 72-300; *Books of Science*, 74-266, 73-278, 72-279; botany, 74-269; cultivation of technology, 72-29; endangered birds, 73-50; endangered plants, *Essay*, 73-11; geoscience projects, 72-321; grizzlies in Yellowstone, 73-35; microbiology, 74-323; oceanography, 74-332; Pawnee Grasslands, 72-81; pesticides, 74-246; technological solutions to pollution problems, 72-283; water pollution and aquaculture, 72-119. See also **Air pollution;** *Education; Environment;* **Metal pollution; Natural resources; Pesticide; Phosphate detergent; Pollution;** *Zoology.*
Ecosystems: ecology, 74-284, 72-300; variables in, 72-91; Yellowstone, 73-35
Eddy, Nathan B., 74-413
Edelman, Gerald M.: Nobel prize, 74-408
Education, 73-301, 72-305; behaviorist techniques, 73-356; China, 73-199; decreasing enrollment in graduate programs, 73-358; medical illustration, 72-11; precision teaching approach, 73-357; programmed instruction, 73-356; science curriculums, 72-305; women in science, *Close-Up*, 73-302, *Special Report*, 74-24. See also **Learning; Science, rejection of.**
EFMU: virus, 74-311

Einstein Award (physics), 73-416
Electric power: *Special Report*, 74-213
Electrical force, 72-207
Electrical nerve stimulation, 74-319
Electricity: energy, 74-292; energy, *Special Report*, 74-52
Electromagnetism: neutrinos, *Special Report*, 74-80
Electron antineutrinos: neutrinos, *Special Report*, 74-80
Electron beam: plasma physics, 73-353, 72-354
Electron microscope: superconducting, 73-382
Electron neutrinos: neutrinos, *Special Report*, 74-80
Electron scattering measurements, 74-343
Electronics, 74-287, 73-304, 72-308; charge-coupled device, 73-304; large-scale integration, 72-308; medicine, 73-305; pain control, 74-319; personal rapid transit, *Special Report*, 74-185; timepieces, 72-309. See also *Astronomy* (high energy); *Communications; Computers; Light; Magnetism; Physics* (solid state); **X rays.**
Electrons: interaction with light, 72-355; magnetism, 73-353; recombination with ions, 73-343
Electrophysiology, 72-138
Elementary particles. See *Physics* (elementary particles).
Elliot Medal, 72-417
Emergency Core Cooling System, 73-309
Emission spectrometry, 73-141
Employment: scientific, *Special Report*, 74-24
Endangered species: grizzly bears, 73-35; osprey, 73-50; plants, *Essay*, 73-11
Endarterectomy, 72-185, 330
Endosseous implants, 73-321
Energy, 74-290, 73-306, 72-311; consumption, 72-30; exchange, 72-300; lasers, *Special Report*, 74-199; neutrinos, *Special Report*, 74-79; new scale of, *Close-Up*, 73-347; physics, 74-338; psychosomatic health, *Special Report*, 74-143; science support, 74-355; *Special Report*, 74-51, 72-213; transportation, 74-362; X rays, *Special Report*, 74-65
Energy crisis: geology, 74-304; *Special Report*, 72-213; transportation, 74-362
Engineering: *Books of Science*, 74-267, 72-279
Entomology: bees, *Special Report*, 74-148; *Books of Science*, 74-267; spiders, 74-284
Entropy increase, law of: aging, 74-122, *il.,* 123
Environment, 74-293, 73-308;

Index

education, 72-305; effect on child
development, 72-359; energy, 74-290;
Environmental Studies, 73-303;
long-range effect of dust on, 72-105;
microbiology, 74-323; oceanography,
74-332; pesticides, 74-246; Russian
science, *Special Report*, 74-240;
science support, 74-355; stream
channelization, *Close-Up*, 74-294. See
also **Air pollution**; *Ecology*; **Natural
resources**; **Pesticide**; **Public health**;
Water pollution.
Environmental geoscience, 72-321
**Environmental Protection Agency
(EPA)**, 74-293, 364, 73-310, 72-34,
364
Environmental Quality, Council on,
72-34, 361, 373
Enzymes: α-chymotrypsin, 73-288;
beta-galactosidase, 72-339; inducible,
72-339; L-asparaginase, cancer
therapy, 72-298; lactate dehydrogenase,
72-202; Nobel prize, 74-404;
peptidase B, 72-202; reverse
transcriptase, 73-94, 313; RNA-directed
DNA polymerase, 72-273; RNA
polymerase, 72-337; sebaceous glands
and, 72-158; synthetic, 72-292;
thymidine kinase, 72-202; trace
element functions, 73-134, *il.*, 138;
trypsin, structure of, 72-290;
trypsin-trypsin inhibitor, 73-288
Epidermis, 72-154, *il.*, 155
Equatorial bulge: on Mars, 73-65
Erythropoietin, 72-330
Eskimos: cheekbones, 73-258;
skeletons, 72-258
Esophageal bleeding: surgery,
74-317
Estrogens: in animal feed, 73-255
Eucaryotic cells, 73-106 *ils.*, 107, 108
**European Center for Nuclear
Research.** See **CERN.**
Eutrophication, 72-285. See also
Ecology; **Water pollution.**
Evans, Herbert M., 72-423
Evans, Ronald E., 74-360, 73-366
Evaporites, 72-323
Evapotranspiration, 73-298
Evolution: botany, 74-269; cell,
Special Report, 73-105; genetics,
74-299; human, *il.*, 73-236;
knuckle-walking, 73-258, *il.*, 379
Exercise: *Special Report*, 73-146
**Experimental allergic
encephalomyelitis**, 72-274
Explorer 42: X-ray satellite, 72-269
Explorer 49: space exploration,
74-361
Extracorporeal liver perfusion: liver
regeneration technique, 73-329
Extragalactic supernova, 74-255

Extraterrestrial civilization:
Close-Up, 73-364
Eye: hologram, *il.*, 73-345; visual
cortex experiments, 72-139, *ils.*, 142, 143

F

Faber, Herman, 72-12, *il.*, 17
Fabing, Howard D., 72-423
Fairchild, Sherman M., 72-423
Fallout pollutants, 73-209
Farming methods: effect on climate,
72-101, *il.*, 103
Farnsworth, Philo T., 72-423
Farrar, Clarence B., 72-423
Fat: as energy source, 73-158
Fatty acids, 72-172
Fault zone: Santa Susana-Sierra
Madre, 72-319
Favaloro, René G., 72-186
FDA. See **Food and Drug
Administration.**
**Federal Communications
Commission (FCC)**, 74-279,
73-291, 72-293
Federal Radiation Council, 72-224
**Federal Water Quality
Administration**, 72-306
Feeding ecology, 73-299
Fermi Award (atomic energy),
74-407, 73-416, 72-416
Fernandez-Moran, Humberto:
biography (*A Man of Science*),
73-382
Ferree, Gertrude Rand, 72-423
Ferromagnets: solid state physics,
73-353
Fertility: animal, 74-166
Fertilization: test-tube, 72-379, *il.*, 376
Feynman, Richard P., 74-338, 73-346,
72-348
Fillings: dentistry, 74-310
Film editing, 72-296
Fingerprint identification, 72-295
Firn, 72-95
Fish: carp, 72-108; catfish, 72-109,
ils., 110; fish antifreeze, 74-366;
flatfish, 72-109; salt-water, 72-109;
selective breeding, 72-107; supertrout,
72-107, *il.*, 108; yellowtail, 72-111
Fission-track analysis: anthropological
dating method, 73-229
Flatworms: memory transfer, 72-145
Flu vaccine: medicine, 74-314
Fluorescent dye: chromosome stain
technique, 73-311; *ils.*, 312, 72-314
Foelling, Ivar A., 74-413
Food: additives, 73-274; additives,
Generally Recognized as Safe (GRAS)
list, 73-295; *Books of Science*, 74-267;
energy source, 73-157; from electricity
and gas, 72-340; labeling, *Close-Up*,
74-331; processed foods and nutrition,
73-134, 145; protein-producing
hydrocarbons, 72-284. See also **Diet**;
Nutrition; **Obesity.**

Food and Agriculture Organization,
72-119
**Food and Drug Administration
(FDA)**: drugs, 74-282, 73-295, 72-298;
environment, 74-297; medicine, 74-312;
metal pollution, *Close-Up*, 72-302;
nutrition, 74-331; restrictions on feed
additives, 73-255
Food chain: concentration of
pollutants, 73-52, 212, 342;
consumption and elimination studies,
73-300. See also *Ecology.*
Foraminifera, microfossils, 72-97,
il., 99
Ford Foundation, 72-306
Forecasting, weather: computer-
produced, 72-336; long-range, 73-295
Forest regeneration, 72-304
Forestry: *Books of Science*, 72-279
Formaldehyde molecules: in the Milky
Way, 72-67
Fornix bundle: axons, 73-338
Forsythe, George E., 73-423
Fossil fuels: energy, *Special Report*,
74-51
Fossil plants: botany, 74-270
Fossils: animal, 73-230; human, 73-226;
ocean-floor sediments, 72-127;
paleontology, 74-308. See also
Anthropology; *Archaeology.*
Founders Medal (engineering), 74-411,
73-421, 72-419
Fowler, William A.: Vetlesen Prize,
74-407
Foxes: foot temperature regulation,
73-375
Fra Mauro highlands, 72-366
Fracture, bone: treated with electricity,
73-330
France: space program, 72-371
Franklin Medal, 74-407, 73-416,
72-416
Friedman, Herbert: Michelson Medal,
74-410
Friends of the Earth, 72-308
Froshe, Franz, 72-12, *il.*, 17
Fruit fly: genetics, 74-297; life span,
74-123
Fuel: energy, *Special Report*, 74-51;
from wastes, 73-283, 72-284
Fungus: blue-green algae, 73-115
Fusion: cell, 72-191; nuclear, 74-81,
207, 72-214; quark, 72-273

G

Gabor, Dennis, 73-416 *il.*, 415
Gairdner Awards, 74-408, 73-418,
72-418
Galaxy: cosmology, 74-260; dwarf,
72-267; expanding universe, 73-272;
quasars, 74-261. See also *Astronomy*;
Milky Way; **Quasar**; **Star.**
Galileo, 72-38
Gallstones, 74-316, 73-327
Ganglionectomy: cervical, 72-182

Index

Garbage: fuel source, *73*-283, *72*-284
Garlic: pesticide, *73*-280
Gas: from coal, *73*-283, 307, *72*-31; natural, *74*-304; noble, synthesized compound, *73*-290; synthetic natural, *74*-290
Gas endarterectomy, *72*-330
Gast, Paul W., *74*-413
Gate theory: pain control, *74*-319
Gell-Mann, Murray, *72*-348, 360
Genetics, *74*-297, *73*-311, *72*-314; aging, *Special Report,* *74*-122; bees, *Special Report,* *74*-157; *Books of Science,* *73*-278; cell evolution, *Special Report,* *73*-105; cell fusion, *Special Report,* *72*-191; counseling, *73*-323, *72*-315; engineering, *72*-202; gene mapping, *72*-192, 202; mRNA, *72*-337; pharmacogenetics, *73*-325; RNA-directed DNA polymerase, *72*-275; sickle cell anemia, *73*-323; structural genes, *73*-311; trace element concentration in nucleic acids, *73*-136, *il.,* 139; transfer of genes, *73*-335. See also *Biochemistry; Cell; DNA; RNA.*
Geochemistry, *74*-299, *73*-313, *72*-315
Geodynamics Project, *72*-324
Geological Society of America, *72*-320
Geology, *74*-302, *73*-316, *72*-318; activity on Mars, *73*-61; *Books of Science,* *74*-267; Day Medal, *74*-407; earth movements, mathematical model, *73*-171; plume theory, *Special Report,* *73*-163; shell deposits as clue to marine life, *72*-342. See also *Oceanography.*
Geophysics, *74*-306, *73*-319, *72*-321; *Books of Science,* *74*-267; earth movements, mathematical model, *73*-171; gravitational variations, mapping of, *73*-164; neutrinos, *Special Report,* *74*-91; plume theory, *Special Report,* *73*-163. See also *Oceanography.*
Geoscience, *74*-299; awards and prizes, *74*-407; geochemistry, *74*-299; geology, *74*-302; geophysics, *74*-306; geothermal energy, *Special Report,* *74*-52; paleontology, *74*-308. See also *Climate; Ecology; Meteorology; Oceanography; Weather.*
Geothermal energy: *Special Report,* *74*-52
Gerasimov, Mikhail M., *72*-423
Germer, Lester H., *73*-423
Gerontology: *Special Report,* *74*-120
Gershon-Cohen, Jacob, *72*-423
Geysers: energy, *Special Report,* *74*-52; *il.,* *73*-174
Gibberellin: and plant growth, *73*-197; synthesized, *73*-289
Gibbon, John H., Jr., *74*-413

Giemsa stain, *73*-311
Gilbreth, Lillian M., *73*-423
Glacier, *72*-102
Glass fiber: optical systems, *72*-295
Glass technology, *72*-286
Glaucoma: electronics, *74*-288
Glenn, John H., *73*-406, *il.,* 401
Glial scar, *73*-336, *il.,* 337
Global Atmospheric Research Program (GARP), *72*-335
Glomar Challenger: ocean drilling, *Special Report,* *72*-121; oceanography, *74*-333
Glucose: hunger regulator, *73*-121; metabolism, *73*-140; repression of enzyme synthesis, *72*-339
Glueck, Nelson, *72*-424
Glycogen: metabolism, *73*-158
Glycoprotein: fish antifreeze, *74*-366
Gneisses: metamorphic rocks, *72*-320
Goddard Award (aerospace), *74*-410 *73*-419
Gofman, John W.: Stouffer Prize, *74*-409
Goiter, *73*-137
Goldfish, *72*-146
Goldstack interferometer, *73*-273
Gondwanaland, *74*-334
Gordian Knot: Martian caldera, *73*-62
Gordon, Robert S., Jr.: Stouffer Prize, *74*-409
Gorillas: zoology, *74*-370
Gould Prize, *74*-410, *72*-420
Grain: synthetic, *72*-255
Grand Tour: space flight, *73*-370
Graphite fiber: used in reinforced plastics, *73*-282
Grassland Biome Program, *72*-81
Gravitational collapse, *73*-77
Gravitational radiation: astronomy, *74*-260; *Special Report,* *72*-52
Gravitational waves, *72*-52
Gravity: irregularities, *73*-164; irregularities on Mars, *73*-66
Grazing, *72*-83
Great Lakes, *73*-332
Great Rift Valley, *73*-224
Greece: archaeology, *Special Report,* *74*-93
Gregory, William K., *72*-424
Grizzly bears: *Special Report,* *73*-35
Gross national product (GNP): and pollution potential, *73*-208
Growth: animals and plants, *73*-214; botany, *74*-269; nutrition, *74*-330
Guanosine (G), *72*-289
Guggenheim Award, *74*-410, *72*-419
Gulf Stream: cold-water eddies, *73*-340
Gum Nebula, *72*-267, *il.,* 268

H

H. erectus. See *Homo erectus.*
Hadley-Apennine region, *73*-363, *72*-367
Hadley Rille, *73*-363

Hair follicles, *72*-156
Halpern, Julius, *73*-424
Hamilton, Walter C., *74*-413
Hamster, hairless, *il.,* *73*-378
Handedness, transfer of, *72*-144
Harappan civilization, *72*-101
Hartley Public Welfare Medal, *73*-421
Hartwell, Leland H.: Lilly Award, *74*-408
Hawk, red-tailed: artificially inseminated, *il.,* *72*-377
Health, Education, and Welfare, Department of: Russian science, *Special Report,* *74*-234; sexism in science, *Close-Up,* *73*-302
Health-testing center, *72*-295
Heart: artificial, *73*-329, *il.,* 330, *72*-331; coronary arteries, *72*-176, 330; mechanical, *72*-179; physiology of, *73*-153, *72*-181
Heart disease: angina pectoris, *72*-176; arteriosclerosis, *72*-181, 330; cardiovascular, *74*-312; coronary thrombosis, *72*-181; diet, *73*-322; myocardial infarct, *72*-181; surgical solutions, *Special Report,* *72*-176; vital statistics, *72*-182, 330
Heart surgery: coronary artery by-pass *73*-329, *72*-176, 330; endarterectomy, *72*-185, 330; infant, *74*-318; revascularization, *72*-182, 330; transplantation, *73*-329, *72*-331
Heat: plume theory, *Special Report,* *73*-163
Hedberg, Hollis D.: Thompson Medal, *74*-407
Heiser, Victor G., *73*-424
Helium: interstellar clouds, *72*-69
Helium-selentium laser, *il.,* *74*-201
Helium 3, *73*-354
Hellas, *73*-68
Hematoporphyrin: biochemistry, *74*-265
Hemoglobin: effect of iron on, *73*-136; gene synthesis, *73*-313; molecule, *72*-290; structural abnormalities in mutants, *73*-288
Herbicides: botany, *74*-269; in Vietnam, *73*-308, *72*-247
Hercules X-1: neutron star, *74*-73, 259
Heredity: child development, *72*-359
Hermit crab, *72*-376
Heroin: drugs, *74*-282
Héroux, Bruno, *72*-12, *il.,* 17
Herpes virus: cancer research, *Special Report,* *73*-95, *il.,* 96; medicine, *74*-311
Herzberg, Gerhard, *73*-414, *il.,* 415
Hesperia, *73*-67
Hexachlorophene, *73*-295
Hi-fi channels, *73*-306
Hibernation: aging, *Special Report,* *74*-123; bat gestation, *73*-376; grizzly bears, *73*-40; squirrels, *74*-366
High energy astronomy, *74*-257
Hill Space Transportation Award, *74*-410, *73*-419, *72*-419

Index

Hippocampus: protein production and learning, 72-145
Hitchcock, Lauren B., 74-413
H.M.S. *Challenger*, 72-122
Hoabinhian people, 72-259
Hoag, David G.: Hill Space Transportation Award, 74-410
Hobbs, Leonard S.: Sperry Award, 74-411
Hodgkin's disease: drugs, 72-298; possibility of contagion, 73-95
Hoffmann, Roald: Cope Award, 74-404
Hog cholera, 73-256
Hogweed, giant: skin disorders, 72-282, *il.*, 281
Holland, James F.: Russian science, *Special Report,* 74-232, *il.*, 235
Holography, *il.*, 73-345
Hominid. See *Anthropology;* and names of specific species.
Homo: anthropology, 74-247
Homo erectus, 73-256
Homo habilis, 73-231, *il.*, 229
Honeybees: *Special Report,* 74-148
Hopewell culture: American Indians, 73-262
Hormone: acne therapy, 72-161; aging, *il.*, 74-130; agriculture, *Close-Up,* 74-246; androgens, 72-156; botany, 74-269; calcitonin, 72-276; dihydroepitestosterone, 72-172; diethylstilbestrol, 74-282; erythropoietin, 72-330; human growth hormone, 72-277; human placental lactogen hormone, 72-277; medicine, 74-313; ovine prolactin, 72-277; parathyroid, 72-276; pest control, 73-255, 72-292; testosterone, 72-172
Horned bladderwort, *il.*, 73-13
Hornykiewicz, Oleh: Gairdner Award, 74-408
Horowitz Prize (biochemistry), 73-417
Horticulture: *Books of Science,* 72-279. See also *Agriculture; Botany; Plant.*
Hot accretion theory: lunar history, *Special Report,* 72-50
Houssay, Bernardo A., 73-424
Hubble, Edwin, 74-262
Hubble constant, 73-272
Human growth hormone, 72-277
Human parathyroid hormone, 74-263
Human placental lactogen hormone, 72-277
Hume, David M., 74-413
Hummingbirds: botany, 74-270
Hunger: appetite and obesity, *Special Report,* 73-117
Hunting camps: Peruvian, 72-262
Hurricane: cloud seeding, 73-333, 72-335; Ginger, 73-333
Hutchinson, John, 74-413

Hydrocarbons: catalytic converters, *Close-Up,* 74-364; environment, 74-293
Hydrodynamic instabilities, 72-222
Hydroelectric power: *Special Report,* 74-213
Hydrogen: interstellar clouds, 72-69; Lamb shift, 73-343; molecular, 72-70; planetary composition, 72-263
Hydrogen cyanide: interstellar molecule, 72-265
Hyperalimentation, intravenous, 74-316, 72-333
Hyperstat: drugs, 74-283
Hypertension: drugs, 74-283
Hypophysation, 72-111
Hypothalamus: lateral areas, feeding centers, 73-120; ventromedial areas, satiety centers, 73-118, *il.*, 120

I

Ice Age: neutrinos, *Special Report,* 74-91
Ice floe, 73-339
Iceland: climate, 72-104
Image enhancement: hurricane study, 72-336
Image sensor: CCD function, 73-304
Immunology: Jonas Salk, 74-389; medicine, 74-315; *Special Report,* 74-108
Inactivated Sendai virus, 72-192
India: dust concentrations, 72-99
Indian, American, 73-36, 72-262
Indiana: archaeology, 74-252
Industrial waste: environment, 74-297. See also **Waste.**
Infantile paralysis: Jonas Salk, 74-389
Influenza: medicine, 74-314
Infrared light: astronomy, 74-255
Infrared radiometer: meteorology, 74-322
Infrared spectroscopy: used to study Martian dust, 73-315
Inhibitor I: plant protein, 73-281
Inland Sea, 73-339
Insecticide: *Close-Up,* 74-246. See also **Pesticide.**
Insects: reaction to chemicals in plants, 73-280
Institute for Primate Studies, 74-40
Institute of Muscle Research, 72-393
Insulin: effect of chromium, 73-139; self-powered infusion pump, *il.*, 72-333
Integrated circuits: surplus, 72-308
Integrated pest control: *Close-Up,* 74-246
Intelsat: communications satellite, 73-291, 72-293
Inter-American Institute of Ecology, 72-300
Inter-Society Commission for Heart Disease Resources, 72-182
Intermediate boson, 73-347
International Association for Dental Research, 72-325

International Atomic Energy Agency, 72-214
International Biological Program, 72-81, 300
International Field Year for the Great Lakes, 73-332
International Maize and Wheat Improvement Center, 72-255
International Telecommunications Satellite. See **Intelsat.**
International Transportation Exposition, 73-370
Intrauterine device (IUD), *il.*, 72-330
Intravenous hyperalimentation, 72-333
Io: Jupiter satellite, 72-264
Iodine: human requirement, 73-137
Ion: crystals, 72-355; implantation, 72-357; storage at constant energy, 73-343
Ionicity, measure of, 72-356
Iraqi paintings, 72-261
Iron: human requirement, 73-136, *il.*, 137
Iron Age: burial site, 73-259, *il.*, 261
Irrigation: prehistoric, 73-263; *Special Report,* 74-213; trickle, 74-244, 72-255
Irwin, James B., 73-363, 72-367
Isaac, Glynn, 73-231, 261, *il.*, 230
Iselin, Columbus O'Donnell, 72-424
Isocyanic acid, 72-73
Isothermal plateau: solar layer, 72-265
Isotopic power sources, 72-312
IUD. See **Intrauterine device.**

J

Japan: environment, 74-295; space program, 72-371; technology, *Special Report,* 72-220
Japanese National Railways, 72-374
Jentschke, Willibald K., 73-251
Jet Propulsion Laboratory, 73-62
Joint Oceanographic Institutions for Deep Earth Sampling, 74-302, 73-319, 72-122
Jones, Chester Morse, 74-413
Jones, Tom, 72-13, *il.*, 20
Jupiter: astronomy, 74-253; giant satellites, 73-264; Io, 72-264; occultation of Beta Scorpii, 73-265; space probe, 73-370
Justice, U. S. Department of, 72-299
Juvenile hormones: pesticides, 73-255, 72-292, *Close-Up,* 74-246

K

Kalinga Prize (science writing), 74-411, 73-421
Karrer, Paul, 73-424
Katchalsky, Aharon Katzir, 73-424
Katz, Sir Bernard, 72-418
Kaunaoa worm: anticancer ingredient, 72-377
Keller, Fred S., 73-357

Index

Kellermann, Kenneth I.: Gould Prize, 74-410
Kelvin, Lord, 72-249
Kendall, Edward C., 73-424
Kennedy, Edward M., 72-363
Kennedy, John F., il., 73-401
Kenward, Rory, 73-148
Kerwin, Joseph P., 74-357, 73-367
Kidney: artificial, 74-314, 72-332; zoology, 74-368
Kilauea volcano, 73-316
Kinetic energy, 72-207
Kistiakowsky, George B., 74-407, 73-414, il., 416
Kittay Award, 74-409
Kow Swamp: anthropology, 74-248
Kraft, Christopher: biography (A Man of Science), 73-398
Krakatoa volcano: climate, Special Report, 72-99; geology, 72-318
Krebs, Hans A., 72-387
Kuffler, Stephen W.: Horwitz Prize, 74-408
Kurchatov, Boris V., 73-424

L

L-asparaginase: cancer therapy, 72-298
Lactase: nutrition, 74-329
Lake Rudolf: anthropological studies, 73-231
Lamb shift, 73-343
Lamont-Doherty Geological Observatory, 72-122
Large-scale integration (LSI), 73-293, 72-308
Lark bunting, 72-85, il, 86
Laser Cane, il., 74-204
Laser ranger, 74-204
Laser-scan photography, il., 74-205
Laserphoto: electronics, 74-289
Lasers: electronics, 74-287; excitation of molecules in chemical reaction, 73-287; fusion reactor, 73-351; gas molecules, excitation of, 72-287; improvements, 72-311; in the far-ultraviolet range, 72-345; molecular fluorescence, laser-induced, 72-287; picosecond pulses, 72-288; satellite tracking stations, 73-319; solid state, 72-311; solid state, optical signal generators, 72-295, il., 293; Special Report, 74-199; spectroscopy, 73-345; ultraviolet, 74-337; used to prevent tooth decay, 73-305
Lasker Award: basic research, 73-418, 72-418; clinical research, 73-419, 72-419
Lava, 73-168. See also Volcano.
Lawrence Memorial Award, 72-416
Lawson criterion, 72-221
LD 50: pollutant concentration, 73-214

Lead poisoning, 72-302
Lead pollution: botany, 74-269; environment, 74-297; in Greenland, 73-209, il., 213
Leakey, Louis S. B., 74-413, 73-228, 257
Leakey, Mary D., 73-226
Leakey, Richard E., 74-247, 73-233, 72-257
Learning: biochemistry, 74-264; brain, Special Report, 72-141; neurology, 74-327; psychology, 74-348. See also Education; Psychology.
Lefschetz, Solomon, 74-414
Lehrman, Daniel S., 74-414
Leloir, Luis F., 72-414, il., 416
Lepier, Erich, il., 72-22
Leptospira: animal disease, 73-256
Leukemia: myeloid, chromosomal identification, 72-314; tuberculosis vaccine (BCG), 72-328
Lewis, Oscar, 72-424
Lewis, Warren K.: Founders Medal, 74-411
Libby, Willard F., 72-229
Lidar: lasers, Special Report, 74-204
Life expectancy: in America, 73-322
Life span: aging, Special Report, 74-120
Light: infrared, 74-255; interaction with electrons, 72-355
Light Amplification by Stimulated Emission of Radiation. See Laser.
Light-emitting diodes, 72-309
Lilbourn Village, 73-263
Lilly Award (microbiology), 74-408, 72-417
Linear induction motor, 73-373
Lipofuscin: age pigment, 74-127
Liquid-crystal device, 72-310
Liquid thread, 72-286
Lister, Joseph, 72-12
Lithium, 72-221
Lithosphere, 72-322
Liver: medicine, 73-328; surgery, 74-318
Livestock: agriculture, 74-245
Lizards, 74-284
Lobsters, 74-367, 72-342
Lodgepole pine, il., 73-15
Long Island Sound: osprey population, 73-50
Los Angeles Man, 72-257
Lothagam Hill, 73-224
Lothagam Mandible, 72-257, il., 258
Lowell, Percival, 73-63
Lunar geophysics, 73-320
Lunar highlands: geochemistry, 74-299
Lunar mapping, 72-403
Lunar rocks: geochemical research, 73-313, il., 314, 72-316; Rock 12013, 72-316, il., 317; Special Report, 72-36
Lunar Rover, 73-363, 72-367
Lunar seismic velocities, 73-320
Lunar seismometer, 73-366
Lunar soil: chemical composition, 72-317; geochemistry, 74-300; plant studies, 72-255

Lung: zoology, 74-368
Lunokhod, 72-312, 369
Lunokhod 2: space exploration, 74-361
Lust, Reimar: Guggenheim Award, 74-410
Lymphocytes: immunology, Special Report, 74-108

M

MacArthur, Robert H., 74-414
Macrocosms, 72-303
Macrophages: biochemistry, 74-265; immunology, Special Report, 74-112
Magnetic bottle, 72-222
Magnetic neutron star: astronomy, 74-259
Magnetic storm, 74-322
Magnetism: nuclear magnetic resonance, 73-353
Magnetohydrodynamic generator, 72-313
Maize agriculture, 72-262
Malnutrition: among the affluent, 73-322. See also Diet; Food; Nutrition.
Marek's disease, 72-255
Marine life, 73-214
Mariner series. See Mars.
Mars: geochemistry, ils., 74-300, 301; Mariner 7, 72-265; Mariner 8, 72-370; Mariner 9, 74-361, 73-61, 264, 315, 368, 72-371; Mariner 77 program, 74-361; space exploration, 73-368, 72-370; Special Report, 73-61; Viking, 74-361. See also Space Exploration.
Martí-báñez, Felix, 73-424
Maslow, Abraham H., 72-424
Mass transportation: Special Report, 74-185; transportation, 74-362
Mathematics: Books of Science, 74-267
Matrixed system: hi-fi channels, 73-306
Mattingly, Thomas K., 73-366, 398
Matulane (procarbazine-hydrochloride): Hodgkin's disease treatment, 72-298
Maturation: learning, 74-350
Mayall telescope, 74-257
Mayer, Maria Goeppert-, 73-424
McCown's longspur, 72-87
McDivitt, James A., 72-366
Mead, Margaret, 73-421, il., 72-246
Meadows, Dennis L., 73-362
Measles: in cattle, 73-256
Meat: bowel cancer, 74-312
Median forebrain bundle, 73-338
Medicine, 74-310, 73-321, 72-325; acupuncture anesthesia, 73-192, 328, il., 194; automated physical examination, 72-295; awards and prizes, 74-408; Books of Science, 74-267, 73-278, 72-279; cancer research, Essay, 73-24, Special Report, 73-90; dentistry, 74-310, 73-321, 72-325; drugs, 74-282;

Index

electronics, *74*-288; home-testing kits, *il.*, *73*-322; in China, *73*-191; internal, *74*-311, *73*-322, *72*-326; medical genetics, *72*-315; medical illustration, *72*-11; neurology, *74*-327, *73*-336; pain control, *74*-319; paraprofessionals in China, *73*-191; prenatal diagnosis, *72*-315; psychosomatic health, *Special Report*, *74*-137; surgery, *74*-315, *73*-327, *72*-330; veterinary, *74*-160; viroids, *74*-326. See also **Childhood disease; Disease.**
Medulla, *72*-144
Megalithic archaeology: *Special Report*, *72*-227
Melanoma: medicine, *74*-315
Memory, *72*-141
Memory technology, *74*-289, *72*-297
Mercury: botany, *74*-269; microbiology, *74*-326; poisoning, *72*-302; pollution, *73*-377, *72*-249, 302
Merrill Award (aerospace), *73*-420
Mesons, *74*-338, *72*-348
Metabolism, rate of: fetal development in bats, *73*-376
Metal-oxide semiconductor (MOS) technology, *73*-293, *72*-297, 308
Metal pollution: *Close-Up*, *72*-302
Metal reclamation, *72*-284
Metallic compounds, *73*-288
Metazoans: paleontology, *74*-309
Meteorology, *74*-320, *73*-331, *72*-333; Improved Tiros Operational Satellite, *72*-336; Nimbus 4, *72*-335; *Special Report*, *74*-220; Stormfury Project, *73*-333, *72*-335. See also **Climate; Weather.**
Methadone: drugs, *74*-282
Methylmercury, *74*-326
Metropolitan Meteorological Experiment (Metromex), *73*-332
Mexico: archaeology, *74*-252
Mice: "double" litter, *72*-378
Michelson Medal, *74*-410
Microbiology, *74*-323, *73*-333, *72*-337; viroids, *74*-326. See also *Biochemistry;* **Biology; Cell; Disease;** *Ecology; Genetics.*
Microcomputers: electronics, *74*-290
Microcosm, *72*-300
Microsurgery, *73*-397
Microwave background radiation: cosmology, *74*-260
Microwave radiometer: meteorology, *74*-322
Microwave signals: interstellar molecules, *Special Report*, *72*-67
Microwave spectrometer: meteorology, *74*-322
Mid-Ocean Dynamics Experiment (MODE), *73*-341

Migraine headaches: self-regulation, *74*-141
Migration: and ocean pollution, *73*-213
Mikoyan, Artem I., *72*-424
Milk: nutrition, *74*-329
Milky Way: cosmology, *74*-260; from moon, *il.*, *73*-269; interstellar molecules, *Special Report*, *72*-67
Millionshchikov, Mikhail Dmitrievich, *74*-414
Mind: *Special Report*, *72*-137. See also **Brain;** *Psychology.*
Mineralogy: *Books of Science*, *73*-278
Mini-computers, *72*-297
Minischool, *73*-301
Minimal brain dysfunction, *74*-348
Minimata Bay (Japan): mercury poisoning, *72*-302
Minnesota Messenia Expedition: archaeology, *Special Report*, *74*-93
Miocene epoch, *72*-131
Mitchell, Edgar D., *73*-403, *72*-366
Mites, *72*-83, 256, *il.*, 91
Mitochondria: and ATP production, *73*-157; cell, *Special Report*, *73*-106; energy production, *73*-274; *il.*, 111
Mitosis, *72*-193
Mohole, Project, *74*-302
Molecular beam chemistry, *73*-285
Molecular fluorescence, *72*-287
Molecule: cancer detection, *72*-327; collision experiments, *74*-274; interstellar, *Special Report*, *72*-67
Molybdenum: deficiency, *74*-312
Momentum order, *73*-391
Mongolism (Down's syndrome), *72*-314
Monocab, *73*-370
Montelius, Oskar, *72*-227
Moon. See headings beginning **Lunar. . . .**
Moore, Stanford, *74*-404
Morphactin, *72*-282
Mosquitoes: anthropology, *74*-249
Moss campion, *il.*, *73*-23
Motion picture: used in obesity studies, *73*-129
Moxibustion, *73*-191
Mu-mesons, *72*-348
Mummies: X-ray analysis, *il.*, *73*-258, *72*-260
Muon antineutrinos: neutrinos, *Special Report*, *74*-80
Muon neutrinos: neutrinos, *Special Report*, *74*-80
Murchison stone meteorite: extraterrestrial amino acids, *72*-315, *il.*, 316
Murphy, Robert Cushman, *74*-414
Muscle, *73*-152, 275, *ils.*, 156, 157, 158, *72*-393
Musk oxen, *il.*, *73*-380
Mussels, *72*-113, *ils.*, 114
Mycenaean civilization, *72*-232
Myelin membrane proteins, *72*-274
Myocardial infarction, *72*-181

Myofibril, *73*-155, *ils.*, 156, 158
Myosin: muscle building, *73*-155, *ils.*, 156, 158; muscle contraction, *73*-275

N

Naphtha: synthetic natural gas, *74*-290
Narcan (naloxone hydrochloride), *73*-297
Narcotics, *74*-282, *72*-299
National Academy of Sciences (NAS), *73*-211
National Academy of Sciences Awards: Day Prize, *73*-417; Microbiology Award, *73*-417; U.S. Steel Foundation Award, *73*-417; Walcott Medal, *73*-417
National Accelerator Laboratory (NAL): physics, *74*-340; Robert Wilson, *74*-372; Russian science, *il.*, *74*-241; *Special Report*, *72*-205
National Acne Association, *72*-175
National Advisory Committee for Aeronautics, *73*-401
National Aeronautics and Space Administration (NASA), *74*-332, *73*-358, 405, *72*-255, 335, 367, 404
National Air Pollution Control Administration, *72*-99
National Association of Independent Schools, *72*-306
National Cancer Act, *73*-27
National Cancer Plan, *73*-30
National Cancer Institute (NCI), *73*-28, 359
National Environmental Policy Act, *73*-360, *72*-34
National Foundation for Infantile Paralysis, *74*-395
National Institutes of Health: cancer research, *72*-363; science support, *74*-353
National Medal of Science, *72*-421
National Meteorological Center, *73*-333
National Oceanic and Atmospheric Administration (NOAA), *74*-320, 332, *73*-331, 333, 339, *72*-334, 341
National Parks and Conservation Association, *73*-23
National Research Council: bees, *74*-153
National Science Foundation: educational grants, *73*-301; Manpower Survey, *73*-302; science support, *74*-351; Very Large Array project, *73*-269
National Science Teachers Association, *72*-305
Natural gas: supply, *74*-304, *73*-306, *72*-30
Natural resources: fossil fuels, *72*-30; geology, *74*-304; oceanography, *74*-332
Navigational aids, *72*-122
Neanderthal Man, *73*-256, *72*-256
Nebula, spiral, *74*-262

Index

Nectar: botany, *74*-269
Néel, Louis E. F., *72*-416
Negev Desert: water management, *72*-255
Neptune: giant satellites, *73*-264
Nervous system: acetylcholine receptor cell, *72*-274; basic operation, *72*-144; nerve-muscle junctions, *73*-275; neurons, *73*-336; pain control, *74*-319; psychosomatic health, *il.*, *74*-141; severance and connection of nerve fibers, *72*-150. See also **Brain**; **Neuroanatomy**; *Neurology*.
Nesting: turtles, *74*-285
Netter, Frank H., *72*-13, *il.*, 23
Neuquense Chopper complex, *72*-262
Neuroanatomy: brain, *Special Report*, *72*-138; psychosomatic health, *Special Report*, *74*-137
Neurology, *74*-327, *73*-336; psychosomatic health, *Special Report*, *74*-138. See also **Brain**; **Nervous system**; **Neuroanatomy**.
Neuromodulation: pain control, *74*-319
Neutrinos: astronomy, *73*-268; physics, *72*-348; *Special Report*, *74*-79
Neutron star: astronomy, *74*-259
New York State Identification and Intelligence System, *72*-295
Nimbus 5: meteorology, *74*-321
Nitrates: and water pollution, *73*-310
Nitrilo triacetic acid: phosphate replacement, *72*-285
Nitrogenase genes, *73*-335
Nitrogen oxides: emission standards, *73*-311; environment, *74*-293
Nitroglycerin: angina pectoris therapy, *72*-176
Nix Olympica, *73*-62, *il.*, 64
Nixon, Richard M., *74*-351, *73*-358, 361, *72*-345, 361, 373
Nobel prizes: chemistry, *74*-404, *73*-414; medicine, *74*-408; peace, *72*-254, 421; physics, *74*-406, *73*-416, *72*-416; physiology and medicine, *73*-418, *72*-418
Noradrenalin, *73*-338
Nordihydroguairetic acid, *73*-274
North American Osprey Research Conference, *73*-52
Nuclear explosives: energy, *73*-307, *72*-313
Nuclear fusion: neutrinos, *Special Report*, *74*-81
Nuclear magnetic resonance (NMR): cancer detection, *72*-327; solid state physics, *73*-353
Nuclear physics. See *Physics (nuclear)*.
Nuclear power: energy, *Special Report*, *74*-51; nuclear fusion, *74*-81, 207, *73*-308. See also **Accelerator, particle**; *Physics*.
Nuclear power plants: arguments against, *72*-32; discharge, *72*-224; energy crisis, *73*-307; environment, *74*-296; nuclear research, *72*-352; safety programs, *73*-309
Nuclear reaction: neutrinos, *Special Report*, *74*-81
Nucleotides: chemical structure, *73*-287
Nucleus: high energy proton probes, *73*-349; magnetism, *73*-353; reaction to added energy, *73*-350
Nutrition, *74*-329; chromium deficiency, *73*-140; dietary supplementation of trace elements, *73*-145; identifying trace elements, *73*-140; iodine deficiency, *73*-137; iron deficiency, *73*-136; silicon deficiency, *73*-144; trace elements, *Special Report*, *73*-133; vitamin synthesis, *74*-262; zinc deficiency, *73*-133. See also **Diet**; **Food**.

O

Obesity: calorie utilization, *73*-125; childhood, *73*-129; "epidemic," *73*-322; genetic, *73*-123, 129, *il.*, 124; metabolic nutrition, *74*-332; *73*-123; regulatory, *73*-123; role of exercise in weight control, *73*-124; *Special Report*, *73*-117
Obsidian: prehistoric trading, *74*-251; stone tools, *72*-262
Occultation, *73*-265
Ocean currents: mathematical model, *73*-341
Ocean Dumping Pact, *74*-332
Ocean-floor drilling: *Special Report*, *72*-121. See also *Oceanography*.
Ocean pollution: mercury in fish, *73*-377; microbiology, *74*-323; oceanography, *74*-332, *73*-341, *72*-343; plant life monitoring device, *73*-342; *Special Report*, *73*-205
Oceanographer: research ship, *73*-339
Oceanography, *74*-332, *73*-339, *72*-341; *Books of Science*, *74*-268, *73*-278, *72*-279; geology, *74*-302. See also **Continental drift theory**; *Geoscience*; **Marine life**; **Ocean pollution**; **Plate tectonics**.
Oersted Medal, *74*-411
Oil: consumption and demand, *74*-304, *72*-30, 321; energy, *74*-290; pollution, *73*-219, 284, *72*-344; spills, *73*-323; supertankers, *Special Report*, *73*-216; transportation, *74*-362
Old age. See **Aging**.
Olduvai Gorge: anthropological studies, *73*-227, *ils.*, 228, 229
Omo Research Expedition, *73*-234, 257
Omo River Basin, *73*-234, *il.*, 232

Oncley, John L.: Stouffer Prize, *74*-409
Open classroom, *72*-307
Oppenheimer, Robert, *73*-80
Optical activity: in amino acids, *72*-315
Optical communications, *74*-280
Optical isomer separation: chemical synthesis, *74*-278
Optical microscope: limitations, *73*-387
Optical transmission: communications, *72*-295, 311
Oral implantology, *73*-321
Orang-utan, *il.*, *73*-379
Orbiting Astronomical Observatory, *74*-257
Orbiting Solar Observatory, *73*-268
Organ transplant. See **Transplantation**.
Organic chemistry: Cope Award, *74*-404
Organic fiber, *73*-282
Orion Nebula, *72*-76, *il.*, 78
Orthopedic surgery, *73*-330
Osprey: *Special Report*, *73*-50
Ovine prolactin, *72*-277
Oxidation, *72*-387
Oyster, *72*-113, *ils.*, 115
Ozone, atmospheric: absorption by plants, *73*-255, *72*-282
Ozone layer (stratosphere): reduction by SST, *74*-322, *72*-333

P

Pacemaker, heart, *73*-305
Pacific Plate, *73*-169, 339, *il.*, 166
Pain: control, *74*-319
Paint: polymer-based, *73*-282
Paleobotanical investigations, *72*-281
Paleontology, *74*-308; *Books of Science*, *72*-279; *Special Report*, *74*-93. See also *Archaeology*.
Palmoxylon: new genus, *72*-281
Palynology: archaeology, *Special Report*, *74*-101
Palytoxin, *72*-376
Panda, *il.*, *73*-380
Pangaea: supercontinent, *73*-318
Paper: plasticized, *72*-286
Paraprofessionals: in China, *73*-191
Parapsychology: psychosomatic health, *Special Report*, *74*-145
Parasexual hybridization, *74*-245
Parasites: used to control wasps, *73*-197
Parathormone, *74*-263
Parathyroid hormone, *72*-276
Pardee, Harold E.B., *74*-414
Parthenocarpy, *72*-282
Particle physics. See *Physics (elementary particles)*.
Partons: physics, *74*-338, 341, *73*-346, *72*-348
Passano Foundation Award, *73*-419
Patsayev, Viktor I., *72*-365, *il.*, 366
Patten, Bernard, *72*-91
Patterson, Bryan, *72*-257

Index

Pauling, Linus: ionicity of solids, 72-356
Pawnee National Grassland: *Special Report,* 72-81
PCBs (polychlorinated biphenyls), 73-53, 208, 341
Peat bogs, 73-297
Peck, Paul, *il.,* 72-22
Pecora, William T., 74-414
Penguin, *il.,* 73-377
Pennsylvania: archaeology, 74-252
Penrose Medal (geology), 74-407, 73-417, 72-417
Penzias, Arno, 72-73
Perception: in infants, 72-357
Periodic table: elements essential for nutrition, 73-144, *il.,* 135
Peristalsis: medicine, 74-314
Perkin Medal (chemistry), 74-406, 72-414
Personal Rapid Transit (PRT) system: *Special Report,* 74-185; transportation, 74-363, 73-370, *il.,* 371
Peru: archaeology, 74-253
Pest control, integrated, 72-256
Pesticide: bacterial pesticide, 73-255; cedarwood extract, 73-280; chemistry, 72-291; *Close-Up,* 74-246; effect on bird population: 73-53; garlic, 73-280; necessity for, 73-214; pheromones, 73-255
Petroleum: consumption and demand, 74-304, 72-30, 321; energy, 74-290; pollution, 73-219, 284, 72-344; supertankers, *Special Report,* 73-216; transportation, 74-362
Pets: dogs, *Special Report,* 74-160
Pharmacogenetics, 73-325
Phase matching: physics, 74-337
Pheromone: pest control, 73-255
Philippine Sea Plate, 73-339
Phillips, George E., 73-424
Phobos: Martian moon, 73-65, *il.,* 66, 72-265
Phonoangiography, 72-329, *il.,* 326
Phonon, 73-355
Phosphate detergent: environment, 73-310, 72-285
Photography: electronics, 74-289; laser-scan, *il.,* 74-205
Photon, 72-345
Photosynthesis: cell evolution, 73-108
Physics: atomic and molecular, 74-337, 73-343, 72-345; *Books of Science,* 74-268, 73-278, 72-279; CERN, 73-244; elementary particles, 74-338, 73-346, 72-347; energy scale, 73-347; in China, 73-198; lasers, *Special Report,* 74-199; Nobel prize, 74-406; nuclear, 74-342, 73-349, 72-351; neutrinos, *Special Report,* 74-79; particle chemistry, *Close-Up,* 74-341; plasma, 74-344, 73-351, 72-353; Russian-

American cooperation, 74-241; solid state, 74-346, 73-353, 72-355; solid state, Buckley Prize, 74-406. See also **Accelerator, particle; Laser; Nuclear power; Plasma.**
Physiology: aging, *Special Report,* 74-120; psychosomatic health, *Special Report,* 74-137. See also **Biology.**
Piaget, Jean: Kittay Award, 74-409
Picosecond pulses, 72-288
Picturephone, 72-293
Piezoelectricity: use in healing fractured bones, 73-330
Pigment: aging, *Special Report,* 74-127
Pigmentation gene, 72-196
Pimple: *Special Report,* 72-152
Pink lady's-slipper, *il.,* 73-20
Pioneer 10, 74-361, 73-292, 364, 370, *il.,* 363
Pioneer 11, 74-361
Pions: nuclear physics, 73-350; neutrinos, *Special Report,* 74-80
Pithecanthropus VIII, 73-256
Plagioclase: lunar soil, 72-42
Planetary astronomy, 74-253
Planetary systems: probing by occultation, 73-265. See also *Astronomy* (planetary).
Plankton, 72-113
Plant: agriculture, 74-245; botany, 74-270; cell fusion, 74-299, 72-203, *il.* 202; chemicals, 73-280; primitive, 73-281; tumors, 73-334. See also *Agriculture; Botany;* **Horticulture.**
Plaque: inhibition of, 73-321
Plasma: generation by laser, 73-351; magnetic confinement, 73-351. See also *Physics* (plasma).
Plastics: dental sealant, 72-325; disposal of, 72-283; paper, 72-286; reinforced, 73-282
Plate tectonics: geology, 74-305; geophysics, 74-306, 73-319, 72-322; oceanography, 73-339; origin of, 73-169; Pacific Plate, 73-169, *il.,* 166; plume theory, *Special Report,* 73-163; volcano, 73-169, 316. See also **Continental drift theory; Sea floor spreading.**
Pleasure: and learning, 74-349
Pleiotropic genes, 74-122
Plutonium: naturally occurring, 73-349; nuclear power plants, 73-307
Plutonium 239, 72-218
Polio: Jonas Salk, 74-389
Pollen: archaeology, *Special Report,* 74-101; in primitive angiosperms, 73-281
Pollination: botany, 74-270
Pollution: atmospheric fallout, 73-209, *il.,* 210; danger to plants, 73-11; determining safe levels for pollutants, 73-214; diffused, 73-205; LD 50, 73-214; point-source, 73-206; pollution potential index, 73-208; transportation, 74-362. See also **Air**

pollution; Automobile; Conservation; *Ecology; Environment;* Ocean pollution; Phosphate detergent; Thermal pollution; Water pollution.
Polychlorinated biphenyls (PCBs), 73-53, 208, 341
Polyps: *Calliactis,* 72-376; limu-make-o-Hana, 72-376
Ponderosa pine, *il.,* 73-15
Pontecorvo, Bruno M., 74-80
Population: animal, 74-167
Positrons: neutrinos, *Special Report,* 74-80
Postglacial vegetation, 73-298
Potassium-argon dating method, 73-229
Powdermaker, Hortense, 72-424
Power plants: geothermal, 74-58
Precambrian rock, 72-320
Precision teaching approach, 73-357
Precognition, 74-145
Pregnancy: T-globulin in blood, 72-326
Premature aging: aging, *Special Report,* 74-128
President's Science Advisory Committee, 74-351
Press, Frank: Day Medal, 74-407
Preventive dentistry, 72-325
Priestley Medal, 74-406, 73-414, 72-414
Prismane: chemical synthesis, 74-279
Progeria, 74-128
Programmed instruction, 73-356
Project Mohole, 74-302
Project Stormfury: meteorology, 74-320
Promethium: stellar radioactivity, 72-265
Prostaglandins: drugs, 74-282
Protein: aging, *Special Report,* 74-128, *il.,* 132; as energy source, 73-158; botany, 74-269; chemical structure, 73-287; fish protein concentrate, 73-283; from cattle manure, 73-283; synthesis, 73-114
Protein-calorie malnutrition: related to chromium deficiency, 73-140
Proton-proton (pp) chain: neutrinos, *Special Report,* 74-81
Proton synchrotron, 73-248, 72-205. See also **Accelerator, particle.**
Protons: head-on collision, 74-338, 73-248, 72-347; nuclear research, 73-349
Psychiatry: amphetamine therapy, 72-299; Kittay Award, 74-409
Psychobiology: molecular, 72-137
Psychokinesis, 74-145
Psychology, 74-348, 73-356, 72-357; *Books of Science,* 73-279, 72-280; building self-esteem in students, 73-303; population and aggression, 72-305; psychosomatic health, *Special Report,* 74-137. See also **Brain;** *Environment;* **Psychiatry.**

Index

Psychosomatic health: *Special Report,* 74-137
Public health: in China, 73-196; in the United States, 73-322. See also **Cancer;** *Medicine;* **Vaccines; Virus.**
Public transportation, 74-362
Pulsars: astronomy, 74-259; X-ray astronomy, 73-270. See also *Astronomy* (high energy).
Pulsed laser: lasers, *Special Report,* 74-208, *il.,* 201
Puromycin: protein synthesis inhibition, 72-145
Pyramid of the Sun, 74-252

Q

Quantum theory: resolution of contradictory theories, 73-349
Quarks: and partons, 73-346; fusion reactions, 72-273; physics, 74-341; theory, 72-348
Quasar: association with galaxies, 73-273, 72-268; astronomy, 74-257, 261; *Close-Up,* 74-258; red shift, 74-257
Quinacrine hydrochloride: chromosome stain, 72-314

R

Rabinowitch, Eugene, 74-414
Race, Robert Russell: Gairdner Award, 74-409
Radar, Doppler, 74-230
Radcliffe Institute, 73-303
Radiation: cosmology, 74-260; far ultraviolet, 74-337; hazards of, 72-32; infrared, 74-321; Raman scattered, 72-347; vitamin D-producing, 72-256
Radio astronomy: astronomy, 74-257; binary stars, 73-269; double stars, 73-271; quasars, 74-259
Radio interferometers, 72-77
Radio telescope: cosmology, 74-260
Radio waves: agriculture, 74-244
Radiocarbon dating, 72-97, 228
Radioisotopic thermoelectric generator, 72-312
Rado, Sandor, 73-424
Railroad: transportation, 73-373, 72-374
Rainfall: acidic, 73-297; measurement of, 72-88
Raja yoga, 74-143
Rama, Swami: psychosomatic health, *Special Report,* 74-142, *il.,* 143
Raman scattered radiation, 72-347
Raman, Sir Chandrasekhara Venkata, 72-424
Ray, Dixy Lee, 74-354
Recommended dietary allowance, 74-331

Recycling, *il.,* 73-310
Red blood cells: regeneration of, 73-378. See also **Blood.**
Red mangrove, *il.,* 73-18
Red Paint Indians, 72-261
Red shift: cosmology, 73-272; galaxy, 72-268; quasar, 74-257; 72-268, *Close-Up,* 258
Regeneration: of damaged axons, 73-336; salamanders, 72-377
Reinforcement: behaviorism, 73-356
Renal disease: medicine, 74-314
Renfrew, Colin, 72-228, *il.,* 238
Reproduction: studies of deer cycles, 73-380
Research, applied, 73-361
Research, basic: in China, 73-198; science support, 73-361
Research and development (R & D): federal funding, 73-359, 72-363, *il.,* 364; science support, 74-351
Respiration: cell evolution, 73-109; effects of exercise, 73-153
Retinoic acid, 72-172
Revascularization: *Special Report,* 72-176; surgery, 72-330. See also **Heart surgery.**
Reverse transcriptase: enzyme, 73-94
Rheumatic fever: immunology, 74-117
Rho factor: mRNA synthesis termination, 72-338
Ribonuclease: Nobel prize, 74-404
Ribonucleic acid. See RNA.
Ribosome: aging, *Special Report,* 74-134; cell identification, 73-110; model, 73-274; protein synthesis, 73-275
Rice: hulls used to control oil spills, 73-284
Rickets: in Neanderthal Man, 72-256
Riemsdyk, Jan Van, *il.,* 72-16
Rifampin (rifampicin), 72-299
RNA (ribonucleic acid): aging, *Special Report,* 74-129; biochemistry, 73-275, 72-273; brain, *Special Report,* 72-144; chemistry, 72-289; *Close-Up,* 72-275; genetics, 74-299, 73-313; messenger RNA (mRNA), 74-129, 73-275, 72-337; microbiology, 73-335; structural chemistry, 74-276; transfer RNA (tRNA), 74-129, 276, 311; viral, 73-94; viroids, 74-326. See also **Cell; DNA;** *Genetics;* **Virus.**
Rolfing, *il.,* 74-146
Roosa, Stuart A., 72-366
Rosenstiel Institute of Marine and Atmospheric Sciences, 72-122
Roseroot, *il.,* 73-23
Rubidium-strontium method: lunar rock dating, 72-41, 316
Ruby laser, *il.,* 74-202
Ruckelshaus, William D.: environment, 74-293

Russia: archaeology, 74-250; communications satellite systems, 73-291; science, *Special Report,* 74-232; space program, 73-75, 367, 72-365

S

Sabin, Albert, 74-398
Saccharin, 73-295
Safety: electronics, 74-287; personal rapid transit, *Special Report,* 74-196
Sailing: cargo ships, transportation, 74-365
Salk, Jonas: biography (*A Man of Science*), 74-389
Salk Institute for Biological Studies, 74-389
Salt: iodization, 73-138
Salt-water fish: domestication, 72-109
Salyut space station, 72-365
San Andreas fault, 73-319, *ils.,* 166, 319
San Fernando earthquake, 72-319
Sandage, Allan R., 73-272
Sanger, Ruth: Gairdner Award, 74-409
Sarabhai, Vikram A., 73-424
Sarepta, *il.,* 73-259
Satellite, communications, 74-280, 72-293
Satellite, meteorological, 74-320, 72-335
Satellite, monitoring: agriculture, 74-247; oceanography, 74-332
Satellite, transponder, 72-293
Satellites: astronomy, 73-264
Satiety: brain functions, 73-118
Saturn: astronomy, 74-253; giant satellites, 73-264
Sauropod: paleontology, 74-308
Scale invariance: physics, 73-346
Schistosomiasis, 73-196
Schlesinger, James R., Jr., 73-360
Schmitt, Harrison H., 74-360, 73-366, 72-409
Schreiffer, J. Robert: Nobel prize, 74-406
Schwarzschild radius, 73-80
Schwarzkopf, Paul, 72-424
Science, general: *Books of Science,* 73-279, 72-280; education, 73-301; ethics, 73-361; rejection of, 72-28, 360; religion, 72-280; technology, 73-358
Science and Technology, Federal Council for, 74-351
Science and Technology, Office of, 74-351
Science Support, 74-351, 73-358, 72-360. See also *Ecology; Education; Space Exploration; Transportation.*
Scorpion: heat experiments, 72-300
Scotophobin: neurology, 74-327; peptide, 72-148; synthesized, 74-264

Index

Scott, David R., *73*-363, *72*-367
Scripps Institution of
Oceanography, *72*-122
Sea-floor nodules: source of ores,
72-342
Sea floor spreading: geology, *74*-305;
geophysics, *72*-322; oceanography,
74-334; plume theory, *Special Report*,
73-163; primitive mammals, *73*-318.
See also Continental drift theory;
Plate tectonics.
Sea snakes: salt glands, *73*-377
Seabed treaty, *72*-345
Seaborg, Glenn T., *73*-360, *72*-248
Seat belts: electronics, *74*-287
Sebaceous glands, *72*-154, *ils*., 155,
157
Sebum, *72*-154
Sediment: ocean-floor, *72*-122
Seed bank, *73*-13
Seedless fruit: cucumber, *73*-254;
cultivated by parthenocarpy, *72*-282
Seegal, David, *74*-414
Seismic profiler, *72*-123
Senses: brain, *Special Report*, *72*-137
Sewage: fish culture, *72*-108
Sex: discrimination and stereotypes,
Special Report, *74*-24
Shadowgram: chromosome
examination, *72*-314
Shapely, Harlow, *74*-262, 414
Sheep: new breeds, *74*-244
Shemyakin, Mikhail M., *72*-424
Shepard, Alan B., Jr., *72*-366
Sheps, Mindel, *74*-414
Shirane, Gen: Buckley Prize, *74*-406
Shoemaker, Eugene: biography
(*A Man of Science*), *72*-398
Sickle cell anemia: *Close-Up*,
73-323; medicine, *72*-328, *il*., 327
Sigma factor: mRNA synthesis,
72-337
Sikorsky, Igor I., *74*-414
Silberman, Charles E., *72*-307
Sioux village, *73*-263
Skin: peeling, acne therapy, *72*-161,
172; structure and function, *72*-154,
il., 155
Skylab: oceanography, *74*-332; space
exploration, *74*-356, *73*-366, *72*-365;
zoology, *74*-370
Slayton, Donald K., *73*-363, *il*., 403
Smell: identification of mother,
72-379
Smith Medal (geology), *72*-420
Snakes: ability to detect warm-blooded
prey, *73*-376; salt glands, *73*-377
Snowmobiler's back, *72*-258
Snowy Mountains Scheme:
Special Report, *74*-213
Soil, lunar. See Lunar soil.
Solar clouds, *73*-268

Solar energy: *Special Report*, *74*-59
Solar farms, *74*-60
Solar flares, *74*-255
Solar heating: dust storms on Mars,
73-67
Solar neutrinos: neutrinos, *Special
Report*, *74*-86
Solar observatory: orbiting, *73*-268;
underground, *73*-268
Solar polar caps, *73*-268
Solid state physics, *74*-346; Buckley
Prize, *74*-406; solid state diode laser,
72-346
South America: archaeology, *74*-253
South Spot: Martian crater, *73*-62
Soybean: male-sterile, *73*-254
Space: geometry of, *72*-55
Space Exploration, *74*-356, *73*-362,
72-365; Apollo program, *73*-360, 398,
72-366; British spacecraft, *72*-371;
Gemini 5, *73*-408; geochemistry,
74-299; Luna 16, *72*-368; Luna 17,
72-369; Luna 18, *73*-368; Luna 19,
73-368; Luna 20, *73*-368; Lunar Rover,
73-363, *72*-367; Lunokhod, *72*-369;
Mariner series, *73*-61, 315, 368,
72-265, 370; pictures from space,
73-306; Pioneer 10, *73*-292, 364, 370, *il*.,
363; Project Mercury, *73*-405; Ranger
Project, *72*-39; Russian-American
cooperation, *74*-239; Russian space
program, *73*-366, *72*-365; Skylab,
73-366, *72*-365; space shuttle, *73*-360,
367; space station, *72*-366; Stratoscope
II, *72*-263; U.S.-Russian manned
mission, *73*-360; Venera 7, *72*-264,
370; Venera 8, *74*-255, *73*-370;
Viking series, *73*-74; Zond 8, *72*-369;
zoology, *74*-368. See also Apollo
program; *Astronomy* (planetary);
Science Support.
Space shuttle. See *Space Exploration*
(space shuttle).
Space station. See *Space Exploration*
(space station).
Spartina: grass, *72*-300
Speedy Bee, The, *il*., *74*-156
Sperry Award, *74*-411
Sphygmometrograph, *il*., *73*-322
Spiders: ecology, *74*-284; zoology,
74-368
Spin-flip laser, *72*-347
Spitzer, Lyman: Bruce Medal, *74*-410
Spitzer, Paul, *73*-50
Split-brain technique, *74*-328,
72-151
Squirrel: feeding ecology, *73*-299;
zoology, *74*-366
Stable fly: *Close-Up*, *74*-246
Stack, John, *74*-414
Stafford, Thomas P., *73*-363
Stanford Linear Accelerator
(SLAC): physics, *74*-338, *73*-346,
72-348
Stanley, Wendell M., *73*-424
Star: binary, *73*-268; changes in mass,

73-268; communication with, *Close-Up*,
73-364; HR 465, *72*-265; infrared,
73-269; X-ray, *72*-268. See also
Astronomy.
Steam: geothermal energy, *74*-53
Steam-powered buses, *73*-374
Stearns, Carl L., *74*-414
Stegosaurs: paleontology, *74*-308
Stein, William H., *74*-404
Stellar astronomy, *74*-255
Stem cells: immunology, *74*-110
Stock market quotations:
computerized, *72*-296
Stomata: potassium content of guard
cells, *73*-281
Stonehenge, *72*-232, *il*., 234
Stormfury Project: meteorology,
74-320
Stouffer Prize (medical research),
74-409, *72*-419
Stratoscope II, *72*-263
Stratosphere: volcanic discharges into,
72-318
Streams: channelization, *74*-294
Structural chemistry, *74*-276
Submarine: geology, *74*-304
Subperiosteal implants, *73*-321
Sucking responses, *72*-359
Sulfur, *73*-285, 306
Sulfur dioxide, *74*-292, *73*-307
Sun: astronomy, *74*-255; climatic
effects, *72*-96; energy, *Special Report*,
74-59; isothermal plateau, *72*-265;
neutrinos, *Special Report*, *74*-79.
See also headings beginning Solar . . .
Sunflower: hybrid, *73*-255
Superaccelerator, *72*-205
Superconducting electron
microscope, *73*-382
Superconductivity: Nobel prize,
74-406
Superconductors, *73*-355
Supernova, extragalactic, *74*-255
Supernova remnant, *74*-260
Supersonic transport (SST) project,
74-322, *72*-333, 360
Supertanker: *Special Report*, *73*-216
Surfactants: detergent ingredients,
72-285
Surgery. See *Medicine* (surgery).
Surveying, *il*., *74*-207
Surveyor: research ship, *73*-339
Svedberg, Theodor H.E., *72*-424
Swami Rama: psychosomatic health,
Special Report, *74*-142, *il*., 143
Sweat glands, *72*-154
Swigert, John L., *73*-363
Swordfish: mercury content, *73*-378,
72-302
Synapse, *73*-336, 337
Synchrotron radiation, *74*-255
Synthetic natural gas: energy, *74*-290
Synthetic suture, *il*., *72*-332
Syphilis: in Neanderthal Man, *72*-256
Szent-Györgyi, Albert: biography (*A
Man of Science*), *72*-382

Index

T

Tamm, Igor Y., 72-424
Taste, sense of, 73-133
Tautavel Man, 73-256, il., 257
Taxi-car pools: transportation, 74-365
Taylor, Edward S.: Goddard Award, 74-410
Teacher education, 72-305
Technological Assessment, Office of, 74-355
Technology: assessment, energy, 74-292; Essay, 72-26; foreign industrial competitors, 72-365; nonmilitary uses for nuclear explosives, 72-313; rejection of, 72-360
Tectonic plate. See Plate tectonics.
Tektite project, 72-341
Tektites: geochemistry, 74-302
Telephone, 74-279, 73-293
Telescope: astronomy, 74-257; cosmology, 74-260
Television: pictures from space, 73-306
Teller, Edward, 72-242
Temin, Howard M., 73-94, 417, 72-273
Teminism, 72-273
Temperature, body: aging, Special Report, 74-124, il., 125
Temperature inversion, 73-172
Teotihuacán (Mex.), 72-262
Termites: microbiology, 74-327
Test-tube embryo, 72-379, il., 376
Thar Desert, 72-99
Theiler, Max, 74-414
Thermal pollution, 72-32, 256. See also Water pollution.
Thermionic converter, 72-311
Thermonuclear fusion, 73-351
Thompson Medal, 74-407
Thorium 232, 72-218
3C-273 quasar, 72-270
3C-279 quasar, 72-271, il., 272
Thrombosis: coronary, 72-181
Thymidine (T), 72-289
Thyroidectomy, 72-182
Tilton program, 72-305
Tiselius, Arne, 73-424
Tissue culture, 72-192
Titan: Saturn satellite, 74-254
Titanium: lunar rocks, 72-40
Tokamak, 73-351, 72-215, 353
Tolbutamide, 72-298
Ton 256, 74-261
Toomey, Bill, 73-148
Tooth, artificial, 74-310
Tooth impressions: used to determine age of bears, 73-39, il., 38
Topography: Mars, 73-267; Venus, 73-267. See also Geoscience.
Torrey Canyon, 73-220
Trace elements: deficiency, 74-312; dietary supplements, 73-145;

experiments in producing deficiency, 73-143, il., 141; Special Report, 73-133. See also Nutrition.
Tracked air cushion vehicles (TACV), 73-373
Traffic: Special Report, 74-185
Transatlantic cable, 73-291
Transcutaneous electrical nerve stimulation, 74-319
Transfer ribonucleic acid (tRNA), 74-311, 72-290. See also RNA (transfer RNA).
Transistor: communications, 74-281
Transplantation: heart, 73-329, 72-331; tissue rejection, 73-378
Transpo 72, 73-370
Transportation, 74-362, 73-370, 72-372; rapid transit systems, 73-370, il., 371; Special Report, 74-185
Transportation, U.S. Department of: air-cushion vehicles, 72-375; meteorology, 74-322; Urban Mass Transportation Administration (UMTA), 73-373, 72-375
Transuranic elements, 72-351
Trapezium, 72-77
Trauma, 73-326
Trees: ecology, 74-286; growth affected by motion, 73-281, il., 280
Trickle irrigation, 74-244
Triticale: synthetic grain species, 72-255
Tritium, 72-218
Trobicin: antibiotic, 73-297
Trout, 72-107
Tuberculosis: BCG vaccine, protection against leukemia, 72-328; drug therapy, 72-299
Tumor angiogenesis factor, 73-328
Tuna: mercury content, 73-377
Tunable laser, 73-345, 72-345
Tunneling: chemical dynamics, 74-276
Turtles, 74-285
2, 3-DPG, 73-155

U

Uhuru satellite, 74-65, 259, 73-270, 72-269
Ultraviolet laser, 74-337, 72-345
Ultraviolet light: used to harden dental sealant, 72-325
Ultraviolet spectroscopy, 74-257
Underdeveloped countries: overpopulation, 71-82
Underground solar observatory, 73-268
Underwater habitat, 74-336
Unemployment: scientists and engineers, 73-358, 72-350
United Nations: seabed treaty, 72-345
United Nations Conference on the Human Environment, 74-295, 73-309
United Nations Educational, Scientific, and Cultural Organization (UNESCO), 73-260

United States Steel Foundation Award, 74-408, 72-418
Universe: expansion, 73-272
Universities Research Association, 72-205
University Group Diabetes Program, 72-298
Upwelling, artificial, 72-343
Uranium-235, 72-214
Uranus, 72-263
Urban dogs: Special Report, 74-160
Urban transportation: Special Report, 74-185
Urea, 73-323, 72-329
Urey, Harold C.: Priestley Medal, 74-406
U.S. Steel Foundation Award, 74-408, 72-418
Uterus: tissue transplants, 73-378

V

Vaccines: animal diseases, 73-256; influenza, 74-314; Jonas Salk, 74-389; low U.S. vaccination rate, 73-324
Vagus nerve, 73-121
Van Slyke, Donald D., 72-424
Vegetation: postglacial, 73-298
Vela: X-ray emission, 74-72
Venera 8: space exploration, 74-361
Venezuela: archaeology, 74-253
Venezuelan equine encephalomyelitis, 73-256
Venezuelan Institute of Neurological and Brain Research, 73-385
Venus: geochemistry, 74-301; space probe, 74-254, 73-370, 72-264, 370. See also Space Exploration.
Vesalius, Andreas, 72-11, il., 15
Veterinary medicine, 74-160
Vetlesen Prize, 74-407, 72-417
Video picture storage, 72-295
Vidicon, 73-304
Vietnam: environment, 73-308
Vietnam War: cloud seeding, 74-320
Vinci, Leonardo da, 72-11, ils., 10, 14
Vineberg, Arthur M., 72-184
Virchow, Rudolf L. K., 72-256
Viroids: Close-Up, 74-326
Virus: B-type, 73-97; breast cancer, 72-326; C-type, 73-100, 72-326; cancer-producing, 73-90, 72-299; herpes virus, 73-95, il., 96; inactivated Sendai virus, 72-192; Lambda H80, 72-340; medicine, 74-311; research, Jonas Salk, 74-389; RNA-directed DNA polymerase content, 72-275; SV-40, 72-201; vaccines, 72-255, 327; viroids, 74-326. See also Cancer; Leukemia; Microbiology; Vaccines.
Visual perception: infants, 72-357
Vitamin A acid, 72-172
Vitamin B$_{12}$: biochemistry, 74-262; chemical synthesis, 74-278
Vitamin C: identification, 72-382;

Index

iron availability, *73*-145; nutrition,
74-329
Vitamin E, *73*-297
Volatile pollutants, *73*-209
Volcanoes: geology, *72*-318; Kilauea,
73-316; located over plumes, *73*-168,
ils., 168, 170, 175
Von Békésy, Georg, *74*-414
Von Braun, Wernher, *73*-362
Von Euler, Ulf S., *72*-418

W

Wankel, Felix, *73*-372
Wankel engine: *Close-Up, 73*-372;
environment, *74*-295
Waste: computerized disposal, *72*-295;
human, *73*-196, 207; industrial,
74-297; radioactive materials, *73*-308,
318; ocean disposal, *74*-333. See also
Air pollution; *Ecology;* **Ocean
pollution; Pollution.**
Water: astronomy, *74*-254; energy
source, *74*-52; irrigation, *74*-244;
on Mars, *73*-71
Water pollution: chemical technology,
72-284; education, *72*-305; fish
farming, *72*-119; metal pollution,
Close-Up, 72-302; oil spills, *Close-Up,*
72-344; phosphates, *73*-310; *Special
Report, 73*-205. See also *Ecology;*
Ocean pollution; Thermal pollution.

Water reclamation, *72*-286
Watson, James D., *72*-362, 417
Watson Medal, *73*-420
Weather: forecasting, *74*-238;
modification, *74*-320, *73*-333, *72*-334;
modification, *Special Report, 74*-220;
prediction, *73*-295, 333, *72*-336.
See also **Climate;** *Meteorology.*
Weed control, *74*-244
Weight control: nutrition, *74*-332.
See also **Obesity.**
Weitz, Paul J.: space exploration,
74-357
Whales: oceanography, *74*-335
Whitby, G. Stafford, *73*-424
Whiteheads, *72*-158
Wilderness Society, *72*-308
Wildlife telemetry system, *73*-37,
il., 39
Wilson, Robert: biography (*A Man
of Science*), *74*-372
Wind: effect on tree growth, *73*-281;
on Mars, *73*-67
Wintersteiner, Oskar P., *73*-424
Witschi, Emil, *73*-424
Wolves, *74*-367, *73*-375
Women: scientists, *Special Report,*
74-24
**Woods Hole Oceanographic
Institution,** *72*-122
Woodward, Robert B.: Cope Award,
74-404
Worden, Alfred M., *73*-366, *72*-367
World Meteorological Organization,
72-334
Wrist watch, electronic, *72*-309

X

X-ray astronomy, *74*-65, *73*-270,
72-269
X rays: acne therapy, *72*-161; analysis
of mummies, *72*-260; angiography,
72-178; astronomy, *74*-259; cosmic,
72-269; Cygnus X-I, *72*-270;
diffraction methods, *72*-289; galactic
emissions, *72*-270; identifying atoms,
72-351; radiation from binary stars,
73-268; *Special Report, 74*-65

Y

Yellowstone National Park, *73*-35,
il., 41
Yoga, *74*-143
Young, John W., *73*-366, *il.,* 369

Z

Zinc: function in humans, *73*-133
Zond (probe) series, *72*-369
Zones of upwelling: marine life
concentrations, *72*-118
Zoology, 74-366, *73*-375, *72*-376;
Books of Science, 74-268, *73*-279,
72-280; chimps, *Special Report,*
74-34; diving sea mammals, *74*-335;
dogs, *Special Report, 74*-160;
endangered birds, *73*-50; grizzly bears,
73-35; social structure among bears,
73-41; whales, *74*-335
Zooplankton, *72*-303
Zworykin Award, *74*-411

Acknowledgments

The publishers of *Science Year* gratefully acknowledge the courtesy of the following artists, photographers, publishers, institutions, agencies, and corporations for the illustrations in this volume. Credits should be read from left to right, top to bottom, on their respective pages. All entries marked with an asterisk (*) denote illustrations created exclusively for *Science Year.* All maps were created by the *World Book* Cartographic Staff.

Cover
Herb Herrick*

Essays
10-23 NASA
24 Carl Purcell*

Special Reports
34-48 Lee Balterman*
50-51 Jackson-Zender*
54 Brian Brake, Rapho Guillumette
55 James Coombs, University of California, Riverside
56 Jackson-Zender*
58 *U.S. News & World Report;* Dr. Georg Gerster, Rapho Guillumette
59 Mas Nakagawa*; George Kew, University of Arizona
60-61 Jackson-Zender*
62 National Research Council, Canada; M. Brigand, French Engineering Bureau
64-65 Herb Herrick*
66-67 NASA
68-73 Herb Herrick*
74-75 Lund Observatory, Sweden
78 Brookhaven National Laboratory
80 California Institute of Technology
81 Brookhaven National Laboratory
82-85 Don Meighan*
86-87 Brookhaven National Laboratory
88 Don Meighan*
92-93 University of Minnesota
94 Duane Bingham
95 William Coulson; University of Minnesota
97 Duane Bingham
99 University of Minnesota; University of Minnesota; Donald Wolberg
100 William Coulson; Duane Bingham
101 Duane Bingham
103 William Coulson; Donald Wolberg; Robert Black
105-106 Duane Bingham
109 Lee Balterman*
110 Dr. Mary Ann South, Baylor College of Medicine
111 Jim Curran*; Dr. Curtis B. Wilson, Jr., Scripps Clinic and Research Foundation
112-117 Jim Curran*
118 Nicola Fabris, University of Pavia, Italy
121 WORLD BOOK photo*; John Launois, Black Star
123 Leonard Hayflick, Stanford University
125 Joseph Erhardt*
127 John Launois, Black Star
129 Bernard Strehler, University of Southern California
130 Daphne J. Osborne, University of Cambridge
132-133 Tak Murakami*
134 Bernard Strehler, University of Southern California
136 Thelma Moss, UCLA
139-144 Don Richards, The Menninger Foundation
145 Gilbert L. Meyers*; Harold Ellithorpe, Black Star
146 Thelma Moss, UCLA

149 G. F. Townsend, University of Guelph, Ontario, Canada; Walter Rothenbuhler, Ohio State University
150 Norman E. Gary, University of California, Davis
151 G. F. Townsend, University of Guelph, Ontario, Canada
152 Tak Murakami*
154 Dr. A. C. Stort, Faculdade de Filosofia, São Paulo, Brazil; G. F. Townsend, University of Guelph, Ontario, Canada
155 G. F. Townsend, University of Guelph, Ontario, Canada
156 Reprinted from *The Speedy Bee*
158 Mike Sheldrick, Washington University
160-161 Alan Beck, Washington University
163-164 Joseph Erhardt*
167 Steve Eagle from Nancy Palmer
168 Calvin Buikema
169 Joseph Erhardt*
170 Alan Beck, Washington University; Michael Fox, Washington University; Michael Fox, Washington University
171 Alan Beck, Washington University
173 Herb Weitman*
175 Michael Fox, Washington University
176 Warren Garst, Tom Stack & Associates; Durward L. Allen, Dept. of Forestry & Conservation, Purdue University; Bruce Coleman, Inc.; Michael Fox, Washington University; Michael Fox, Washington University
179 Jan Van Wormer, Bruce Coleman, Inc.; R. Austing, Bruce Coleman, Inc.; Michael Fox, Washington University; Rod Allin, Bruce Coleman, Inc.; S. C. Bissert, Bruce Coleman, Inc.; Michael Fox, Washington University
180-183 David Cunningham*
184-185 George Jones*
187-188 Gilbert L. Meyers*
189 George Jones*
190 George Jones*; West Virginia University
191-192 Gilbert L. Meyers*
193 Peter Britton, *The Lamp,* Exxon Corporation; Transportation Technology, Inc.
194-195 George Jones*
198-199 Fritz Goro
201 Fritz Goro; Avco Everett Research Laboratory
202 Mas Nakagawa*; Fritz Goro
204 Bionic Instruments, Inc.; Fritz Goro
205 Howard Sochurek, Woodfin Camp, Inc.
206 National Bureau of Standards; Howard Sochurek, Woodfin Camp, Inc.
207 National Bureau of Standards
208 Hughes Aircraft Company
209 Avco Everett Research Laboratory; Howard Sochurek, Woodfin Camp, Inc.; Avco Everett Research Laboratory
210 Mas Nakagawa*; University of Rochester
212 Photographic Library of Australia
214 David Moore, Black Star; Snowy Mountain Hydro-Electric Authority
215 Snowy Mountain Hydro-Electric Authority; Photographic Library of Australia; David Moore, Black Star
216-217 Dorothy Louise Nelson*
218-219 David Moore

Typography

Display–Univers
Monsen Typographers, Inc., Chicago
Text–Baskerville Linofilm
Total Typography, Inc., Chicago
Text–Baskerville monotype (modified)
Total Typography, Inc., Chicago
Monsen Typographers, Inc., Chicago

Offset Positives

Capper, Inc., Knoxville, Tenn.
Kingsport Press, Inc., Kingsport, Tenn.
Process Color Plate Company, Chicago

Printing

Kingsport Press, Inc., Kingsport, Tenn.

Binding

Kingsport Press, Inc., Kingsport, Tenn.

Paper

Text
Childcraft Text, Web Offset (basis 60 pound)
Mead, Escanaba, Mich.

Cover Material

Flax Lexotone
White Offset Blubak
Holliston Mills, Inc., Kingsport, Tenn.